DEMOCRATIC ECO-SOCIALISM AS A REAL UTOPIA

Democratic Eco-socialism as a Real Utopia
Transitioning to an Alternative World System

Hans A. Baer

First published in 2018 by
Berghahn Books
www.berghahnbooks.com

© 2018, 2020 Hans A. Baer
First paperback edition published in 2020

All rights reserved. Except for the quotation of short passages for the purposes of criticism and review, no part of this book may be reproduced in any form or by any means, electronic or mechanical, including photocopying, recording, or any information storage and retrieval system now known or to be invented, without written permission of the publisher.

Library of Congress Cataloging-in-Publication Data

Names: Baer, Hans A., 1944– author.
Title: Democratic eco-socialism as a real utopia : transitioning to an alternative world system / Hans A. Baer.
Description: New York : Berghahn Books, 2018. | Includes bibliographical references and index.
Identifiers: LCCN 2017037763 (print) | LCCN 2017042507 (ebook) | ISBN 9781785336966 (ebook) | ISBN 9781785336959 (hardback : alk. paper)
Subjects: LCSH: Socialism. | Capitalism. | Democracy. | Social ecology.
Classification: LCC HX73 (ebook) | LCC HX73 .B3235 2018 (print) | DDC 335—dc23
LC record available at https://lccn.loc.gov/2017037763

British Library Cataloguing in Publication Data

A catalogue record for this book is available from the British Library

ISBN 978-1-78533-695-9 hardback
ISBN 978-1-78920-533-6 paperback
ISBN 978-1-78533-696-6 ebook

CONTENTS

List of Illustrations	vi
Acknowledgments	vii
Introduction	1
Chapter 1 The Contradictions of the Capitalist World System at the Beginning of the Twenty-First Century	15
Chapter 2 Twentieth-Century Attempts to Create Socialism: Successes and Failures	43
Chapter 3 Technoliberal and Countercultural Visions of the Future	103
Chapter 4 Efforts to Reconceptualize Socialism	127
Chapter 5 The Role of Anti-systemic Movements in Creating a Socio-ecological Revolution	173
Chapter 6 Transitional System-Challenging Reforms	201
Chapter 7 Conclusion: The Future in the Balance	254
References	267
Index	300

ILLUSTRATIONS

Figures

1.1	Four Potential Climate Change–Induced Social Formations	40
2.1	Types of "Socialist" Systems	79
3.1	Possible Future Scenarios	104
5.1	Types and Subtypes of Anti-corporate Social Movements	178

Tables

1.1	New GDP (US$) per Capita Compared to the Old (World Development Indicators), for the Thirteen Most Populous Countries in the World	20
1.2	Projected Demographic Trends during the Twenty-First Century	24
2.1	Postrevolutionary Societies, ca. mid 1980s	47
2.2	Five Dimensions of Various Types of Leninist Societies	48
2.3	The End of Postrevolutionary or "Leninist" Regimes in the Soviet Union and Eastern Europe	49
2.4	U.S. and Soviet Strategic Arsenal, 1985	59
2.5	Economic Growth in the Capitalist Core and Postrevolutionary Societies, 1961–1968	87
2.6	Regions of the Soviet Union Suffering Serious Environmental Problems	93
5.1	Eco-villages in Karen Litfin's Study	194
6.1	Die Linke.PDS/DIE LINKE Vote Share in Bundestag Elections, 2005–2009	209
6.2	Voting Patterns in the 2011 Berlin Land Election— West and East	210

ACKNOWLEDGMENTS

My interest in communitarian societies, postrevolutionary societies, alternatives to capitalism, and democratic eco-socialism began in the 1970s, when I began my training in anthropology and the social sciences. In 1975 I met Merrill Singer, who currently is based in the Department of Anthropology at the University of Connecticut, at the University of Utah and we have maintained a close friendship and collaboration that has lasted over four decades. We have published seven books and numerous book chapters and articles together that have touched on critical medical or health anthropology, African-American religion, and the critical anthropology of climate change. Along with Ida Susser (Department of Anthropology, Hunter College of the City University of New York), we employed the concept of *democratic eco-socialism* in three editions of our textbook titled *Medical Anthropology and the World System: A Critical Perspective* (1997, 2003, 2013). Many scholars and social activists have inspired me over the course of my long academic career as a scholar-activist. While there are too many to list all of them, I particularly would like to acknowledge Robert Anderson, Alan Klein, Vicente Navarro, Ray Elling, David Hakken, John Johnsen, Arachu Castro, John Bellamy Foster, Verity Burgmann, Jim Falk, Ariel Salleh, Ted Trainer, Debbi Long, Arnaud Gallois, David Holmes, Andrea Bunting, Sue Bolton, David Spratt, John Ebel, Brendon Gleeson, and Adam Bandt. I would like to give a special acknowledgment to Jack Roberts, a Ph.D. student in the School of Social and Political Sciences at the University of Melbourne, for his assistance in formatting this book. I would also like to acknowledge that Next System Project put its invitation to publish a 9,000-word essay titled "Democratic Eco-socialism as the Next World System" on its website (thenextsystem.project.com) in April 2016 and that an associated video link inspired by my essay appeared on the website later in the year. Last but not least, I would like to acknowledge various colleges and universities where I taught over the years, either on a regular or a visiting basis. These include, in chronological order, Kearney State College (Nebraska), George Peabody College for Teachers, St. John's University, the University of Southern Mississippi, the University of Arkansas at Little Rock, Humboldt University of Berlin, University of California–Berkeley, Arizona State University, George Washington University, the Australian National University, and the University of Melbourne. I was also a postdoctoral fellow in the medical anthropology program at Michigan State University in 1980–1981.

INTRODUCTION

This book is guided by the recognition that social systems, whether they exist at the local, regional, or global level, do not last forever. Capitalism as a globalizing political economic system that has produced numerous impressive technological innovations, some beneficial and others destructive, is a system fraught with contradictions, including an incessant drive for profit making and economic expansion; growing social disparities; authoritarian and militarist practices; depletion of natural resources; and environmental degradation resulting in global warming and associated climatic changes. Even more so than in earlier stages of capitalism, transnational corporations make or break governments and politicians around the world. Capitalism has been around for about five hundred years but manifests so many contradictions that it has been increasingly clear to progressive thinkers that it must be replaced by an alternative world system—one committed to social parity and justice, democratic processes, and environmental sustainability, which includes a safe climate.

As delineated in this book, I term the vision for an alternative world system democratic eco-socialism. Due to the shortcomings of efforts to create socialism in the twentieth century, the notion of socialism has been discredited in many quarters, an unfortunate reality that has prompted various scholars and social activists who have sought, in their efforts to preserve the ideals of socialism, such as collective ownership, social equality, and participatory democracy, to utilize terms such as radical democracy, economic democracy, global democracy, and Earth democracy, or to turn to one or another form of anarchism. Nevertheless, it is important for progressive people to come to terms with the historical discrepancies between the ideals of socialism and the realities of what passed for socialism and to reconstruct socialism, to create a global socialist system with manifestations at regional, national, provincial, and local levels that is highly democratic rather than authoritarian, that ensures that all people have access to basic resources, and that is at the same time environmentally sustainable and recognizes that we live on a fragile planet with limited resources. Democratic eco-socialism constitutes, in the terminology of sociologist Erik Olin Wright (2010), a *real utopia*, a utopian vision that is theoretically achievable but with much reconceptualization and social experimentation. As the existing capitalist world system continues to self-destruct due to its socially unjust and environmentally unsustainable practices, democratic eco-socialism provides a vision to mobilize human beings around the world, albeit in different ways, to prevent ongoing human socioeconomic disparities, environmental devastation, and catastrophic climate change. This vision is not utopian in the sense that it imagines a perfect world free of problems or conflicts, but rather is utopian in the sense that it is aimed at making the best world possible within existing constraints.

Although I have a Ph.D. in anthropology, I have over the course of my career come to view myself as an engaged historical social scientist committed to creating a better world. In this regard, my work is guided by the insights of two scholars: the first is John Bodley (2012) who in the six editions of his book *Anthropology and Contemporary Human Problems* envisions a *sustainable planetary society*. The second is comparative sociologist Immanuel Wallerstein (1998: 1), who coined the term *utopistics*, which he defines as "serious assessment of historical alternatives, the exercise of our judgment as to the substantive rationality of alternative possible historical systems."

I have become convinced that humanity faces two imperatives, an assertion that I often have expressed to my students at the University of Melbourne, where I have taught since January 2006. The first imperative is to learn how we humans from many different countries and cultures can live in relative harmony with each other, to ensure that we all have access to certain basic resources, such as food, clothing, shelter, health care, education, and meaningful and satisfying work and social relationships. The second imperative is to learn to live in relative harmony with nature or the planet in achieving the former and I add that these two imperatives are intricately interwoven. Of course, the idea of humans living in complete harmony with each other and nature is completely utopian because even in more or less pristine hunting and gathering or foraging societies, which were highly egalitarian, communal, and consensual, social conflict did exist, more at the individual level than the collective level. The real utopian vision is the hope that we achieve relatively harmonious relationships with each other, despite our national, ethnic, and cultural differences, on a fragile planet that has provided us—as protohumans in the form of various lines of hominines and eventually more or less full-blown humans—with sustenance for several million years.

At least in theory, these two imperatives are embedded in the concept of *democratic eco-socialism* that Merrill Singer, Ida Susser, and I coined in our vision for a healthy planet in a textbook titled *Medical Anthropology and the World System: A Critical Perspective,* which we first published in 1997 with subsequent editions in 2003 and 2013 (Baer, Singer, and Susser 1997, 2003, 2013). In adopting this term, we drew from two other terms that had been floating around at the time, namely, *democratic socialism* and *eco-socialism.* Various scholars have devised the term "democratic socialism" to flag those regimes that often have been termed "socialist" or "communist" were or are neither of these designations because they were or are not democratic. By design, socialism should be far more democratic than capitalism.

Various scholars proposed alternative terms such as *radical democracy* and *economic democracy* in order to disassociate their conception of the good society from what had occurred and was occurring in authoritarian regimes that generally called themselves socialist such as the Soviet Union, China, Cuba, North Korea, etc. However, when one delineates the basic dimensions of socialism, such as public and social ownership of the means of production, a high degree of social parity, and workers' democracy, many people will say that this sounds like socialism and add that history has shown over the course of the twentieth century that socialism was a failure and continues to be a failure in the few countries that still use that designation to refer to themselves, such as China, Vietnam, North Korea, and Cuba. Ultimately, self-proclaimed socialists or Marxists must come to terms with the discrepancies between the ideals and realities of what has historically gone under the label of socialism. It is my assertion that what I term *postrevolutionary societies* or *socialist-oriented societies,* which some scholars term *actually existing socialist societies,* exhibited and in some cases still exhibit positive features. They also exhibit notable negative features. Unfortunately, all too many of the negative features have been tragic and horrific, to the point that they have discredited the notion of socialism in the minds of many people.

Eco-socialism is a more recent term that has emerged as various scholars and activists have come to realize that an authentic socialist world must be based both on principles of environmental sustainability and a disavowal of the growth paradigm that is an integral component of global capitalism and that all too often has been accepted by people who called themselves socialists or Marxists. While eco-socialism has had precursors, including even Marx (Foster 2000), it began to take off in the early 1980s.

My Personal Journey in Becoming a Critical Anthropologist and a Democratic Eco-Socialist

People sometimes ask me how I came to define myself as a "socialist" or a "democratic socialist" or, more recently, a "democratic eco-socialist." This entailed a somewhat slow and convoluted process. I often tell my students and other people that I am a "leftover from the 60s." However, during the 1960s my politicization occurred not so much in the university but in the bowels of the corporate world. Between 1963 and 1966 I studied engineering mechanics at the Pennsylvania State University or "Penn State," situated in the bucolic Nittany Valley of the Appalachian Mountains of central Pennsylvania. I opted to study engineering largely due to the influence of my father, a mechanical engineer, and because for better or worse I did relatively well in my mathematics and physics courses in high school, thus prompting my guidance counselors to reinforce my father's insistence that I study engineering so that I could secure some semblance of job security. Furthermore, in the post-Sputnik era of the late 1950s and 1960s, young Americans, particularly boys, were encouraged to study some technological field lest the United States fall behind the Soviet Union in the "space race." However, I was an oddball engineering student in that I enjoyed taking liberal arts courses, which many of the engineering students regarded to be a complete waste of time, referring to them as "gut courses" and "Mickey Mouse courses."

Upon graduating from Penn State with a B.S. in Engineering Mechanics in early September 1966, I went to work for United Aircraft in Connecticut, where I spent most of my time doing stress analysis on the JT9D engine that propelled Boeing 747 "jumbo" jets. While there I became restless and followed Horace Greeley's advice, "Go West young man," and took a job at the Boeing Corporation in the Seattle area, where most of my time was spent doing stress analysis on the 747. However, I found myself increasingly questioning the corporate emphasis on profit making and neglect of the social and environmental consequences of the military and the commercial airplanes that the Boeing Company was manufacturing. The late 1960s as well as the early 1970s was a period of social ferment and I found myself being influenced by various social movements, including the peace, environmental, countercultural, civil rights, and feminist movements. I came to the conclusion that the world was very screwed up and I wanted to understand why it was screwed up and wanted in some small way to contribute to the process of making a better world for humanity and the ecosystem. I believe that in large part over the course of the past forty-six or so years I have achieved the first objective, but I leave it to others to determine the extent to which I have achieved the latter objective.

In order to achieve these two objectives, I ended up studying anthropology, initially at the University of Nebraska–Lincoln, where I obtained an M.A. in 1971, and following a short teaching stint at Kearney State College (now the University of Nebraska at Kearney) in south-central Nebraska, I continued to study anthropology at the University of Utah, where I obtained a Ph.D. in 1976. While in Nebraska, I conducted ethnographic research on a Hutterite colony in South Dakota, which piqued my interest in communitarian societies (Baer 1976a). I did my doctoral dissertation on the Levites of Utah, a Mormon schismatic group (Baer 1976b; Baer 1987). Both the Hutterites and Levites interested me because they sought to create egalitarian and communal societies, one steeped in the Anabaptist tradition and the other in the Mormon tradition. As a Ph.D. student at the University of Utah, I taught a course on Communitarian Societies as part of my effort to unravel the complexities of efforts to create egalitarian and communitarian lifestyles.

Over the course of my long academic career, I have examined a wide variety of topics, including African-American religion; medical pluralism in the United States, Great Britain, and Australia; critical health or medical anthropology; socio-political and religious life in East Germany before and after unification; the political economy of higher education; and Australian climate politics. Much of my work has focused on issues of power, domination, social (class, racial/ethnic/gender) stratification, the contradictions of global capitalism, and efforts to create a more socially just, democratic, and environmentally sustainable world. In these efforts, I have often collaborated with Merrill Singer, whom I met in 1975 while we were both graduate students in the anthropology department at the University of Utah. While at the University of Utah, I was particularly influenced by Robert Anderson (1976), who reportedly was Leslie A. White's first Ph.D. student and described himself, like his mentor, as a "culturologist." I still remember in a seminar Anderson telling us there were two important questions in anthropology. The second most important was how did social stratification emerge, noting that while we do not have all the answers to that question, there were social structural, demographic, and ecological factors that contributed to the shift from foraging to state societies and social stratification. When Anderson asked us what we thought the first most important question in anthropology was, none of us in the seminar knew exactly what to say, and he said, "How the hell do we get rid of it [namely, social stratification]"? I had the honor of being a teaching assistant for Anderson in his introductory cultural anthropology class. He perhaps more than anyone prompted me to shift from my early flirtation with psychological anthropology to a materialist perspective, although ultimately more of a historical materialist one than a cultural materialist one.

I began to call myself a "socialist" in my early thirties during my stint as an assistant professor at George Peabody College for Teachers (now part of Vanderbilt University) in Nashville, Tennessee. I told my students there that I was a socialist and have been doing so ever since, even though I have had many colleagues in various institutions who have said that I should have been more neutral about stating my political orientation. Given that much of my teaching career was spent in the American South, namely, Tennessee, Mississippi, and Arkansas, or the "Bible Belt," as it is known, my admission sometimes met with horror among my students. While openly declaring that one is a socialist or an anthropologist who draws on a historical materialist or neo-Marxian perspective probably does not raise as many eyebrows at the University of Melbourne as it did at the various U.S. Southern tertiary institutions where I had taught, even at Melbourne I have felt somewhat isolated in my theoretical and political stances. Ironically, as an academic, I felt the most comfortable during my stint in 1988–1989 as a Fulbright Lecturer at Humboldt University in East Berlin in what was then still the German Democratic Republic (Baer 1998). It is not that my colleagues were all enamored of the Socialist Unity Party of Germany to which many of them belonged or that they believed that the GDR had even achieved an authentic socialism. Indeed, many of them hoped that the winds of *glasnost* coming from the east would somehow translate into a drive to pave the way for a democratic socialism in the most technologically advanced of the Soviet bloc countries. The opening of the Berlin Wall in November 1989 and the reunification of the two Germanys in December 1990 short-circuited such hopes. In reality, however, the reunification was an *Anschluss*, or annexation, in which the Federal Republic of Germany absorbed the GDR. Indeed, after the reunification, many East Germans spoke of the process as a form of *Coloniserung*, or colonization.

The Implications of Climate Change for the Human Condition

In mid 2005, while I was working on an introductory medical anthropology textbook with Merrill Singer, the penny dropped for me in terms of the devastating impact of climate change on human societies and the role of global capitalism as the principal driver of anthropogenic climate change (Singer and Baer 2007, 2012). I have come to the conclusion that humanity is in an ever-growing crisis as it lurches ever deeper into the twenty-first century, probably the most critical century in our history as a species over the course of the past five to six million years. Various radical environmentalists, eco-socialists, eco-anarchists, and certain critical social scientists view climate change as yet one more manifestation of the contradictions, perhaps the most profound contradiction, of the capitalist world system. While humans indeed have been emitting greenhouse gases for many centuries, the mid-nineteenth-century Industrial Revolution, with its heavy reliance on fossil fuels, contributed to anthropogenic climate change, which some scholars trace back to the Agricultural Revolution. William Ruddiman (2005: 171) contends that "beginning in the late 1800s, use of fossil fuels (first coal, and later oil and natural gas) rapidly increased, eventually replacing deforestation as the primary source of CO_2 emissions by humans."

It has become increasingly apparent that climate change constitutes a major threat to human well-being and even survival for many people. The overwhelming majority of climate scientists have come to the conclusion that the warming of the planet and associated climatic events that the planet has been experiencing is largely anthropogenic or the result of human activities, particularly since the Industrial Revolution. Over the course of the past two decades or so, a slow but gradual interest in climate change has emerged as well in the social sciences, including my discipline of anthropology (Crate and Nuttall 2009, 2016; Dove 2014). As a critical anthropologist who has come to view myself more and more as a historical social scientist who happens to have a Ph.D. in anthropology, I have attempted to contribute to the anthropology of climate change, particularly the critical anthropology of climate change, and to contribute to a dialogue between anthropologists working on climate change and other social scientists and natural scientists working on the same topic (Baer and Singer 2009; Baer 2012; Burgmann and Baer 2012; Baer and Singer 2014).

The work of natural and social scientists across multiple disciplines has demonstrated that the earth is steadily warming; that human activities, especially since the Industrial Revolution, have been the dominant driver of this process; that the pace and effects of warming have been increasing; and that this change in the world we inhabit threatens significant, if not severe, consequences for human well-being on the planet. Despite these momentous and potentially dire developments, the governments of the world, as a whole, have been slow to respond to this pending threat, as seen in the failure of a series of international climate conferences designed to generate such a response. Moreover, while manufacturing and agrobusiness producers of greenhouse gases have developed a public discourse of *green capitalism*, continued emphasis on unceasing growth contradicts assertions that the world economic system can achieve sustainability. Indeed, as John S. Dryzek, Richard B. Norgaard, and David Schlosberg eloquently acknowledge, climate change has proven to be an extremely difficult challenge because "[t]he size of the threat calls into question received ideas about the inevitability of human progress: if progress requires continued economic growth based on ever-increasing emissions of greenhouse gases, then that kind of progress is clearly no longer sustainable. The economic stakes could not be higher, calling into question the future of industries such as coal and cars, and leading to deep political conflicts as those whose industries, profits, employment, and lifestyles feel threatened resist the necessary changes" (Dryzek, Norgaard, and Scholosberg 2013: 1).

At the same time, a corporate-sponsored climate change denial campaign has succeeded in sowing confusion, which, in turn, has contributed to lowering of public concern about climate change despite ever mounting scientific evidence that anthropogenic climate change is real and pressing fact. All of these events have produced a significant challenge for anthropological relevance and for Sidney Mintz's (1985: xxvii) vision of crafting an "anthropology of the present" that entails detailed examinations and critiques of "societies that lack the features conventionally associated with the so-called primitive." In that anthropogenic climate change will manifest profound impacts on human societies as they proceed further and further into the twenty-first century, the still evolving anthropology of the climate change can make an important contribution to the anthropology of the future.

Historically, anthropologists have concerned themselves with human societies of the distant past—the domain of archaeology—and of the recent past or present—the domain of sociocultural anthropology. In contrast to the age of the earth, the period that we humans have lived on this planet, roughly five to six million years, has been a blip in time. When we consider how long we have lived in farming and herding societies, some ten thousand years, or how long we have lived in state societies, some six thousand years, marked by different power relations and social stratification, our presence in such as social arrangements is a tiny fraction even of the already brief timeline of our species. The Intergovernmental Panel on Climate Change and other scenario setters often speak of what the state of humanity on this planet may be like in 2050 or 2100 but generally not beyond. As Sheila Jasanoff (2010: 241) observes, "[c]limate change invites humanity to play with time," including with the mind's eye into the future as we might imagine it will unfold. Over the past several decades, along with other social scientists, anthropologists have often alluded to a cavalcade of "posts": postcolonialism, postindustrialism, post-Fordism, postsocialism, postmodernism, poststructuralism, and postfeminism. Anthropologists might, indeed should, entertain the possibilities of two other "posts," namely, postcapitalism and postanthropogenic climate change. White anticipated the anthropology of the future quite early on when he wrote in a letter to Elsie Clews Parsons dated 13 June 1931: "It is possible to discern trends in the development of civilization, trends that are living in the present, and which will push into the future. It seems to me that no one is so well equipped as the anthropologist to understand and to foresee the future. And I think he [or she] should render this service to society as best he [or she] can" (quoted in Peace 2004: 77).

Bronislaw Malinowski (1941: 166) placed hope in the emergence of a "superstate, or, better, a federation or union of nations," an antidote to international wars, a vision that was partly fulfilled with the creation of the United Nations, but not one that successfully ended regional wars. White (1949: 164) predicted the eventual formation of a world state or something of that sort in his assertion: "Social evolution has been moving toward larger political units ever since the first human grouping was established, and the goal of a single world organization embracing the whole planet and the entire human race is now almost in sight," a prediction that in retrospect was premature. Conversely, he did not regard the United Nations as an instrument for creating a world state, noting this task would not be achieved by "frock-coated diplomats in a United Nations *opera bouffe*." On the issue of the "growth of great human communities," Raoul Naroll (1967: 85) had higher hopes for the United Nations, arguing that "behind a reformed and strengthened United Nations, the two superpowers will establish a world order through negotiation rather than conquest." Obviously he did not anticipate that less than a quarter of a century later, one of the superpowers as such would cease to exist, although its leading republic, namely, Russia, like the United States, continued to hold nuclear warheads. Thus, the threat of a nuclear holocaust still looms in the background, despite the end of the Cold War. Ruth Benedict (1971: 236) also commented on growing political centralization and stated that "we can project this upward curve into the future and recognize that someday mankind will organize the whole world."

In the 1970s and 1980s, various anthropologists grappled with future scenarios for humanity. Margaret Mead (1978) argued that anthropology has the potential to contribute to the "science of the future." Roger W. Wescott (1978: 513) argued that anthropology "has much to offer futuristics" and that the anthropology of the future constitutes a sub-discipline that could serve as a component of the emerging discipline of future study.

Eugene L. Ruyle (1978) delineated some of the features of a "socialist alternative for the future," including a participatory economy in which workers elect and control management; operate factories, farms, schools, hospitals, and other publicly owned enterprises; the replacement of money by socialist credit card system; and the popular election of all legislative, judicial, and higher administrative officials, subject to instant recall and paid at the same level as ordinary workers. *Cultural Futures Research* served as the official journal of future interest groups in the American Anthropological Association, the Society for Applied Anthropology, and the Futurology Commission of the International Union of Anthropological and Ethnographic Sciences (Riner 1998).

The demise of the Soviet bloc countries and the disillusionment with grand theory under the guise of postmodernism appear to have predisposed a younger generation of anthropologists to steer away from seemingly grandiose projects of attaining a better world based on both social justice and environmental sustainability. Yet a revival of the anthropology of the future strikes me as long overdue and appears to be re-emerging (Pels 2015; Hannerz 2015). A notable effort in this regard is *All Tomorrow's Cultures: Anthropological Engagements with the Future*, in which Samuel Gerald Collins (2008) argues that activist anthropology is involved in creating future change and offering alternative future scenarios to those defined by corporations and governments. Arjun Appadurai (2013) argues that anthropologists need to construct an understanding of the future by examining imagination, anticipation, and aspiration as important components shaping the future as a cultural fact and that the anthropology of the future is significant for the future of anthropology itself. Another recent effort to revive the anthropology of the future is an anthology edited by Australian anthropologists Jonathan Paul Marshall and Linda H. Connor (2015) titled *Environmental Change and the World's Futures*, for which I have contributed a chapter (Baer 2015a). Anthropologists need to recognize that despite the historical association of their discipline with colonial powers and imperialism, anthropology has a long tradition of fostering progressive causes. The search for an authentically progressive stage in social evolution harkens back to Lewis Henry Morgan's (1964: 467; original 1877) prediction of a "next higher plane of society," which will involve not only the "termination of a career of which property is the end and aim" but also a "revival, in a higher form, of liberty, equality and fraternity of the ancient gentes." Almost 100 years later, White made a similar prediction by arguing that: "if civilization is not destroyed by warfare ... there will undoubtedly be another political revolution and an entirely new type of society will be established. This new type will most likely be one in which the church-state will be nonexistent: a society in which there will be an administration of *things* rather than the governing of men: and again society will be organized on the basis of personal relations rather than property relations" (White and Dillingham 1973: 65).

In seeking to assess possible future scenarios with respect to climate change, one must consider the possibility of a dystopian future, a grim possibility that I explore in chapter 3, with the hope that this will contribute to the realization that serious mitigation efforts will require an alternative to the capitalist world system, one that is based on both social parity and environmental sustainability and that will allow humanity to reach a steady state for itself and other forms of biological life, both large and small. Tariq Ali (2013: 3) argues that history "rolls along at its own unpredictable pace" and may entail a revolutionary acceleration that moves humanity or a portion of humanity forward, but that this movement may be followed by reversals.

Wallerstein (1984) envisioned a world socialist system with a world government that would require a transition of about 100–150 years. Later Wallerstein (1998) asserted in *Utopistics* that capitalism has only perhaps fifty years or so left and that humanity may be heading toward a great historical transition that will probably be culminated in some type of socialist world government. Wallerstein (2007: 382) more recently has asserted that the present historical period is in a terminal structural crisis that will culminate in a "chaotic transition to some other system (or systems)" within the next twenty-five to fifty years, with the possibility of being, on the one hand, more equitable, or, on the other hand, more inequitable.

Overview of Remaining Chapters in this Book

While the contradictions of global capitalism are numerous, chapter 1 (The Contradictions of the Capitalist World System at the Beginning of the Twenty-First Century) focuses on what might be its principal contradictions in terms of social justice and environmental sustainability, namely: (1) profit making, economic growth, and the treadmill of production and consumption; (2) growing social inequality within and between nation-states; (3) population growth as a by-product of poverty; (4) depletion of natural resources and environmental degradation; (5) climate change; and (6) resource wars. Given that climate change scenarios prompt us to imagine dystopian visions of the future, I explore several mainstream and radical worst case scenarios that humanity must avoid in order to preserve itself as a species along with other species disappearing from planet.

In chapter 2 (Twentieth-Century Attempts to Create Socialism: Successes and Failures), I focus on the discrepancies between the ideals and realities of socialism as they played out during the twentieth century. While others have grappled with these discrepancies, I seek to provide a concise overview of them by examining efforts to create socialism in five contrasting countries, namely, Russia and the Soviet Union, China, the German Democratic Republic, North Korea, and Cuba. I also examine various interpretations that seek to determine the nature of postrevolutionary societies, whether they were instances of (1) "actually existing socialism" or some form of state socialism; (2) aborted transitions between capitalism and socialism; (3) state capitalism; or (4) new class societies. This chapter also examines selected positive and negative features of postrevolutionary societies, particularly in terms of the economy and workplace, social stratification, and environmental problems. The history of postrevolutionary or socialist-oriented societies over the course of the twentieth century proved to be a very mixed record. However, when one considers not only issues of social inequality but also the limits to growth, even a reformed and supposedly more environmentally friendly capitalism may spell the end of much of humanity. This strongly suggests that the concept of socialism must be rejuvenated to ensure social parity, democratic processes, and environmental sustainability for humanity.

The growing realization of the gravity of the global ecological crisis and anthropogenic climate change has prompted the development of numerous mainstream visions as well as countercultural visions of the future, which I explore in chapter 3 (Technoliberal and Countercultural Visions of the Future). While by no means exhaustive, this chapter examines future scenarios devised by the Global Scenario Group; Plan B devised by Lester R. Brown and Plan C developed by Pat Murphy; cosmopolitanism devised by sociologist Ulrich Beck; the future trends model devised by sociologist Stephen K. Sanderson; the Green New Deal model; various postgrowth or low-growth models; the climate emergency mobilization model; various counter-countercultural visions of the future; and Al Gore's views on the future. Ultimately, a shortcoming of some of these future scenarios is that they are premised primarily on ecological modernization, which advocates a shift to renewable energy sources and energy efficiency but does not adequately address issues of social parity. A shortcoming of the Green New Deal and postgrowth models is that they assume that some version of capitalism can function as a steady-state or zero-growth economy, when history tells us that capitalism is inherently committed to continual economic expansion as part and parcel of its pursuit of profits.

Chapter 4 (Efforts to Reconceptualize Socialism) argues that socialism remains very much a vision, one with which various individuals and groups continue to grapple, often by seeking to frame it in new guises. As humanity enters into an era of significant climate change accompanied by tumultuous environmental and social consequences, it will have to consider alternatives that will circumvent the dystopian scenarios depicted earlier. After briefly reviewing several Marxian-inspired future scenarios, this chapter seeks to reconceptualize socialism by examining the notions of democratic socialism, eco-socialism, and democratic eco-socialism. This chapter also critically examines efforts to create socialism for the twenty-first century in Venezuela, Bolivia, Ecuador, and Cuba, and examines the pros and cons of Amin's notion of *delinking* as a strategy for escaping the clutches of global capitalism.

Chapter 5 (The Role of Anti-systemic Movements in Creating a Socio-ecological Revolution) acknowledges that anti-systemic movements are sure to be a permanent feature of the world's political landscape so long as capitalism remains a hegemonic political-economic system. This chapter examines the role of specific anti-systemic movements, namely, the labor, ethnic and indigenous rights, women's, anti-corporate globalization, peace, and environmental and climate movements in creating a socio-ecological revolution committed to both social justice and environmental sustainability. Anti-systemic movements are a crucial component of moving humanity to an alternative world system but the process is a tedious and convoluted one with no guarantees, especially in light of the disparate nature of these movements.

While not seeking to create a blueprint per se for creating an alternative world system that will be manifested in different ways in the many societies around the world, chapter 6 (Transitional System-Challenging Reforms) proposes several system-challenging reforms that potentially could facilitate a transition from the present existing capitalist world system to a democratic eco-socialist world system. These include: (1) the creation of new left parties designed to capture the state; (2) emissions taxes at the sites of production; (3) public and social ownership of the means of production; (4) increasing social equality and achieving a sustainable population size; (5) workers' democracy; (6) meaningful work and shortening the work week; (7) challenging or rethinking the growth paradigm (8) energy efficiency, renewable energy sources, appropriate technology, and green jobs; (9) sustainable public transportation and travel; (10) sustainable food production and forestry; (11) resisting the culture of consumption and adopting sustainable and meaningful consumption patterns; (12) sustainable trade; and (13) sustainable settlement patterns and local communities.

Finally, in chapter 7 (Conclusion), I argue that as humanity proceeds into the twenty-first century, our survival as a species appears to be more and more precarious, particularly given that the impact of climate change in a multiplicity of ways looms on the horizon. More so than has ever been the case, it is essential for critical scholars, including anthropologists and other social scientists, to envision possible future scenarios and strategies for achieving an alternative world system based on principles of social justice, democracy, and environmental sustainability, regardless of how we term our real utopian vision. Perhaps more important is developing strategies to shift from the existing system of globalized capitalism to an alternative that transcends its numerous contradictions and limitations.

CHAPTER 1

The Contradictions of the Capitalist World System at the Beginning of the Twenty-First Century

Capitalism as an economic system oriented to profit making or the creation of surplus value emerged in Western Europe some five hundred years ago but has since then diffused to virtually every humanly inhabitable part of the world; thus, Wallerstein (2004) and other world systems theorists refer to it in the modern context as a "world system" and is often referred to as "global capitalism." Its drive for profits requires ongoing accumulation, expansion, and economic growth. Global capitalism systematically exploits human beings and the natural environment in pursuing its aims, despite the rhetoric that it contributes to the prosperity and well-being of all human beings, albeit some more than others. It fosters an incessant treadmill of production and consumption primarily for the purpose of generating profits for a few and, in the process, because they are rated of lesser importance relative to profit making, sacrifices basic needs and environmental sustainability. Machines of all sorts have played a central role in sustaining particularly industrial capitalism. Anthropologist Alf Hornborg (2001: 2) contends that machine "power" entails "power to conduct work, power over other people, and power over our minds." Conventional proponents of capitalism laud its technological feats and proclaim that they eventually will result in material prosperity.

Capitalism as a global economic system contains numerous contradictions. The term contradiction generally refers to a logical incompatibility or inconsistency between two or more propositions. In Marxian terms, a contradiction refers to a tension among social forces, such as a tension between the forces of production or the techno-economic apparatus and the relations of production, such as in who owns and controls the means of production, namely, the capitalist class or the bourgeoisie, and who does not but nevertheless works within them, namely, the working class, simply to subsist and survive. For Marx, this contradiction would cease in part with the culmination of socialism, when the working class would become the dominant class (thus, the "dictatorship of the proletariat") and would cease eventually with the culmination of communism. Like previous social formations, capitalism has numerous contradictions. Anthropologist John H. Bodley (2012) contends that capitalism has produced a global crisis that has resulted in many social problems, including ongoing population growth, overconsumption, social stratification, environmental degradation, militarism, crime, and many personal crises, both physical and mental.

For purposes of this book, I focus on the following contradictions emanating from global capitalism: (1) profit making, economic growth, and the treadmill of production and consumption; (2) the growing socioeconomic gap; (3) poverty and population growth; (4) depletion of natural resources and environmental degradation; (5) climate change; (6) and resource wars.

Beginning in the late 1970s and particularly after the collapse of the Soviet bloc in 1989–1991, a late stage of capitalism, namely, neoliberalism, took hold around the world. Neoliberalism undermined much of the welfare state in developed societies but also developing societies by undercutting government regulation of capitalist economies and pushing for reductions in public social services. Ganti states: "An institution deeply connected to the transformed nature of the state, specifically its role in social welfare and development is the NGO [non-government organization]. The growth of NGOs has been an essential feature of the decentralized and privatizing political-economic landscape associated with neoliberalism. NGOs are a critical feature of the global political economy as more development aid is channelled to the Global South through NGOs than through the World Bank and the IMF combined" (Ganti 2014: 97).

Despite the ascendancy of neoliberalism, in reality capitalist corporations and enterprises and the state are so intricately interwoven that one could argue that around the world we see numerous national variants of state capitalism, including an American state capitalism, a Japanese state capitalism, an Australian state capitalism, and even a Chinese state capitalism, a point to which I will return in chapter 3. As Boswell and Chase-Dunn (2000: 86–87) observe, "all societies are 'state capitalist,' if what is meant by this elusive term is using the state to promote economic development within the capitalist world-system." While multinational or transnational corporations are at the center of the capitalist world system in today's world and their primary objective is profit making, which they increasingly achieve through globalization of their operations, each multinational corporation remains closely tied to its country of origin, which generally is a Western country, but may also be a non-Western country, such as Japan, China, or India.

Profit Making, Economic Growth, and the Treadmill of Production and Consumption

Capitalism is a global economic system that in its drive for profits requires ongoing accumulation and economic growth or expansion. According to Harvey (2014: 222), "[c]apital is always about growth and it necessarily grows at a compound rate." Various technological optimists and environmentalists maintain that economic growth can be decoupled from pollution and greenhouse gas emissions, a claim that has not been borne out to date (Hawken, Lovins, and Lovins 1999; Brown 2008).

Capitalism systematically exploits human beings, some more than others, and nature or the natural environment in pursuing its aims. According to Minqi Li, "Even if is technically feasible to make climate stabilizations compatible with economic growth, efforts aimed at climate stabilization are likely to face insurmountable political obstacles under the capitalist world system. Capitalism is a system based on inter-state competition, which is essential to secure favourable political conditions for capital accumulation.... Within the world system, states are under constant pressure to compete against each other economically and militarily" (Li 2009: 1043).

Global capitalism also fosters a treadmill of production and consumption primarily for the purpose of creating profits for a few, while at the same time sacrificing the well-being of billions and the sustainability of the natural environment. Capitalism relies heavily on advertising to convince people around the world that they need a wide array of products that they often do not need per se. It creates what Richard Hofrichter (2000: 1) terms a *toxic culture*, which includes the "unquestioned production of hazardous technologies, substandard housing, chronic stress, and exploitative working conditions." The treadmill of production and consumption associated with global capitalism that results in environmental degradation, greenhouse gas emissions, and climate change contributes to the toxic culture of capitalism at the macro level.

Consumer capitalism began to take off in a profound way beginning in the 1950s when households in developed societies were flooded with energy-intensive appliances and devices, including electric cookers, washing and drying machines, refrigerators, toasters, electric irons, microwave ovens, electric toothbrushes, electric razors, televisions, computers, printers, electric tools, power lawn mowers, cassette players, leaf blowers, elaborate lighting systems, and so on. Capitalism, with its predilection for built-in obsolescence, encourages people to update older models of appliances and entertainment devices, such as regularly occurs with new models of plasma televisions, CD and DVD players, mobile and smart phones, computers, and many other items. For a moment I will single out one consumer item, namely, the computer that I have heavily relied on in the writing of this book. Richard H. Robbins, who has written a provocative anthropological textbook that focuses on the global problems of capitalism as a cultural system, observes that the construction of a computer starts with the silicon chip: "There are 220 billion chips manufactured each year that contain highly corrosive hydrochloric acid: metals such as arsenic, cadmium, and lead; volatile solvents such as methyl chloroform, benzene, acetone, and trichloroethylene (TCE); and a number of super-toxic gases. A single batch of chips requires on average twenty-seven pounds of chemicals, nine pounds of hazardous waste, and 3,787 gallons of water requiring extensive chemical treatment" (Robbins 2011: 338).

Furthermore, in 2007, there were more than 500 million obsolete computers in the United States, many of which would end up in landfills, incinerators, or hazardous waste sites in developing countries.

Although developed countries constitute the leading cultures of consumption, various developing countries, such as China, South Korea, India, Brazil, and Mexico, have quickly been joining the pack, thus demonstrating that capitalism promotes growth around the world. Capitalism fosters not only the production of material commodities such as large dwelling units, private automobiles, televisions, and clothing, but also immaterial goods, such as communication, entertainment, information, and travel, including to distant places.

The Growing Socioeconomic Gap

Marx was probably, more than anyone, cognizant of the fact that capitalism produces incredible social equalities. He wrote: "It is true the labour produces wonderful things for the rich—but for the worker it produces privation. It produces palaces—but for the worker, hovels. It produces beauty—but for the worker, deformity. It replaces labour by machines, but it throws one section of the workers back into barbarous types of labour and it turns the other section into a machine. It produces intelligence—but for the worker, stupidity, cretinism" (Marx 1977: 65).

Capitalism subscribes to the "trickle-down theory," which asserts that the wealth generated at the top will eventually trickle down to people at the bottom, thereby lifting them out of their dismal condition. In reality, while some people in lower classes have seen their overall standard of living improve under capitalism, widespread poverty continues to persist, at least in relative if not absolute terms. Apologists for capitalism often assert that that "everyone's boat is rising," although some boats are rising faster than others. The *Human Development Report 2013* claims that on the basis of the Human Development Index (HDI), a "composite measure of indicators along three dimensions: life expectancy, educational attainment and command over the resources needed for a decent living," all countries and regions "have seen notable improvement in all HDI components, with faster progress in low and medium HDI countries" (United Nations Development Programme 2013: 1). Thus, at a certain level, this fact would appear to bear out the frequent assertion about "rising boats." Conversely, reality is generally much more complex, as the report itself observes: "[N]ational averages hide large variations in human experience. Wide disparities remain within countries of both the North and the South, and income inequality within and between many countries has been rising" (United Nations Development Programme 2013: 1).

The report, hardly a radical document, goes onto state: "[O]ne can ... state that there is a 'south' in the North and a 'north' in the South. Elites, whether from the North or the South, are increasingly global and connected, and they benefit the most from the enormous wealth generation over the past decade, in part due to accelerating globalization. They are educated at the same universities and share similar lifestyles and perhaps values" (United Nations Development Programme 2013: 2).

The report also mentions some other grim realities, including that an "estimated 1.57 billion people, or more than 30% of the population of the 104 countries studied for this Report, live in multidimensional poverty, a measure of both the number and the intensity of overlapping human deprivations in health, education and standard of living" (United Nations Development Programme 2013: 13–14). Furthermore, out of 132 countries, almost a quarter increased in inequality as measured in terms of inequality-adjusted Human Development Index and that sixty-six countries experienced only a marginal decline in overall inequality, "because declining inequality in health and education was offset by rising inequality in income" (United Nations Development Programme 2013: 14).

Branko Milanovic, a renowned World Bank economist, conducted a sophisticated analysis in which he found: "The recalculations of international and global inequalities, using the new PPPs [purchasing power parity rates], show that inequalities are substantially higher than previously thought. Inequality between global citizens is estimated at 70 Gini points rather than 65 as before" (Milanovic 2012: 1).

Table 1.1 below indicates GDP per capita based on the old figures and new ones for the thirteen most populous countries in the world.

Table 1.1. New GDP (US$) per Capita Compared to the Old (World Development Indicators) for the Thirteen Most Populous Countries in the World*

Country	GDP per capita based old PPPs	GDP per capita based on new PPPs
Vietnam	3,106	2,143
Philippines	4,991	2,956
Mexico	10,356	11,387
Japan	31,262	30,290
Nigeria	1,200	1,520
Bangladesh	2,025	1,068
Russia	11,053	11,858
Pakistan	2,437	2,184
Brazil	8,854	8,474
Indonesia	3,898	3,209
USA	42,454	41,813
India	3,556	2,222
China	6,666	4,088

Source: Adapted from Milanovic (2012: 6).

These new figures reveal that the upward social mobility often touted by neoliberal economists for China, India, Bangladesh, and Vietnam were grossly inflated.

An Internal Revenue Service analysis reports that income tax returns in New York City in 2012 revealed "that the average income of the top 1 per cent in that year was $3.57 million, which half of the population in this extremely high-rent and high-cost-of-living city were trying to get by on $30,000 a year or less" (Harvey 2014: 164). Relative poverty has been on the rise in many developed countries. In the United States, the population with an income below the poverty level increased from 11.1 percent in 1973 to 15 percent in 2011 (National Public Radio 2012). The percentage of the population "at risk of poverty" in seventeen EU countries between 2005 and 2011 increased from 12.2 to 16.2 overall; from 12.2 to 15.8 percent in Germany, from 19.7 to 21.8 percent in Spain; and from 13 to 14 percent in France (Eurostat 2012). An OECD (2011: 24) study indicated that the Gini coefficient increased in seventeen out of twenty-two developed countries between 1985 and 2008, remained roughly the same in three, and decreased in only two, Greece and Turkey.

An Oxfam media briefing reported that the share of national income earned by the top 1 percent doubled from 10 percent in 1980 to 20 percent recently and that there are now some 1,200 billionaires in world, who are not only found in the developed regions and countries in North America, Europe, Japan, and Australia, but also in developing countries, such as China, India, Brazil, and Mexico (Harvey 2014: 169).

In reality, income inequality only reveals a partial picture about social inequality, whereas total wealth inequality presents a far grimmer picture. In March 2011 the *Wall Street Journal* reported that the total net wealth of richest four hundred Americans accounted for 40 percent of all U.S. wealth (Rothkopf 2012: 8). Robinson reports:

> The annual *World Wealth Report* published by Merrill Lynch and Capgemini identifies what it terms High-Net-Worth Individuals, or HNWIs, those people who have more than $1 million in free cash, not including property and pensions. The 2011 report identified some ten million of these HNWIs in 2010, concentrated in North America, Europe, and Japan, with the most rapid growth among the group taking place in Asia-Pacific, Latin America, Eastern Europe, Africa, and the Middle East. The collective wealth of the HNWIs surpassed $42 trillion in that year. Well over double of what it had been ten years earlier, and 10 percent higher than the previous year (Robinson 2014: 61).

The richest one thousand people in the world have assets roughly equal to the wealth of the two billion poorest (Rothkopf 2012: 19). In 2013 *Forbes* magazine reported that Bill Gates was the richest person in the world with a net worth of some $76 billion, with Carlos Slim Helu a close second with a net worth of about $72 billion, most of which he acquired as a result of the privatization of Mexico's telecommunication industry (Sayer 2016: 8). Christy Walton, who had inherited part of the Walmart wealth, was the richest woman with a net worth of $52.5 billion. Rupert Murdoch, the global mass media magnate, came in at seventy-eighth place, with a net worth of $13.5 billion. The rise of the superrich over the past four decades has been closely linked with the "shift from those whose money comes primarily from control of the production of goods and services to those who get most of their income from control of existing assets that yield rent, interest, or capital gains," including financial speculation (Sayer 2016: 18).

High economic growth rates in Brazil, Russia, India, and particularly China (the BRIC countries) have released millions from poverty, although at a heavy cost in terms of pollution and greenhouse gas emissions resulting from of the manufacture of products, many of which are exported to the developed countries. While on the surface it appears that disparities in global distribution of wealth and income between countries have been reducing due to rising per capita income in many developing parts of the world, except for Africa, at the same time most countries in the world have been witnessing increases in income and wealth between the rich and poor. Extreme wealth does have its downsides for the privileged of the world. According to Danny Dorling, "The multimillionaires' biggest fear is not a slowdown to economic growth or a rise in terrorism, other than personal risk of kidnap. As inequalities within certain countries grow, it is threats to their private property claims that keep the rich awake at night" (Dorling 2013: 172).

Oxfam (2015: 2), hardly a radical NGO, reports that "In 2014, the richest 1% of the people in the world owned 48% of global wealth" and that the wealth of the eighty people listed by *Forbes* as the richest people in world had a total wealth of $1.9 trillion in 2014 and had as much wealth as 50 percent of the world's population. The following year Oxfam reported the following grim statistics:

- "In 2015, just 62 individuals had the same wealth as 3.6 billion people—the bottom half of humanity. This figure is down from 388 individuals as recently as 2010"
- "Since the turn of the century, the poorest half of the world's population has received just 1% of the total increase in global wealth, while half of that increase has gone to the top 1%."
- "The average annual income of the poorest 10% of the people in the world has risen by less than $3 each year in almost a quarter of a century" (Oxfam 2016: 2).

On a more positive note, there have been some reductions in social inequality in various Latin-American countries due to progressive state policies. In Venezuela, the poverty and extreme-poverty rates fell from 48.6 percent in 2002 to 28.5 percent and 8.5 percent by 2007 under the guise of the Bolivarian Revolution, which is committed to "socialism for the 21st century" (Webber 2011: 322). However, this sort of trend has occurred under less radical or center-left regimes such as Argentina, Chile, Costa Rica, and Uruguay as well.

Poverty and Population Growth

According to Daniela Danna (2014: 213), "The capitalist world-economy ... has generally welcomed population growth ever since mercantilist states' need for expanding population was openly theorized." Europeans migrated to newly conquered lands in the Americas beginning in the sixteenth century and then to Australasia beginning in the late eighteenth century. The population of humanity from its dawn some five to six million years ago and since the dawn of modern humans around 200,000 years ago gradually grew, but particularly with the Agricultural Revolution and the beginnings of the Industrial Revolution. Oil-dependent or industrial agriculture contributed to population growth as did the Green Revolution since "the need for more hand labor grew because only jobs performed by men were mechanized, counting on unrewarded women's and children's labor for the labor-intensive tasks such as weeding in the fertilized crop fields" (Danna 2014: 216).

It is estimated that humanity hit its first billion people in 1820, its second billion in 1926, its third billion in 1960, its fourth billion in 1975, its fifth billion in in 1988, and its sixth billion in 2000 (Dorling 2013: 104–108). In the period during which humanity increased its population from six billion to seven billion people by December 2011, as Dorling (2013: 153) observes, "a global war of terror had begun, there were two economic crashes, the banking system almost collapsed from the latter; world food prices soared; hundreds of millions more people went hungry; the poor became absolutely poorer; there was again absolute immiseration; and the world's super-rich continued their advancement up the upper slopes of Mount Olympus to take on what initially appeared to be a god-like status."

The capitalist world system has required a labor force to sustain its drive for profits and economic expansion. Workers respond to the dictates of capital by having more children, which is also a means of economic survival and to address poverty that emanates from their exploitation. Capitalist economies encourage population growth because a growing population means more production of whatever: houses, furniture and appliances, cars, and on and on. As Murdoch (1980: 307) argues, "Rapid population growth and inadequate food supply are but the symptoms of poverty." Furthermore, population growth is a consequence of a lack of women's rights, patriarchy as exemplified in traditional religions, including Judaism, Christianity, Islam, Hinduism, and Buddhism, and restricted access to contraception and information about birth control. Table 1.2 below depicts UN-projected population trends for world regions between 2010 and 2050.

While Asia will remain the most populated region, Africa, the poorest region in the world, will undergo the greatest increase in population percentagewise, thus illustrating the role of poverty in stimulating population growth. The UN predicts a peak in Asia's population at 5.2

Table 1.2. Projected Demographic Trends during the Twenty-First Century

	Total population 2010 (millions)	% of total 2010	Projected population 250 (millions)	% of total
World	6,908.7	100.0	9,150.0	100.0
Africa	1,033.0	14.9	1,998.5	21.8
Asia	4,166.7	60.3	5,231.5	57.2
Europe	732.8	10.6	691.0	7.6
Latin America & Caribbean	588.6	8.5	729.2	8.0
North America	351.7	5.1	448.5	4.9
Oceania	38.8	0.6	51.3	0.5

Source: United Nations Population Fund (2010).
(Wainwright and Mann 2012: 9)

billion in 2052, while it predicts that Africa's population will rise to 3.6 billion by 2100. The case of sub-Saharan Africa illustrates the observation of demographers and other scholars who assert that the greatest rate of population growth tends to occur among the poorest people on the planet. Poston and Bouvier state: "An ever-growing proportion of [the global] population resides in the developing regions of the world. More than 95 percent of the projected growth in the world population between 2010 and 2050 is expected to occur in the developing countries. By 2050, 85 percent of the world's projected population will hail from developing countries. This population growth will come disproportionately from people on the margins, those with limited resources and life opportunities" (Poston and Bouvier 2010: 368).

While upper- and middle-class people worldwide frequently are perplexed as to why the poor have more children—because the affluent realize that financially supporting children to adulthood and often beyond constitutes a massive expenditure—poor people in developing countries, both in rural and urban areas, commonly view children as breadwinners who add to the family coffers. As Sachs (2008: 16) observes, "[p]overty contributes to high fertility rates, while high fertility rates prolong poverty." Peasants in India and other developing countries particularly welcome additional sons because they can help work the land. Even in urban areas, children are viewed as an asset among the poor. Boys in Mexican cities, for example, often work as shoe shiners, and both young boys and girls sell items such as chewing gum, tortillas, and other foods prepared by their mothers. Poor children often sell handicrafts produced by their parents on the street to tourists from developed societies. Among the Yoruba of northern Nigeria, young children carry water, collect fuel, transmit messages, sweep, tend younger siblings, care for animals and older children take care of aging parents (Robbins 2011: 148). Furthermore, "[c]ultural factors encourage people to see children as an affirmation of family values, a guarantee of family continue, and the expression of religious principles" (Robbins 2011: 148). Islam and Mormonism are only two examples of religions that encourage their adherents to have large families.

Historically, capitalist penetration, starting out with European colonialism, of indigenous societies and precapitalist state societies contributed to population increases, as local households struggled to meet the demand for taxes imposed by colonial powers. In this context, having more laborers and wage earners became an important household strategy for survival in a globalizing capitalist system. Rural household financial difficulties often pushed males, sometimes pushed females, and even pushed entire families to migrate to urban areas and foreign countries to find employment. As Angus and Butler (2011: 212) observe, "High birth rates aren't the cause of third world poverty—they are the effect of poverty, and building birth clinics, however important that is for other reasons, won't eliminate the underlying causes." Containing population growth ultimately will entail creating a much more equal playing field in terms of access to resources of all sorts. Indeed, Bodley (2008: 193), who views increasing population growth as "one problematic symptom by a world dominated by commercially organized societies," maintains that instead of "focusing on economic growth, the world needs to focus on socio-cultural development and the politics of wealth distribution to meet existing human needs."

While population growth does place pressure on various resources and does contribute to environmental degradation and greenhouse gas emissions, it would undoubtedly level out or actually be reversed in a world where the vast majority of people had access to sufficient basic resources and where a high degree of social equality had been achieved. Indeed, this is what has happened in Japan and to various northern European countries, the latter of which have in part offset their declines in population by allowing immigrants to work and even settle in them. If it were not for a large influx of immigrants to Australia, its population would decline. Studies have repeatedly indicated that when people achieve a reasonable degree of material well-being, the birth rate declines in all countries around the world. Better nutrition contributes to a lower infant and child mortality rates, thus reducing the need for "insurance children" who will look after parents in their old age. Fred Magdoff reports: "The poorest

40 percent of people on Earth are estimated to consume less than 5 percent of natural resources. The poorest 20 percent, about 1.4 billion people, use less than 2 percent of natural resources. If somehow the poorest billion people disappeared tomorrow, it would have barely noticeable effect on global natural resource use and pollution. (It is the poor countries, with high population growth, that have low per capita greenhouse gas emissions). However, resource use and pollution could be cut in half if the richest 700 million lived at an average global standard of living" (Magdoff 2014: 24).

Depletion of Natural Resources and Environmental Degradation

While to a greater or lesser extent all human societies have historically encroached on the natural environment, global capitalism has done so at a much more pronounced level. In essence, capitalism lacks a concept of sufficiency, of individuals having enough, and demands perpetual economic growth. It operates with the assumption that natural resources are infinite, when in reality many of them are finite. Capitalism contributes to the depletion of natural resources and environmental degradation, including climate change. While in their thorough and critical analysis of capitalism Marx and Engels did not focus on ecological issues, they were certainly cognizant of the dialectical relationship between sociocultural systems and the natural environment. In *The Condition of the Working Class in England,* Engels (1969; original 1845) described the devastating impact of industrialization on the natural environment. As Merchant (2005: 146) observes, "Marx gave numerous examples of capitalist pollution: chemical by-products from industrial production; iron filings from machine tool industry; flax, silk, wool, and cotton washes in the clothing industry; rags and discarded clothing industry; rags and discarded clothing from consumers; and the contamination of London's River Thames with human waste." In essence, global capitalism has created an ecological disruption or a "metabolic rift" between society and nature, which is "alienating us from the material-natural conditions of our existence and from succeeding generations" (Foster, Clark, and York 2010: 7). Capitalism has been reliant on what Moore (2015: 155) terms the "Four Cheaps," namely, cheapened raw materials, cheap coal, cheap iron, and cheap steel, all of which exist in limited supply.

Johan Rockstroem and his colleagues at the Stockholm Resilience Center identify nine "planetary boundaries" that are essential to maintaining a global ecosystem in which humanity can survive and thrive safely:

(1) climate change, which is on the verge of reaching a tipping point; (2) ocean acidification, also on the verge of reaching a tipping point; (3) stratospheric ozone depletion, also close to a tipping point; (4) disruption of natural nitrogen and phosphorus cycles; (5) human pressure on global freshwater availability; (6) conversion of forests, wetlands, and other vegetation types to agricultural lands; (7) biodiversity loss; (8) atmospheric aerosol loading; and (9) toxic chemical pollution (Rockstroem et al. 2009). These boundaries constitute the "safe operating space for humanity with respect to the Earth system" but exceeding them would "push the Earth outside the stable environmental state of the Holocene, with consequences that are detrimental or even catastrophic for large parts of the world" (Rockstroem et al. 2009: 472). For climate change, the boundary is 350 ppm (parts per million) of CO_2 in the atmosphere, but this boundary has been surpassed at over 400 ppm of CO_2 and continues to rise because serious mitigation efforts have not been in put in place, despite the fact that the Kyoto Protocol was passed in 1997. According to Dryzek et al. (2013: 116), "Just as the *Limits to Growth* argument of the 1970s had little or no impact on the policies of governments, so arguments for planetary boundaries to human activity have thus far had little effect in modifying responses to climate change at any let alone global level," including the 2012 United Nations Conference on Sustainable Development in Rio de Janeiro. Foster, Clark, and York maintain that:

> The development of capitalism, whether though colonialism, imperialism, or market forces, expanded the metabolic rift to the global level, as distant regions across the oceans were brought into production to serve the interests of capitalists in core nations. While incorporating distant lands into the global economy—a form of geographical displacement—helped relieve some of the demands placed on agriculture production in core nations, it did not serve as a remedy to the metabolic rift. The systematic expansion of production on a large scale subjected more of the natural world to the dictates of capital. The consequence of this is that "it disturbs the metabolic interaction between man and the earth" (Foster, Clark, and York 2010: 78).

Foster provides a long list of "warning signs" signaling the environmental unsustainability of the capitalist world system, which I summarize in condensed form below:

- There is virtual certainty that the conventionally critical threshold of 2°C will be crossed in the near future.
- Various studies suggest that with each 1°C increase in global temperature, rice, wheat, and corn yields could drop 10 percent.

- With the world approaching peak oil production, a global crisis and mounting resource wars appear to be likely (Foster 2009: 254–255).
- Global freshwater supplies are diminishing due to the drawing down of aquifers in many locations.
- Two-thirds of the global fish stocks are being depleted.
- Species extinction is at the highest level that it has been in 65 million years.
- The National Academy of Sciences in the United States released a study in 2002 that indicated that humanity had exceeded the earth's regenerative capacity in 1980 and by 1999 exceeded it by as much as 20 percent.
- Just as various earlier civilizations and societies have experienced ecological collapse, the capitalist world system faces a similar fate.

Capitalist production has played a central role in a dramatic increase in greenhouse gas emissions as well the creation of a global water crisis. As for the first of these developments, the Working Group I component of the 2013–2014 Intergovernmental Panel on Climate Change Fifth Assessment reports the following grim scenario regarding the increase of greenhouse gas emissions: "The atmospheric concentrations of carbon dioxide, methane, and nitrous oxide have increased to levels unprecedented in at least the last 800,000 years. Carbon dioxide concentrations have increased by 40% since pre-industrial times, primarily from fossil fuel emissions and secondarily from net land use change emissions. The ocean has absorbed about 30% of the emitted anthropogenic carbon dioxide, causing ocean acidification" (Alexander et al. 2013: 11).

Whereas only 8 percent of freshwater supplies are utilized for domestic use, industry consumes 22 percent of the world's water and irrigation agriculture uses 70 percent (Parr 2013: 65). In contrast to a global average water footprint of 1,243 cubic meters per capita per year, "[t]he per capita annual footprint of China is 702 cubic meters, with only 7 percent of Chinese water consumption coming from offshore. The water footprint of somebody living in the United States is 2,483 cubic meters per year, with 19 percent of that consumption coming from outside the country. Meanwhile, in Kenya the average per capita water footprint per year is just 714 cubic meters, with 10 percent coming from outside the country" (Parr 2013: 65–86).

In addition to wide discrepancies of water use varying from country to country, there is tremendous class variation in access to freshwater, varying from households that have swimming pools and large lawns to people who have limited access to water, particularly clean water.

Capitalism and Climate Change

Marx's metabolic analysis can be extended to an examination of global climate change involving the linkage of three dimensions: (1) the impact of capitalism on the global carbon cycle; (2) the Jevons paradox, in which technological improvements in energy efficiency lead not to decreased but to increased utilization of natural resources, and thereby contribute to environmental degradation; and (3) the role of capitalism in the destruction of carbon sinks (Foster et al. 2010: 122). As is evident from the imperiled list of planetary boundaries, climate change is not happening in a vacuum but is growing at a time of multiple and other anthropogenic eco-crises, some of which, like ocean acidification and stratospheric ozone depletion, are verging on their own respective tipping points.

The 2013 Intergovernmental Panel on Climate Change Working Group 1 report projects a rise in global temperature anywhere from 0.4°C to 2.6°C by 2046–2065 and from 0.3°C to 4.8° C by 2081–2100, depending on what strategies that humanity takes to mitigate anthropogenic climate change. Reports of the World Bank (2012) and Price Waterhouse Coopers (2012), hardly radical organizations, indicate that the world is presently heading to 4°C or more of warming this century. This global average could mean around 6°C of warming over land, and perhaps 7–8°C in far northerly latitudes, mind-boggling possibilities to say the least. NASA reports the warming trend of the 2000s occurred despite the fact that there was a solar output decline, with a minimum in 2007–2009. Sixteen of the seventeen warmest years on record have occurred since 2001, with 2016 being the warmest years on record. NASA (2017) reports: "Earth's 2016 surface temperatures were the warmest since modern recordkeeping began in 1880, according to independent analyses by NASA and the National Oceanic and Atmospheric Administration (NOAA) ... "2016 is remarkably the third record year in a row in this series," said GISS Director Gavin Schmidt. "We don't expect record years every year, but the ongoing long-term warming trend is clear."

Global warming and associated climatic events are direct effects of an increase of greenhouse gas emissions since the Industrial Revolution, which kicked off reliance on fossil fuels, initially coal but later petroleum and natural gas. Reportedly, 556 billion metric tons of CO_2 emissions have been added to the atmosphere since 1750 as a result of fossil fuel consumption and land cover change due to increased agricultural production and deforestation. While climate scientists have debated for a long time whether recent climate change is primarily a natural phenomenon rather than an anthropogenic one, the vast majority of them now agree that it is has been largely created by the emission of various greenhouse gases, particularly carbon dioxide, nitrous oxide, and methane. Doran and Zimmerman (2009) conducted a survey in which they found that 97.4 percent of the climatologists and 82 percent of the earth scientists in their sample maintain that human-related activities are a significant factor in increasing global temperatures. Despite the existence of many climate denialists or skeptics, particularly in the United States and Australia, they argue that the "debate on the authenticity of global warming and role played by human activity is largely non-existent among those who understand the nuances and scientific bases of long-term climate processes" (Doran and Zimmerman 2009: 23).

Atmospheric CO_2 alone has increased from 280 ppm at the time of the Industrial Revolution to 406.17 ppm in April 2017 (NASA 2017). Whereas during the 1990s acceleration of CO_2 emissions stood at about 1.1 percent per annum, since 2000 the acceleration of CO_2 emissions has been about 3 percent per annum (more than 2 ppm per year) (Rogner et al. 2007). In a similar vein, Friedlington et al. report: "In the past decade (2004–2013) global CO_2 emissions have had continued strong growth of 2.5% yr. This growth rate was below the 3.35% yr averaged over 2000–2009 because of the lower 2.4% yr growth rate since 2010" (Friedlington et al. 2014: 710).

China in particular has become the factory of the world, a place where labor costs are much cheaper than developed countries. According to Malm (2016: 329), "[b]etween 2000 and 2006, 55 percent of the global growth of CO_2 emissions happened there: in 2007, the figure stood at *two-thirds*." An increasing percentage of China's emissions are "export emissions" generated in the process of largely exporting manufactured products to developed countries in North America, Europe, and Australia, which do not generally count "import emissions" in their totals of emissions for which they are responsible. Malm reports: "In 2001, China entered the WTO, dismantled the remaining barriers to investment, abolished restrictions on foreign ownership, relaxed requirements on local cooperation and, in general, flung the gates wide open: then the real explosion began. While a third of the increase in Chinese CO_2 between 1990 and 2002 could be directly attributed to export, the share rose to *half* in the following three years; moreover, according to one estimate, as 48 percent of the country's *total* emissions between 2002 and 2008 were generated in the export sphere" (Malm 2016: 331).

Multinational corporations and state and joint-venture companies in postrevoluntionary societies, particularly China, have created a global factory and a new global ecosystem that includes industrial and motor vehicle pollution, toxic and radioactive wastes, deforestation, desertification, and climate change. As a result of its emphasis on ever-expanding production, global capitalism has been having adverse impacts on large sections of much of humanity and a fragile ecosystem. Global capitalism is heavily reliant on fossil fuels, which prompted Lewis Mumford (1967) to refer to the world system as a form of "carboniferous" capitalism. The "industrial-fossil revolution" or "fossil capitalism" relied first on coal, then oil, and finally natural gas. As Altvater observes, "Without the continuous supply and massive use of fossil energy modern capitalism would be locked into boundaries of biotic energy (wind, water, bio-mass, muscle-power, etc.). Although something like capitalist social forms occasionally could be found in ancient societies (in Latin America as well as in Europe), they could not grow and flourish with fossil energy" (Altvater 2006: 6–7).

Although coal consumption dates back to the China as early as 1,000 B.C., the Industrial Revolution relied very heavily on coal. As Edwin Black observes, "For mankind, coal was both a blessing and a curse. This black combustible was plentiful in the ground. It was also poison in the air" (Black 2007: 19). Not only does coal contribute to air and water pollution, but it is a major source of CO_2 and contributor to climate change.

Anthropologist Leslie A. White (2008: 118) maintained: "Coal might well be called the king, or the father, of the Fuel Revolution but two other fossil fuels have played important roles in this drama, also: petroleum and natural gas." The invention of the internal combustion engine in particular contributed to the significance of petroleum and the motor vehicle in the modern era. It also came to fuel train locomotives, ships, and airplanes. In addition, oil became an essential ingredient in the production of plastics and in many cases replaced coal in the heating of dwelling units, offices, and factories. Of all the countries in the world, as none other than George W. Bush quipped, "America is addicted to oil." Oil serves as an important energy source in transportation, asphalt production, mechanized farm equipment, and the production of pesticides. Despite much discussion of the world approaching "peak oil," or even having already reached it, the oil industry continues to find new sources of oil, regardless of the grim reality of oil spills in bodies of water. Imperial Oil has been exploring the extraction of oil from the tar sands in northern Alberta, which reportedly "releases at least three times the CO_2 emissions of regular oil production procedures and will likely become North America's single industrial contributor to climate change" (Jami 2010: 16).

Although natural gas was used as early as 1821 to heat homes in Fredonia, New York, it began to take off as a major source of fuel in the 1880s, starting out with its use for lighting and industrial health in Pittsburgh (White 2008: 120). Many energy analysts view natural gas as a cleaner energy resource because it produces lower emissions than either oil or coal. It can be converted into many other products, such as liquid fuels, artificial fertilizers, and hydrogen for use in fuel cells.

In addition to making enormous profits, the "fossil fuel complex" is the recipient of massive government subsidies, yet another indicator of the close nexus between corporations and the state. Princen, Manno, and Martin report: "Worldwide, governments subsidize the fossil fuel industry to the tune of some $300-500 billion per year. In the United States in 2008, the petroleum industry paid $23 billion in royalties to the U.S. Treasury. In Saudi Arabia, the world's largest oil producer, oil and gas account for 90 percent of the gross domestic product while employing only 1.6 percent of the active labor force" (Princen, Manno, and Martin 2013: 161).

Greenhouse emissions have particularly increased during the recent late capitalist phase or neoliberalism that started around the 1980s and particularly intensified after the collapse of the Soviet bloc in around 1989–1991. Half of emissions from fossil fuels combustion in the period 1751–2010 occurred after 1986 (Malm 2016: 328) and global CO_2 emissions were 58 percent higher in 2012 than they were in 1990 (Malm 2016: 3).

The manufacture of products obviously too numerous to list contributes to greenhouse gas emissions, particularly carbon dioxide. Steel and aluminum are the most common metals that are utilized in the manufacture of numerous products, including motor vehicles, rail rolling stock, military and commercial aircraft, and ships, along with factories, office and residential buildings, appliances, and electronic equipment. Cement and concrete are important components of many building materials and road construction and result in large amounts of CO_2 emissions. According to Little (2009: 319), "Buildings alone account for nearly 40 percent of all energy use and contribute nearly 40 percent of the world's annual greenhouse gas emissions." Dwelling units in developed societies, particularly in North America, Australia, and New Zealand, have become larger and larger over time. These larger dwelling units generally require more and more energy to heat or cool and encourage their occupants to purchase more and more consumer items, such as wide-screen televisions, computers, large barbeque grills, and so on.

Paul Baran and Paul Sweezy (1966) identified the steam engine, the railroad, and the automobile as having had "epoch-making" impacts on capitalist development in their respective eras. They maintain that the oil industry in large part has been a creation of the motor vehicle. The products associated with the global automobile industry account for nearly half of global oil consumption (Dauvergne 2005: 43). Research drawing on government-reported registrations and historical motor vehicle population trends counted 1,015,000 motor vehicles, namely, cars, light- and medium- and heavy-duty trucks, and buses registered worldwide in 2010 (Sousanis 2011). Sperling and Gordon (2010: 4) project that by 2020 there will be over 2 million motor vehicles, including trucks, buses, motorcycles, scooters, and electric bicycles. Globally, nearly 4,000 square kilometers of land, much of it farmland, are being transformed annually for motor vehicle use in the form of roads, highways, and parking lots (Dilworth 2010: 362). Vandana Shiva reports: "The car has seriously divided India. People can no longer walk on the streets. Neighbors have turned into enemies over car parking. It has cut up rural India through land grabs for factories and highways" (Shiva 2008: 51).

Air travel possibly contributes about 2.5 percent of CO_2 emissions globally, has been increasing at about 5 percent per annum, and now constitutes the fastest single source of greenhouse gas emissions (Goodal 2010: 175). Airplanes also emit nitrous oxide and other contrail or exhaust fumes, meaning that a "factor between two and three is normally applied to the CO_2 emissions from aviation to account for the additional warming impact" (Tickell 2008: 41). As in many other areas of a stratified world system, the affluent, particularly businesspeople but also politicians and even academics, contribute much more overall to greenhouse gas emissions from flights than working-class people and particularly the poor around the globe.

Marine transportation is sometimes posited as a reasonably sustainable sector for moving freight compared with airplanes, trucks, and railways (Black 2007: 209). However, it results in considerably greater sulphur dioxide emissions than these other transportation modes and the shipping of oil in tankers has also resulted in numerous oil spills since 1967 (Black 2007: 21). Sea shipping functions as the "circulatory system of the global economy" in which about "90 percent of the world trade products are carried at some point" (Paskal 2010: 80). Sea shipping increased between the early 1960s and 2006, rising from less than 6 trillion ton-miles to 33 trillion ton-miles. The cost of shipping has decreased by about 80 percent due to "containerization, bigger ships, and computer-assisted resource allocation" (Paskal 2010: 81).

The effort to move production in developing countries to an industrial model based on oil-powered machinery, petrochemical fertilizers, pesticides, and herbicides under the guise of the Green Revolution and industrial agriculture in developed countries are heavily reliant on fossil fuels. An estimated three calories of fossil fuel energy are required to produce one calorie of food energy (Bello 2009). Danielle Nierenberg maintains:

The Green Revolution technologies of the past, although effective at increasing yields in the short term, tended to focus narrowly on yields and very little on biological interaction. Nearly 2 billion hectares and 2.6 billion people have been affected by significant land degradation resulting from large-scale agricultural practices associated with the Green Revolution. Today, 70 percent of freshwater withdrawals are for agricultural irrigation, causing salinization of water in industrial and developing countries alike. The overuse and misuse of artificial fertilizers and pesticides has produced toxic runoff that has created coastal dead zones and reduce biodiversity (Nierenberg 2013: 193).

Various studies indicate that livestock production is responsible for about 18 percent of the world's total greenhouse gas emissions, which includes its role in deforestation and pasture degradation, 9 percent of CO_2 emissions (not including respiration), 37 percent of methane emissions, and 65 percent of nitrous oxide (including feed crops) (Steinfeld et al. 2006; Hertsgaard 2011: 180). What Tony Weis (2013) terms the "industrial grain-oilseed-livestock complex" has emerged as the dominant agricultural system across much of the temperate countries of the world and has been spreading to large portions of the tropics, such as the Amazon Basin, where there has been an intensification of cattle production. Many cattle are not reared on the open range as was the case in the past but in feedlots where they are fed grain and oilseeds that are grown on land that could be used to raise a wide variety of crops that could directly feed humans.

Deforestation, intensive tillage, and overgrazing release CO_2 from living or recently live plants and organic matter in soil. According to Gore (2009: 175), "Brazil is destroying twice as much forestland each year as Indonesia—primarily because the carbon-rich peatlands from which the Indonesian forests being cleared dry up when the tree cover is gone and burn much longer when set ablaze, emitting far larger quantities of CO_2 into the atmosphere."

Many environmentalists and climate activists whom I have personally met as a result of my research on the Australian climate movement often cite population growth as a major driver of resource depletion, environmental degradation, and anthropogenic climate change (Burgmann and Baer 2012; Baer 2014). A seminal study conducted by David Satterthwaite proved that this assertion is overstated. His study indicates that between 1980 and 2005:

- Sub-Saharan Africa underwent 18.5 percent of the world's population growth but only 2.4 percent of the increase in CO_2 emissions.
- The United States experienced a 3.4 percent of the world's population growth but 12.6 percent of the increase in CO_2 emissions.

- China underwent 15.3 percent of the world's population growth but 44.5 percent of the increase in CO_2 emissions, in large part due to its new role as the "factory of the world."
- Most of the countries with the highest population growth rate had low increases in CO_2 emissions, whereas many of the countries with the lowest population growth rates had high increases in CO_2 emissions (Satterthwaite 2009).

Resource Wars

While warfare is essentially nonexistent in most foraging or hunting-and-gathering societies, as opposed to occasional feuds, it often becomes endemic in many horticultural and pastoral societies. States and empires have long engaged in "resource wars" (Klare 2008). War generates huge corporate profits and nation-states to a greater or less degree participate in a global "military-industrial complex." Gabriel Kolko argues: "The destructive potential of weaponry has increased exponentially, and many more people and nations have access to it ... The world has reached the most dangerous point in recent, or perhaps all of, history. There are threats to war and instability unlike anything that prevailed when a Soviet-led bloc existed" (Kolko 2006: 127).

Since the beginning of the twentieth century, oil has played a central role in both world and regional wars. McQuaig (2004: 3) observes, "Even as the competition over dwindling reserves heats up and threatens to cause international conflict, we are faced with a still more devastating consequence of our addiction to oil—global warming." While war is contributing to climate change vis-à-vis greenhouse gas emissions, the latter in turn may already be contributing to conflicts in drought-stricken regions of sub-Saharan Africa and threatens to pose large-scale conflict as the twenty-first century unfolds. The need for more and more oil drives a global war machine that both pushes countries to and is the force that allows countries to go after this scarce resource along with other scarce resources. Eventually, resource wars erupt in various regions, thereby creating scenarios in which the carbon-dependent military machine pollutes the environment, contributes to climate change, creates diseases, and kills and maims people. Environmental degradation, which includes climate change, in turn creates greater competition for resources and raises the specter of even more violent conflict, including war.

Tensions in the Middle East, much of it resulting from competition over access to oil and the Israeli-Palestinian conflict, have dominated international conflict since the collapse of the USSR. Warfare also contributes to environmental degradation, including chemical contamination and devastation of landscapes, and requires enormous amounts of fossil fuels and nuclear fuels for tanks, jet fighters, and cargo planes, and uranium for the construction of nuclear weapons (Jorgenson, Clark, and Kentor 2010: 9). Needless to say, these activities contribute to greenhouse gas emissions that in turn contribute to climate change.

In 1986 global military expenditure stood at 1 trillion dollars, a cost that undoubtedly contributed to the collapse of the Soviet Union as it tried to maintain arms parity with the United States (Dilworth 2010: 384). The end of the Cold War resulted in a reduction of global military spending to about $750 billion per year. The events of 9/11 appear to have given the United States the impetus to militarize global capitalism. U.S. military spending alone increased from $325 billion in 1997 to $902 billion in 2012 (Robinson 2014: 148). In 2008 a total of $1.47 trillion was spent on military expenditures worldwide; the United States accounted for $771 billion (48 percent of it); Europe $289 billion (20 percent); China $122 billion (8 percent); East Asia/Australasia 120 billion (8 percent); the Middle East/North Africa $82 billion (5 percent); Russia $70 billion (5 percent); Latin America $39 billion (3 percent); Central/South Asia $30 billion (2 percent); and sub-Saharan Africa $10 billion (1 percent) (Giddens 2014: 192). In 2010 military spending exceeded $1.4 trillion, "more than the GDP of the world's 50 poorest countries combined" (United Nations Development Programme 2013: 21–22). Dilworth observes: "In 2004 high-income countries, containing only 16 per cent of the world's people, accounted for 75 per cent of these expenditures. In a world where billions of people survive on one to two dollars a day, governments spend on average $166 per person on weapons and soldiers. At a time when endemic poverty, health epidemics, climate change and mass unemployment cry out for attention, the continued growth in military budgets reflects a troubling set of priorities and a failure to address the underlying reasons for much of the world's instability" (Dilworth 2010: 384).

Corporate hegemony over the world system is ensured by the military forces of the United States and its allies, particularly in Europe, Japan, Canada, and Australia. Under this scenario, warfare to a large extent has been privatized in that the "state organizes and directs warfare from above yet doles out the distinct activities directed and indirectly associated with warfare and with the global 'security' establishment to transnational corporations" (Robinson 2014: 151). In 2010, the U.S. government subcontracted $202 billion to mercenary companies (Robinson 2014: 153). The United States and its various allies, including the U.K. and Australia, have been involved in ongoing conflicts in Afghanistan since 2001 and in Iraq since 2003. While wars obviously predate capitalism, under capitalism war has evolved into a massive killing machine and this is most obvious in today's world for the United States, a country:

> where the agenda of global domination has become an article of consensus within [its] ruling class.... The specific expression of this agenda is the self-proclaimed prerogative of the U.S. government to intervene militarily, at its own discretion, in any country at any time. ... What unites the interventions, rather, is a pair of preoccupations central to the rule of capital, namely, (a) maximising the sphere of corporate economic operations (now focusing especially on oil) and (b) blocking, punishing, and ultimately destroying any attempt to chart an independent—especially if socialist—course of development (Wallis 2010: 56).

Welzer (2012) identifies three broad tendencies driving interstate wars: (1) international commodities markets and supply infrastructure, such as pipelines; (2) conflicts over basic resources such as water and potentially new resources being uncovered as a result of the melting of Arctic and Antarctic ice cover; and (3) climate change as a factor contributing to social system breakdown, including civil war or genocide. He delineates a number of social conflicts, wars, and security measures resulting from environmental and climatic factors. These include:

- Soil degradation, flooding, water shortages, or hurricanes resulting from climate change, which place serious constraints on subsistence.
- Violence in societies most adversely impacted by climate change, which will increase the "flow of refugees and migrants, both within and between countries."
- "The terrorism that has grown in step with global modernization and processes is legitimated and strengthened by inequalities and injustices resulting from climate change." (Welzer 2012: 160–161).

The Road to Dystopia

Climate change scenarios prompt us to imagine dystopian visions of the future, if for no other reason than to forewarn us to take serious measures to counteract possible doomsday events. In his book *Six Degrees*, journalist Mark Lynas (2007), based on his perusal of numerous climate scientific reports, vividly portrays climate change scenarios at 1°C to 6°C increases in the global temperature, most of which will have negative impacts on human population. Given that an increasing number of climate scientists are envisioning a 4-degree world by 2100 (see Christoff 2014) if humanity does not drastically begin to curtail its greenhouse gas emissions, Lynas envisions a 4-degree world that would entail the following conditions:

- Loss of one-third of Bangladesh's land area, resulting in the displacement of millions from the Meghna Delta
- Flooding of low-lying islands and deltaic cities such as Shanghai, Mumbai, Alexandria, Boston, New York, New Orleans, London, and Venice
- Massive shrinking of Greenland's ice sheet into the center of its landmass
- Slowing and shutdown of the North Atlantic Conveyor Belt
- Spreading of new deserts in southern Europe
- Possible July and August temperatures of 48°C in Switzerland, accompanied by wildfires and diminished water supplies
- A completely ice-free summer in the Arctic Ocean
- Release of CO_2 contained in frozen Arctic soils

In a 5-degree world, human populations would be greatly restricted in terms of habitable areas due to drought and flooding, with northern Europe possibly constituting a crowded refugee area. Patagonia, Tierra de Fuego, Tasmania, the South Island of New Zealand, and the ice-free Antarctic Peninsula could serve as other refugee areas. In a 6-degree world, the eruption of oceanic methane might result in massive human extinction.

Peter D. Ward (2010), a professor of biology and earth and space sciences, has written a book that presents various future scenarios based on rising sea levels. He maintains that even if humanity were to curtail all of its CO_2 emissions at the present time, the seas will rise three feet by 2050 and nine feet by 2100, the latter resulting in catastrophic impacts on humanity. These would include incursions of salt into the water table, which would destroy the best agricultural land in the world, including the Central Valley of California and the possible abandonment of many coastal cities. In contrast to most climate scientists, Ward even speculates about future scenarios for both humanity and the planet well beyond 2100.

James Lovelock (2006: 180–181), the inventor of the Gaia hypothesis, recommends that humanity stabilize its population at 500 million to 1 billion and warns that Gaia will cull those who break the rules. He argues that future society will be tribal and even more fractionated than presently between the privileged and the poor (Lovelock 2006: 171). In *The Vanishing Face of Gaia*, Lovelock (2009: 11) identified portions of the earth that still may be inhabitable in a dystopian future. These include the northern regions of the United States and Russia, Canada, Scandinavia, Siberia, Patagonia, southern Chile, and island nations or states, such as Japan, Tasmania, New Zealand, and the British Isles. Lovelock (2009: 61) predicts that the summer heat of continental Europe will become increasingly unbearable, even with the use of air conditioning. Lovelock (2009: 56) contends that a 4°C hotter planet may only be able to sustain a population of "as little as 100 million if the carrying capacity of the land surface of a hot Earth falls to 10 percent of what we have now."

More recently, Lovelock has qualified some of his earlier dystopian projections. He argues that the "best course of action may not be sustainable development but a sustainable retreat" (Lovelock 2014: 3). Lovelock asserts that global average temperature and sea level have risen less than climate scientists had been predicting, possibly for a number of reasons including a modest increase in global economic activity, global dimming resulting from industrialization in China and other developing countries, an increased output of aerosols in the atmosphere due to forest clearance, and the redistribution of heat in the ocean (Lovelock 2014: 85–88). Lovelock (2014: 117) advocates the creation of high-rise, high-density, climate-controlled cities like Singapore where people can escape the hot world outside of their artificial environment, but such a development will require the suspension of democracy. In the event that humanity does not curtail climate change, he predicts that: "[I]t is most unlikely that the number of humans surviving will be less than a million: this is more than enough needed for the survival of our species.... [O]ur present choice of city life could be a powerful, benign force moving us to a future existence in city super-organisms" (Lovelock 2014: 122–123).

Perhaps frustrated by the cumbersome nature of global and national governance processes, including in liberal democracies, various scholars have argued that democratic processes are moving too slowly to contain climate change and they suggest that eco-authoritarian, or even eco-fascist, regimes are needed to do so. James Anderson (2006: 245) argues that the "radical changes necessary to sustain capitalism could indeed turn out to be an extremely authoritarian *counter-revolution.*" Shearman and Smith (2007) maintain that "democratic states" are too dominated by special interest groups and materialism to create effective climate change mitigation policies. They assert that liberal democracies need to be replaced by authoritarian states, such as Singapore, which will be governed by "natural elites" who have been socialized from childhood to address complex problems such as climate change. Shearman and Smith (2007: 134) assert that climate change will create an economic and ecological disaster that will require a future government led by "specially trained philosopher/ecologists" committed to environmental sustainability.

In a somewhat similar vein, Lovelock (2009: 61) maintains that "orderly survival ... may require, as in war, the suspension of democratic government for the duration of the survival emergency." Lieven De Cauter (2008: 111) suggests that climate change may contribute to a future world that "looks like some version of Mad Max, a trash sci-fi movie in which oil scarcity has turned the planet into a low-tech, chaotic, neo-medieval society run by gangs." She argues that environmental disasters, such as Hurricane Katrina in New Orleans in 2005 and the tsunami in Indonesia in 2004, are contributing to what Naomi Klein has termed *disaster capitalism*, under which the affluent sequester themselves from the victims of disasters in gated communities and green zones, or a "sort of security stronghold as well as an ecological safe haven" (De Cauter 2008: 115). Prompted in part by Bertram Gross's (1980) classic, *Friendly Fascism*, William Robinson (2014: 163) has raised the specter of the emergence of a twenty-first-century fascism that would allow "dominant groups around the world to assure widespread, organized mass social control of the world's surplus population and of rebellious forces from below." While he does not specifically mention pervasive social unrest resulting from a global climate crisis, such a development potentially could serve to ensure the transnational capitalist class and its political allies to create a "fortress world" to ensure their privileged position.

In contrast to dystopian climate change scenarios envisioned by various conservative commentators, critical geographers Joel Wainwright and Geoff Mann (2012) delineate four potential "social formations" that may emerge as a result of the climate crisis, which they term Climate Leviathan, Climate Behemoth, Climate Mao, and Climate X. These are depicted in Figure 1.1 below.

Wainwright and Mann posit that: "[T]he future of the world will be defined by Leviathan, Behemouth, Mao, and X, and the conflicts between them. This is not to say that all future politics will be simply be determined by climate ..., but rather that the challenge of climate change is so fundamental to the global order that the complex and manifold reactions to climate change will restructure the world along of one these paths" (Wainwright and Mann 2012: 5).

Figure 1.1. Four Potential Climate Change–Induced Social Formations

	Planetary Sovereignty	*Anti-Planetary Sovereignty*
Capitalist	Climate Leviathan	Climate Behemoth
Non-Capitalist	Climate Mao	Climate X

Source: Adapted from Wainwright and Mann (2012: 5).

Under Climate Leviathan, humanity will be governed by a "regulatory authority armed with popular legitimacy, a panopticon-like capacity to monitor and discipline carbon production, consumption, and exchange, and binding technical authority on scientific issues" (Wainwright and Mann 2012: 6). In some ways, the UN Conference of the Parties process has served as a precursor to such a centralized structure in that the former never questioned capitalism as the primary driver of climate change, but has accepted the premise that capitalism could be reformed in such a way that it would ultimately mitigate climate change. Climate Leviathan would create the first global sovereign, under U.S. domination, that will call a climate emergency and create order in the world system. Wainwright and Mann (2012: 8) view the notion of the Green New Deal, which takes various forms, as "perhaps the most sophisticated call for capitalist Leviathan yet." Climate Mao constitutes another version of Climate Leviathan, which is situated at the "end of the red thread running from Robespierre to Lenin to Mao" in that it "asserts the necessity of a just terror in the interests of the future of the collective" (Wainwright and Mann 2012: 9). Climate Behemouth opposes both visions of a Climate Leviathan and entails two contrasting grassroots responses, the first one a type of reactionary populism that rallies to the defense of certain sectors of capital, such as the fossil fuels and the motor vehicle industries, and the second one a "revolutionary anti-state democracy," which is manifested in the more anarchistic strains of the environmental and climate movements. Climate X constitutes a second version of Climate Behemouth, one that seeks to transcend global capitalism through an ecological revolution that results in a rejuvenated socialist agenda.

Conclusion

Given that global capitalism has so many glaring contradictions, can we safely predict that it will end and, if so, anytime soon? Wolfgang Streeck observes: "[A]ll of the main theorists of capitalism have predicted its impending expiry, ever since the concept came into use in the mid-1800s. This includes not just radical critics like Marx or Polanyi, but also bourgeois theorists such as Weber, Schumpeter, Sombart and Keynes" (Streeck 2014: 46).

As history has repeatedly shown us, capitalism both within specific countries and globally has had a profound capacity to recover from a series of crises. The question is, how much longer can this be the case? Streeck (2016) maintains that capitalism will end through a steady stream of small and large imperfections which a variety of management reforms will fail due to the massive weight of contradictions in the system. Only time will tell if this assessment is true but given the gravity of the current world situation socioeconomically, socio-politically, and ecologically, it seems that humanity needs to both revisit earlier attempts to transcend global capitalism, assessing where they succeeded and where they failed, and envision possible pathways for the future.

In contrast to previous indigenous and precapitalist state societies, as Burkett (2014: xx–xxi) observes, "[c]apitalism has the ability, because of its global scope and more basically because of its treatment of labor power and its material conditions as separable inputs, to take a much more 'slash and burn' attitude toward particular natural conditions, to continue accumulating on a global scale despite (and often because of) the damage it causes to nature." Ironically, capitalism has a capacity to turn tragedies of all sorts in sources into profit-making opportunities, thus prompting Canadian journalist Naomi Klein (2007) to write about *disaster capitalism* (see also Loewenstein 2015) and, in the case of climate change, Peter Newell and Matthew Paterson (2010) to coin the term *climate capitalism*. Numerous carbon trading or emissions trading markets, such as the Chicago Climate Exchange and Carbon Markets and Investors Association, and carbon offsetting schemes, such the Carbon Neutral Company, have emerged around the world.

As a globalized system that relies heavily on fossil fuels, it has played a significant role in the emissions of greenhouse gases into the atmosphere. Existing climate regimes, ranging from the UN Framework Convention on Climate Change to the EU's Emissions Trading Scheme to various national emissions trading schemes, have proven ineffective in significantly cutting back on greenhouse emissions. Efforts to create a green capitalism also have proven wanting because they all fail to address the growth paradigm as well as social justice issues.

CHAPTER 2

Twentieth-Century Attempts to Create Socialism
Successes and Failures

Given the authoritarian nature of communist regimes in the Soviet Union, its satellites in Eastern Europe, China, North Korea, Vietnam, Cuba, and other postrevolutionary societies, many scholars and other observers immediately conjure up negative images of the word socialism or find its association with the concept of democracy contradictory. Numerous observers have viewed the collapse of communist regimes in the former Soviet Union and Eastern Europe in particular as evidence that capitalism constitutes the end of history and that socialism was a bankrupt experiment that led to totalitarianism, rapid industrialization, inefficient centralized planned economies, forced collectivization, ruthless political purges, and gulags. Unfortunately, what these commentators often forget is that efforts to create socialist-oriented societies occurred by and large in economically underdeveloped countries. Russia, for instance, was an agrarian nation ruled by an absolutist monarchy upon the eve of the Bolshevik Revolution of 1917. Indeed, the czarist regime did not abolish serfdom until the 1860s, as part of an effort to stabilize imperial rule in the wake of having lost the Crimean War to Britain. The efforts of Lenin, Trotsky, and other Bolsheviks to develop the beginnings of the process that they hoped would result in socialism occurred under very adverse circumstances.

Although I eschew a ritualistic reading of Marx and Engels as ultimate authorities on questions under dispute in political economic studies, unavoidably consideration of the nature of socialism must begin with an examination of what socialism meant to these seminal thinkers. In fact, Marx and Engels refrained from drawing up a detailed blueprint of what form or forms socialism and communism would take. Yet, here and there in their writings they delineated the essential dimensions of the "lower phase of communism" (socialism) and the "higher phase of communism" (communism per se). Socialism, as Navarro observes, "is not a mode of production: it is a social formation in transition from the capitalist mode of production to the communist mode of production.... Socialism as a social formation has several modes of production, including capitalist and communist modes" (Navarro 1982: 85).

A perusal of Marx and Engels's writings indicates that they believed that socialism would exhibit four basic components: (1) social or public ownership of the means of production, (2) increasing social equality, (3) proletarian or workers' democracy, and (4) developed productive forces. Social or public ownership of the means of production by the state or other collective bodies would be a necessary condition for achieving an economic system in which production is oriented toward meeting social needs rather than the creation of profit that enriches a small ruling class. As opposed to communism, which would be guided by the dictum "from each according to his ability, to each according to his needs," socialism, as a transitional phase between capitalism and communism, is based on the dictum "from each according to his abilities, to each according to his work." Income under socialism would be calculated according to the work performed because of the persistence of bourgeois values. Under socialism, individuals who work longer, have greater abilities and skills, and take on more responsibilities are rewarded more than others. Since a select few could no longer derive income or wealth from the private ownership of property, income differentials would diminish greatly. Nevertheless, socialists have over the years engaged in intense debates about what sort of wage differentials would exist in a socialist society. Stilwell (2000: 130) argues that a 3:1 ratio of the highest to the lowest incomes would be a tolerable standard for a socialist society. Other socialists, however, argue for a somewhat large income differential on the grounds that some individuals may choose to work many more hours in order to obtain certain expensive consumer items, and others may choose to work a very short period so as to pursue various avocations. Conversely, it is important to note that there are other compensations for meaningful work than material rewards, such as the intrinsic rewards of intellectual and even physical stimulation and the sense that they have contributed to the greater good. The distinction between manual and intellectual labor would continue under socialism but would theoretically be abolished under communism in the sense that most workers would engage in both.

Marx, who was committed to the ideal of direct democracy, from his earliest writings, referred to his conception of socialist democracy as the "dictatorship of the proletariat." The term "dictatorship" did not refer to an authoritarian government, but to the domination of the working class over productive forces and the institutions of society. Just as the capitalist state, despite the presence of voting rights and parliaments, is governed by a "democracy of all factions of the capitalist class, but a dictatorship over the working class, [Marx] expected a democracy of the working class to exert a dictatorship over their former masters, the capitalists, until the capitalists would disappear completely" (Sherman 1972: 292). Although he never explicitly described how collective decision making might occur under socialist democracy, Marx believed that various features of the Paris Commune of 1871, referred to in *The Civil War in France*, foreshadowed the proletarian state:

> The Commune was formed of the municipal councillors, chosen by universal suffrage in various wards of towns, responsible and revocable at short terms. The majority of its members were naturally working men, or acknowledged representatives of the working class. The Commune was to be a working, not a parliamentary body, executive and legislative at the same time.... From the members of the Commune downwards, the public service had to be done at workmen's wages. The vested interests and the representation allowances of the high dignitaries of State disappeared along with the high dignitaries themselves. (quoted in Tucker 1978: 632)

It is important to add that Marx extended his concept of democracy to include workers' control over the means of production. For example, in *Capital*, he referred to "associations of working-men" who will freely regulate their own labor and determine the terms of their intercourse with nature. For Navarro (1982: 80), the "key criterion in defining a social formation as socialist is whether there is control by the working class and its allied forces of the political instance in that formation."

Marx argued that the attainment of socialism would occur after capitalism had "exhausted" its potential for developing the productive forces. In the *Preface to a Contribution to the Critique of Political Economy*, he stated: "No social order ever perishes before all the productive forces for which there is room in it have developed; and new, higher relations of production never appear before the material conditions of their existence have matured in the womb of the old society" (quoted in Tucker 1978: 5).

In later years, however, both Marx and Engels relaxed the strict determinism of this statement by conceding that revolution in certain underdeveloped countries, particularly Russia, might provide the impetus for proletarian revolution in Western Europe. In 1882, in the preface to the Russian edition of *The Communist Manifesto*, they concluded: "If the Russian revolution becomes the signal for a proletarian revolution in the West, so that both complement each other, the present Russian common ownership of land may serve as the starting point of a communist development" (quoted in Tucker 1978: 472).

An Overview of Selected Postrevolutionary Societies

In this section, I examine efforts to create socialism in various countries, particularly the Soviet Union, China, the German Democratic Republic, North Korea, and Cuba, and explore various theories on the nature of *postrevolutionary* or *socialist-oriented societies*, particularly those that for better or worse have been characterized as Marxist-Leninist or simply Leninist regimes in that they were organized the basis of bureaucratic centralism; and positive and negative features of postrevolutionary societies. Table 2.1 below presents a listing of postrevolutionary or socialist-oriented societies that existed in the mid-1980s, a few years before the opening of the Berlin Wall in 1989 and the collapse of the Soviet Union in 1991.

Gottlieb (1992: 79–80) delineates three key elements in Lenin's political stance: (1) the central significance of a vanguard party that would lead the working class; (2) an alliance between the working class and the peasants in which the former would lead the way to first a bourgeois democratic revolution and then a socialist-oriented revolution; and (3) an internationalist perspective that regards capitalism as an imperialistic world system rather than as simply a system embedded in more or less sovereign states. Lenin viewed the party as acting as the vanguard for the working class because the majority of workers held a "trade union" consciousness that emphasized their own immediate economic interests. Leninist regimes that are not listed in Table 2.1 according to some sources would have included South Yemen, Ethiopia, Mozambique, Angola, Zimbabwe, and even Nicaragua, the latter of which operated as a multiparty state during the period of 1979–1990 when the Sandinistas governed. Six African regimes reportedly had embraced the principles of "scientific socialism" and Marxism-Leninism by 1986, including the Congo beginning in 1969, Benin in 1974, Madagascar in 1975, Angola and Mozambique in 1977, and Ethiopia in 1987 (Keller 1987: 8). In reality, Afro-Marxist regimes merged Leninist and populist elements, whereas Tanzania under the leadership of Julian K. Nyerere constituted an example of what came to be known as "African socialism" or "populist socialism." African socialism drew on traditional notions of communal ownership of land and natural resources and downplayed the notion of class struggle, asserting now that social classes had been part of indigenous African societies, a belief that overlooked the existence of state societies in precolonial Af-

Table 2.1. Postrevolutionary Societies, ca. mid 1980s

Country	Date Leninist regime established	Process by which Leninist took power	Population in 1985 (in millions)
Soviet Union	1917	Revolution	277
Mongolia	1924	Revolution with Soviet aid	1.9
Yugoslavia	1945	Revolution	23
Albania	1945	Revolution with Yugoslav aid	3.0
	1945–1948	Occupation by Soviet army after World War II	
Poland			17
German Democratic Republic			16
Czechoslovakia			11
Hungary			23
Romania			2.3
Bulgaria			9.0
Democratic People's Republic of Korea	1945	Occupation by Soviet army	20
China	1949	Revolution	1,092
Vietnam	1954, 1975	Revolutions	59
Cuba	1959	Revolution	10
Laos	1975	Revolution with Vietnamese aid	3.8
Kampuchea	1975, 1979	Revolution with Vietnamese aid, then with new regime installed by Vietnamese invasion	6.2
Afghanistan	1979	Soviet-backed revolution	12

Total population, 1985 (33 percent of world total) = 1,621 billion

Source: Adapted from Chirot (1986: 264).

rica. In part, Afro-Marxism drew on Marxian class analysis and became popular in the late 1960s, partly as a response to the limitations of the African socialism model. It also drew inspiration from the Cuban Revolution, which was perceived as an example of how developing countries could adopt a Soviet-style development trajectory.

As depicted in Table 2.2 below, Chirot (1986: 268) identifies five dimensions of various types of "Leninist" societies. He characterizes the USSR between the 1930s and the early 1950s as having been Stalinist whereas the Eastern European countries (except for Yugoslavia) fit this characterization in the late 1940s and early 1950s, as did Romania in the 1970s and 1980s and North Korea from the 1950s apparently to the present day. Liberalized Stalinist regimes included the USSR from the late 1950s to the 1980s, most of the Eastern European countries during the same period, and Cuba between the 1960s and the 1980s. Titoism refers to the policies promoted by Marshall Tito in which his regime implemented an experiment in "market socialism" and "economic decentralization, opening Yugoslavia to the West and allowing his citizens more personal freedom," including to take jobs in Western Europe (Chirot 1986: 267). In addition to Yugoslavia from the late 1950s to the 1980s, Titoist regimes in Chirot's view included possibly China from the late 1970s to the 1980s, Czechoslovakia during the mid 1960s but ending with the Prague Spring era in 1968, and Poland in the late 1970s. Finally, Maoist regimes, under which Mao Tse Sung theoretically prioritized peasant agriculture development over industrialization, included China from the mid 1960s to the mid 1970s, Kampuchea in 1975–1979 during the Pol Pot period, and possibly Albania from the 1960s to the 1980s.

Table 2.3 below depicts the fate of postrevolutionary societies in the Soviet Union and Eastern Europe.

Space does not allow a historical account of all the postrevolutionary societies that existed during the twentieth century, some of which have persisted into the twenty-first century. However, for illustrative purposes, I focus on five such societies, namely, the Soviet Union, China, the German Democratic Republic, North Korea, and Cuba. Selection of the Soviet Union and China are obvious in that the former constituted the first

Table 2.2. Five Dimensions of Various Types of Leninist Societies

Regime type	Commitment to equality	Commitment to heavy industrialization	Centralization	Repression	Openness to outside world
Stalinist	low	high	high	high	low
Maoist	high	low	high	low	high
Titoist	low	low	low	low	high
Liberalized Stalinist	low to medium	medium	medium	medium	medium

Source: Adapted from Chirot (1986: 268).

Table 2.3. The End of Postrevolutionary or "Leninist" Regimes in the Soviet Union and Eastern Europe

Country	Year of transition to Leninsm	Current status
Soviet Union	1917	Abandoned Leninism and fragmented into numerous other states after 1991
Albania	1944	Abandoned Leninism after 1989
Yugoslavia	1945	Abandoned Leninism after 1989 and fragmented into Bosnia-Herzegovina, Croatia, Macedonia, Slovenia, and Yugoslavia (Serbia and Montenegro)
Bulgaria	1947	Abandoned Leninism after 1989
Czechoslovakia	1948	Abandoned Leninism after 1989 and split into the Czech Republic and Slovakia in 1993
Hungary	1948	Abandoned Leninism after 1989
Poland	1948	Abandoned Leninism after 1989
Romania	1948	Abandoned Leninism after 1989
German Democratic Republic	1949	Abandoned Leninism in 1989 and became absorbed into the Federal Republic of Germany in 1990

Source: Adapted from Sanderson (2010: 108).

effort historically to create a socialist society and the second because China with its revolution in 1949 became the first and continues to be the largest postrevolutionary society in the world. I have chosen the German Democratic Republic because in its time, it was the most technologically advanced of any of the postrevolutionary societies and because I lived there from July 1988 to February 1989 in two capacities, first as student in a summer course on "Social Life in the German Democratic Republic" at the teachers' college in Erfurt in Thuringia and subsequently as a Fulbright Lecturer in the Department of British and American Studies at Humboldt University in East Berlin (Baer 1998). During the time of my Fulbright Lectureship, I seriously contemplated the possibility of becoming a GDR specialist but, alas, "my country" disappeared on me. Ironically, while occasionally there has been discussion of uniting the two Koreas, the Democratic People's Republic of Korea continues to function as a separate country from the Republic of Korea or South Korea. It remains the most perplexing and secretive of postrevolutionary societies that have existed in the past or presently exist, one that has been ruled for three generations by a dynastic family. Finally, Cuba has persisted for over half a century as a postrevolutionary society, despite an economic embargo placed on it by the United States in 1961 and persistent predictions of its allegedly imminent collapse.

Soviet Union

In the early twentieth century during the era of the 1905 and 1917 revolutions, Russia was largely an agrarian country that was only in the beginnings of industrialization, a process that had started in Western Europe about a century earlier. Isaac Deutscher (1967: 11) describes Russia as "half empire and half colony" during the reign of the last Romanovs in that Western shareholders owned about 90 percent of the mines, over 40 percent of the engineering plants, and 42 percent of the banking stock in the country and it relied heavily on foreign capital for industrial expansion. The Social Democrats who had been inspired by Marx's thought were divided into two factions, the Bolsheviks who pushed for a socialist-oriented revolution, which they hoped would be followed by similar revolutions in the West, and the Mensheviks, who did not think that the Russia was ready for a socialist revolution but only one that would transform it into a developed capitalist society. Until a few weeks before the outbreak of the February Revolution in 1917, Lenin, leader of the Bolsheviks, maintained that a revolution in Russia would not occur until sometime in the distant future (Kuran 1995: 1530). In order to ensure the victory of the subsequent October Revolution, the Bolsheviks created the Cheka in December 1917 in order to suppress counterrevolutionary activities, an organization that participated in a period of political suppression of dissident elements following a failed attempt to assassinate Lenin in May 1918.

There has been considerable debate on the left as to whether Leninism could have moved the Russia and USSR forward if Lenin had lived beyond 1924 or whether Leninism was the inevitable precursor of Stalinism. Paul Le Blanc (2006: 97–98) views Lenin as a genuine democrat who believed that the working class "must be interested not only in wages, hours, and working conditions, but also in broader social and political questions—especially questions of democracy, opposing the opposition of racial and national and religious minorities, opposing the oppression of women, opposing violations of academic free and of liberties, and so on." He asserts that the Bolshevik Party during 1912–1917 "blended democratic functioning with activist coherence, a unified political orientation with substantial autonomy for activists in various locales" (Le Blanc 2006: 150). The party could be described as "a negotiated federation between groups, groupings, factions, and 'tendencies'" (quoted in Cohen 1973: 5). The soviets, which were workers' councils, appeared in Russia during the 1905 Revolution but resurfaced during the February Revolution in 1917 at which time industrial workers, soldiers, sailors, and peasants elected delegates who were subject to recall by them. Reportedly the some four hundred soviets in May 1917 grew to some six hundred by August and some nine hundred by October of the same year (Haynes 2002: 9). They functioned as a parallel power to the Provisional Government and included representatives from various political parties, including the Social Revolutionaries, the Mensheviks, Bolsheviks, anarchists, and smaller socialist groups. The Bolsheviks briefly considered a composite state consisting of the Constituent Assembly and the soviets. Lenin took advantage of the strong support among workers in St. Petersburg in October 1917 and abolished the dual power of the Constituent Assembly and the soviets. The Communist Party came to dominate the Russia's political and social life, abolished other political parties, and co-opted workers' councils, trade unions, and peasant cooperatives. It justified its policies as being necessary given Russia's economic backwardness and internal and external threats to the revolutionary cause. Indeed, this was not without grounds as some of Lenin's opponents had attempted to kill him and others allied themselves with belligerent foreign countries.

Trotsky reportedly initially maintained that Lenin's notion of democratic centralism would eventually lead to a dictatorship over the party and the proletariat. He stated: "The Party organisation substitutes itself for the Party, the central committee substitutes itself for the organisation, and finally a 'dictator' substitutes himself for the central committee" (quoted in McLellan 1983: 69).

In *Organizational Questions of Russian Democracy*, Rosa Luxemburg expressed alarm that democratic centralism rested on "blind subordination ... of all organs to the centre" and accused Lenin of "ultra-leftism" and suppressing the revolutionary impulse of the masses. She believed that Lenin privileged intellectuals over working-class people. However, Trotsky shifted his initial position to one that viewed the party as being the appropriate locus of revolutionary action, not the parliament or the soviets (Thatcher 2007: 40). Luxemburg rejected this stance and insisted on the priority of workers' councils in decision making.

In recent years, scholars on the left have engaged in a debate, particularly in response to new materials released from the Soviet archives since the collapse of the USSR, as to how democratic or authoritarian Lenin may have been, with some viewing him in a more negative light, such as Bronner (2011: 88), who maintains that Lenin distrusted the soviets as vehicles for democratic governance, and others, such as Alan Badiou

(2007: 7), asserting that he recognized the importance of elections, viewing him in a more positive light (see Anderson 2014: 143–147).

In addition to economic underdevelopment and the presence of a tiny trained working class, the new Soviet republic faced a civil war and the military intervention of fifteen foreign powers, including the United States, during the period from 1918 to 1920. Furthermore, Russia at best had only rudimentary experience with parliamentary democracy along the lines of what existed in Western Europe and North America. Certainly, in the aftermath of the civil war, by 1921, while Lenin and Trotsky were in positions of leadership, the Russian Communist Party exhibited authoritarian tendencies (Le Blanc 2006: 107). Isaac Deutscher contended that by 1921–1922 for the first time since 1917 "the bulk of the working class unmistakably turned against the Bolsheviks.... If the Bolsheviks had now permitted free elections to the soviets they would almost certainly have been swept from power" (quoted in Chattopadhyay 2010: 35). The Bolshevik-dominated Central Executive Committee of Soviets excluded the Mensheviks and right-wing Social Revolutionaries and ordered the soviets to follow the same policy. Lenin did not mention the 1905 or 1917 soviets in *State and Revolution*, apparently indicating his preference for the party as a more important revolutionary instrument.

Factors contributing to the decimation of a strong working class in the postrevolutionary Russia included the civil war, foreign invasion in support of the Whites, who opposed the Bolshevik Revolution and efforts to create socialism, and a trade embargo on the part of Western capitalist powers (Roper 2013: 272). In addition to strengthening centralized bureaucratic authority, the civil war experience of 1918–1920, according to Cohen (1973: 60), "brought about a pervasive militarization of Soviet political life, implanting what one Bolshevik called a 'military-soviet'." Ten expeditionary forces, included ones from Britain, France, the United States, Canada, Serbia, Finland, Romania, Turkey, Greece, and Japan, assisted the Whites against the Reds in the civil war (Anderson 2010: 61). The civil war decimated the Soviet working class, reducing it from 2.6 million workers in 1917 to 1.2 million in 1920 (Zimbalist and Sherman 2015: 137). Much of Russia by 1920 was devastated, with abandoned factories in the cities and towns and famine in the villages. The Bolsheviks had hoped that revolutions in developed capitalist European countries would assist them in their efforts to create socialism in an economically underdeveloped country.

The Bolsheviks lacked a clearly thought-out economic plan before they assumed power. In 1918 Lenin himself viewed the new Soviet economy as consisting of five segments: (1) a patriarchal one based on rudimentary agriculture, (2) small commodity production in which most of the peasants participated in trading their grain, (3) private capitalism, (4) state capitalism, and (5) a socialist sector consisting of state-operated enterprises (Resnick and Wolff 2002: 108). Cohen (1973: 57) asserts that this shortcoming "set the stage for the party's twelve-year search for viable economics policies commensurate with its revolutionary ambitions and socialist faith" and "assured that the search would be bitterly divisive." In the wake of the civil war, Lenin initiated the New Economic Plan (NEP) as a supposedly short-term strategy for propping up the devastated Russian economy that was designed to increase peasant agrarian productivity, develop small industry, and combat bureaucratic excesses. While it resulted in increased industrial production, it also allowed the kulaks and middle peasants to produce for the market and the poor peasants to at least subsist. Although the peasants were permitted to make profits during the NEP period of 1918–1921, the Communist Party did not grant them the right to give any serious political input to the NEP.

Theoretically, the soviets were intended to function as exemplars of workers' democracy. They consisted of elected delegates from factories and workplaces and formed in Moscow, Petrograd and other major Russian cities. The All-Russian Congress of Soviets consisted of delegates from various political tendencies, including the Bolsheviks, the Mensheviks, the Internationalists, the Left Social Revolutionaries, other socialists and some anarchists (Roper 2013: 267). However, according to Boswell and Chase-Dunn (2000: 106), the Bolshevik suppression of the Kronstadt rebellion in 1921 "ended hope for renewed direct worker control," with the NEP perpetuating the "trend of adopting capitalist management practices."

The Soviet Union or as it was formally known, the Union of Soviet Socialist Republics (USSR) was formally created in 1922 but did not initially consist of the fifteen republics. The Baltic states of Estonia, Latvia, and Lithuania were annexed in 1940 in the aftermath of the Hitler-Stalin Pact of 1939. The USSR annexed Moldavia, which had been part of the Russian empire, as well in 1940 when Germany allowed for its annexation. Until its demise in 1991, the USSR constituted the largest country in the world in terms of area, about two and a half times larger than the second-largest country, namely, Canada.

The USSR's first Five-Year Plan went into effect in October 1928. Under very draconian and coercive measures, Stalin directed an enormous industrialization campaign and the forced collectivization of the peasantry. The late 1920s saw significant social development in the Soviet Union, with an appreciable increase in the literacy rate, a doubling of enrollment in primary and secondary schools over the pre–World War I level, a 26 percent decrease in the death rate, and an almost 30 percent decrease in the infant mortality rate (Cohen 1973: 273). Furthermore, the average workday had decreased from ten hours to seven and a half hours, real wages increased about 11 percent over the 1913 level; and the "factory worker, like the peasant, was eating better than before the revolution" (Cohen 1973: 274). On the negative side, the NEP period saw a revival of prostitution, gambling, drug trafficking, corruption, and profit making. According to Lewin (2005: 9), although the NEP was "quite successful in restoring the country to minimal levels of physical (biological) and political viability, [it was] still left ... short of what was required to confront the internal, and especially external, challenges that were looming on the horizon."

Both prior to Lenin's death in 1924 and particularly afterward, the notion of democratic centralism devolved into bureaucratic centralism, which essentially "stifles political and theoretical initiative, fosters a need-to-know climate which gives the highest party bodies a monopoly on information regarding party affairs, and opens the door to petty corruption" (Silber 1994: 85). All in all, as Silber so astutely observes: "Lenin's conception of the vanguard party can hardly be separated from the conditions out of which it arose: a working class itself only a decade or two removed from the countryside and constituting a small percentage of the population in a predominantly peasant country; czarist repression; the illegality of most socialist organizations; a rotting political structure holding back the country's economic development; and a people clearly readying for revolution" (Silber 1994: 251).

With Stalin's rise to power, Trotsky as the second leading Bolshevik leader of the Revolution was sidelined for his assertion that the Soviet state had entered into a phase of bureaucratic degeneration. Stalin and his faction prevailed by taking a middle position between Trotsky's "left-wing" approach and Bukharin's "right-wing" approach by following a course of action urged on him by economic planners, managers of heavy industry, military leaders, and the Cheka or secret police (Lichtheim 1970: 280–281). The Left Opposition under Trotsky's leadership lost its political influence by 1928, essentially ending interparty conflicts. Trotsky had been the leader of the Red Army during the civil war while Stalin essentially functioned as a shrewd party hack. In 1928 he supported reform of the party "thorough increased worker control and return to political power for the soviets" (Gottlieb 1992: 95). As Gottlieb observes, "Justifying its dictatorship by appeals to security, the CP eliminated all opposition, modernized agriculture, and initiated the fastest industrialization in world history. It also created a new class of privileged bureaucrats and technical experts" (Gottlieb 1992: 84).

Stalin's theory of "socialism in one country" asserted that the USSR could develop socialism single-handedly and without the assistance of a "permanent revolution" in other capitalist countries (Molyneux 1978: 117). Stalin unleashed one of most gruesome crimes in modern history: terror against workers, peasants, intellectuals, and communists; more than a million communists became victims of a murder machine (Mandel 1992: 49). Mass purges in USSR civil war–like conditions essentially ended any possibility for serious opposition to the Stalinist dictatorship.

Stalin's "Great Turn" resulted in the rapid and massive industrialization of the Soviet Union during the late 1920s and 1930s, which provided the country with the military wherewithal to counter the Nazi military assault launched in 1939 during World War II. Although Stalin had initially favored the NEP along with Bukharin, he reversed his position in the late 1920s and concurrently with rapid industrialization, embarked on a program of "forcing the country's 125 million peasants into state-run collective farms" (Cohen 2009: 3). He drew on Yevgeni Preobrazhenky's (who had been a Trotsky supporter) "fundamental law of primitive socialist accumulation," which meant the people in the cities would need to appropriate their food resources from the peasants in order to build the infrastructure for a modern society. Ironically, Trotsky had been the initial promoter of rapid industrialization in the Soviet Union and had believed that agrarian collectivization would bring about a form of agricultural production that was superior over traditional farming methods of the Russian peasantry (Deutscher 1971: 51). Stalin, however, exiled Trotsky as the leader of the Left Opposition and orchestrated his assassination in Mexico City in 1940.

The collectivized and mechanized agricultural system produced a surplus that sustained industrial growth but one under which the peasants performed their required tasks sluggishly, often reserving their greatest efforts to tend their private plots. Furthermore, Stalin made a "decision to intentionally starve out an estimated five to fifteen million middle and wealthy peasants who resisted" (Gottlieb 1992: 92). Nevertheless, whereas in 1929 around 3.6 percent of agricultural land was collectivized, by 1935 this figure stood at close to 100 percent (Silber 1994: 98). During this period, Soviet political life became not only much more constrained but also repressed with workers being faced with the threat of being forced to labor in Gulag camps or being imprisoned or executed for supposed violations. According to Lewin (2005: 70), the period of 1928–1939 consisted of the following features: "urbanization, industrialization, collectivization, purges and show trials, the spread of education, an often demagogic depreciation of culture, the mobilization of energies and people, increasing criminalization of many aspects of life, hectic creation of administrative structures, and so on." Collectivization, which had depressed rural living conditions through coercive food collections at low prices, prompted millions of peasants to move to the cities where many of them became industrial workers (Zimbalist and Sherman 2015: 156). The party elites did not define the USSR as a socialist society until 1936 when it was claimed that the country consisted of "two classes—workers and collectivised peasants—and a stratum of intelligentsia with these entities living and working in friendly collaboration" (Sarkar 1999: 58).

A variety of external factors, such as World War II and the nuclear arms race associated with the Cold War, and internal forces, such as a centralized command economy and a political system of one-party rule, prevented development of socialist democracy in the Soviet Union. According to Christopher Chase-Dunn (2010: 50), "Building and supporting a Soviet Army that was capable of halting the advance of Germany in World War II meant further concentration of power in the Communist Party, the complete elimination of democracy within the party, and the use of the Communist International as purely the instrument of Russian international interests." Ironically, on the eve of World War II, Stalin decided to purge one-third of the Soviet Union's generals.

Unfortunately, under Stalin's leadership, the USSR constituted a highly centralized regime that relied on massive violence to suppress political opposition. Figures on the number of executions during the Stalinist era vary, and in part reflect the political stance of the source. Various historians who had access to secret Soviet police archives discovered that the total Gulag population in January 1939 was 2,022,976 (Parenti 1997: 79). Nearly a million Gulag prisoners were released during World War II in order serve in the military. While due to the ravages inflicted by World War II the population of the Gulag, a system of forced labor, had declined to some 800,000, it rose to three million by the eve of Stalin's death in 1953 (Lewin 2005: 154). The total number of executions from 1921 to 1953 according to Parenti (1997: 80) came to 799,455. There is a tendency to view the Stalinist era as one during which a madman with paranoid tendencies ruled over a totally subjugated populace. The atrocities that Stalin committed were supported by the actions of party bureaucrats, police officers, and even ordinary citizens who complied with his policies out of sense of duty or fear or both. The Soviet Union during the Stalinist era, as it had been before his Stalin's ascent to the heights of power in the late 1920s, during the civil war of 1918–1920, and after Stalin's death in 1953, was a country besieged by hostile external forces, ranging from Nazi Germany to the United States.

Factors that contributed to the development of a Stalinist dictatorship in the USSR include:

- The autocratic tradition of Czarist Russia, which resulted in a low level of consciousness of even bourgeois democracy in the new Soviet Union
- The underground training of the Bolshevik leadership necessitated by the repressive stance of the Czarist government toward its subversive activities
- The adverse impact of the civil war of 1918–1920, which reinforced authoritarian tendencies and stifled democratic decision making
- The military intervention of numerous capitalist countries, including the United States, Britain, France, and Japan, during 1918–1921 in support of counterrevolutionary forces, the invasion by Nazi Germany in 1941, and the Cold War
- The economic underdevelopment of the Soviet Union, which impelled Stalin to orchestrate a draconian program of industrialization and collectivization (Sherman and Wood 1989: 340–344).

Sherman and Wood (1989) maintain that the emergence of a political dictatorship in the Soviet Union "was *not* caused by socialism, but arose in spite of socialism."

While the use of violence diminished appreciably after the death of Stalin, a collective dictatorship under party elites continued until around 1988 when Gorbachev introduced some democratic reforms. Nikita Khrushchev assumed leadership of the Soviet Union after Stalin's death and despite being one of Stalin's trusted collaborators, chose to reveal in 1956 the atrocities that had been committed under the Stalinist dictatorship, much to the shock of supporters of the USSR around the world. Under Khrushchev, the USSR underwent a drastic reduction in the use of force both in economic and political life but also a reduction in economic growth (Deutscher 1971: 22). He launched the Virgin Lands Program, which sought to boost agricultural production by dramatically increasing the amount of land under cultivation. Khrushchev transformed most of the Moscow-based ministries into local agencies in an effort to improve the economy. He managed to improve the tensions with Yugoslavia, but not to the point that the latter joined the Council for Mutual Economic Access (COMECON). Conversely, tensions with China deteriorated due to his de-Stalinization campaign. Ironically, as part of his effort to learn from the West, particularly in the area of scientific and technological innovations, Khrushchev brashly predicted that the material standard of living in the USSR would exceed that in the United States by 1970.

In the aftermath of the Cuban Missile Crisis, Khrushchev, who fell from power in 1964, was replaced by the "spiritless bureaucrat Leonid Brezhnev whose 18-year rule from 1964 to 1982 was known as the 'period of stagnation'" (Bronner 2014: 117). Brezhnev's rule was followed by short-term general secretaries of the party, namely, Yuri Andropov, who died in February 1984, and Konstantin Chernenko, who only remained in the leadership role until March 1985. According Sherman (1995: 195), the Soviet Union functioned as a dictatorship essentially for most of its duration, first as a "personal dictatorship from about 1928 to 1953, then a collective dictatorship under Party leaders until about 1988 when democratic reforms began" under Gorbachev. Anthropologist Marvin Harris (1992: 299) asserts that the Soviet command economy acted as an impediment to technological innovation because of an underdeveloped incentive system and was essentially "incompatible with the transition to high-tech industrialism with its devices that create, store, retrieve, copy, and transmit information at high speeds over national and international networks." However, he does not romanticize capitalist societies; he asserts that, like the Soviet system, they exhibit numerous dysfunctional dimensions, including the depletion of natural resources, environmental degradation, severe housing shortages, racial and ethnic tensions, bureaucratic inertia, and corruption in the upper echelons of society (Harris 1992: 301).

Under Perestroika, Gorbachev had hoped to transform the Soviet command economy into a mixed economy, which would have incorporated market mechanisms while guaranteeing social benefits. His policy of *glasnost* or transparency probably achieved more popularity outside the USSR than inside of it. In March 1990 the Congress of the Communist Party elected Gorbachev the president of the Soviet Union. By this point, however, the internal and external contradictions of the Soviet Union had reached such a crisis point that its collapse proved to be inevitable, although most analysts did not foresee this. Following its monumental victory in World War II, it faced a new type of war—the Cold War—during which it attempted to match dollar for dollar, ruble for ruble, what the United States was spending on the arms race. The Soviet Union spent a disproportionate amount of its budget on military spending, up to 20 percent of its GDP at times—and in times of war even more (Collins 1995: 1567). Military spending accounted for 33 percent of the state budget in 1940 and 43 percent in 1941 (Flaherty 1991: 142). Table 2.4 below illustrates the U.S. and Soviet strategic arsenals in 1985 and indicates that in terms of at least the number of warheads, the USSR still lagged behind the United States, despite concerted efforts to achieve parity.

Over the period 1945–1990, the United States had had about seventy thousand nuclear warheads and had tested over a thousand of them, while the USSR had had about forty-five thousand nuclear warheads and tested at least 715 (McNeill and Engelke 2014: 157). The U.K. joined the

Table 2.4. U.S. and Soviet Strategic Arsenal, 1985

Weapon carrier	No. of warheads	% of strategic aresenal
USA		
1,025 intercontinental ballistic missiles	2,125	19
36 submarines with 640 missiles	5,728	51
263 B-52 bombers	3,072	27
61 FB-111 bombers	366	3
Total	11,291	100
USSR		
1,398 intercontinental ballistic missiles	6,420	65
62 submarines with 924 missiles	2,688	27
173 bombers	792	8
Total	9,900	100

Source: Adapted from McWilliams and Piotrowski (2009: 517).

nuclear warheads club after 1952, France after 1960, and China after 1964.

Thus, in terms of military spending the USSR was overextended. The Soviet arms industry functioned as the "axis of the super-centralized redistributive command administrative system," which functioned virtually over the course of the Soviet Union's existence (quoted in Flaherty 1991: 143). In essence, the Soviet Union was never able to develop a peacetime economy over the course of its existence. Furthermore, the arms race also prompted the Soviet Union to pressure its Warsaw Pact allies into increasing their military expenditures, "particularly so in the late 1970s, when member states of the WTO were expected to increase defence expenditure by some 3 per cent per annum," thus also adversely impacting their economies (Holmes 1997: 201).

The contradictory nature of Leninist regimes imploded first in Eastern Europe in 1989 and then in the Soviet Union in 1991. Social scientists and other scholars have had a rather poor record at predicting *revolutionary surprises* such as the ones that occurred in the Soviet bloc in the late 1980s and early 1990s. Sociologist Randall Collins had predicted in the 1980s that the Soviet Union would disintegrate sometime within thirty to fifty years (Kuran 1995: 1547). Tariq Ali (2009: 86) expected that Perestroika and *glasnost* would transform the Soviet Union from a "social dictatorship" into a social democratic society, but observes that in actuality, contrary to Gorbachev's intentions, a "strong layer in the Soviet bureaucracy was bent on the total restoration of capitalism," and history tells us benefitted from this development. Gorbachev enjoyed popularity for about five years but began to lose it in the latter half of 1990, and saw it plummet in June 1991 when Boris Yeltsin was elected president of the Russian republic, making Gorbachev's status as president of the USSR somewhat irrelevant (Cohen 2009: 113). Although Gorbachev and Yeltsin had once been allies, political developments turned them into bitter rivals. By July 1991, Yeltsin resigned from the Communist Party, as did the mayors of Moscow and Leningrad. In August 1991, a small group of hardliners attempted a coup against Gorbachev in which he was taken hostage, but the coup collapsed after a few days. Nevertheless, various republics, including Ukraine, Estonia, Lithuania, and Latvia, had made moves to secede from the Soviet Union, events contributing to Gorbachev's decision to resign from the presidency on Christmas Day 1991, an event that resulted in the dissolution of the Soviet Union.

Ultimately, in taking stock of the relatively short history of the Soviet Union (1917–1991), the "Russian Revolution remains a progressive event despite its failure as a socialist event" (Boswell and Chase-Dunn 2000: 90). Despite all of its flaws, the USSR provided space for progressive social programs in core or developed capitalist countries and nation-liberation movements in peripheral capitalist countries. Yeltsin, in the midst of poor health and scandals, managed to serve two terms as president of Russia. Under Yeltsin, Russia underwent two phases of neoliberal reform: (1) the period of 1991–1994, when privatization occurred through a voucher system and accompanying hyperinflation, and (2) the period of 1994–1998, which ended with financial stabilization and almost complete privatization (Kagarlitsky 2002: 188). He managed to form an alliance of three groups: (1) the market-oriented segments of the old ruling class, (2) Soviet free market theorists, and (3) Western advisors, such as Jeffrey Sachs, who advocated "shock therapy," which entailed drastic reductions of social and health services (Haynes 2002: 2004). Russians were provided with vouchers that they theoretically could invest as shares in new private enterprises, but the value of the vouchers in the end was highly deflated, with the "oligarchs" taking advantage of the cheap price of the former state enterprises. The Yeltsin government placed priority on servicing Russia's foreign debt over the course of the 1990s in compliance with International Monetary Fund policies (Kagarlitsky 2008: 304). Russia's GDP fell 29 percent between 1991 and 1992 and another 12 percent by 1993, and agricultural and industrial production fell approximately 50 percent between 1991 and 1998 (Sanderson 2015: 49). The oil and gas sector provided about 20 percent of Russia's GDP by the late 1990s, accounting for about 45 percent of its exports and 60 percent of its foreign currency earnings (Kagarlitsky 2008: 307). Because so many of the state enterprises had been shut down, millions of Russian industrial workers either became unemployed or became petty traders, security guards, and servants of the "new Russians" who had managed to become rich under an increasingly capitalist Russia. These "new Russians" came from the ranks of the old Soviet bureaucracy and came to control the largest oil, gas, and metallurgical companies that exported raw materials and semi-finished industrial products to the world market. At the other end of the social ladder, some Russians, particularly women, became vendors for international direct market companies, such as Tupperware, Avon, Herbalife, and Mary Kay, which began to flow into Russia in the early 1990s and continue to operate there (Patico 2016). Nevertheless, the Russian share of the global GDP had plummeted from 7.08 percent in 1980 to 2.10 percent in 2000 (Kagarlitsky 2008: 311).

Vladimir Putin, the former strong-willed head of the KGB, the Soviet security apparatus, replaced Yeltsin as acting president until he was formally elected president in March 2000. After serving two five-year terms as president, Putin exchanged positions with Prime Minister Dmitri Medvedev, but in essence continued as Russia's principal leader. Under Putin's reign, Russia enjoyed a massive inflow of petrodollars, but this currency reserve in the Russian Central Bank was not used to solve the country's massive social problems. Instead, "in 2004 the Moscow government was cutting social spending and launching a new attack on the remaining elements of the welfare state," with many of the funds in the so-called Stabilization being "invested in US government bonds" (Kagarlitsky 2004: 271). Furthermore, under Putin, the bureaucratic bourgeoisie has displaced the oligarchs and has become "deeply dependent on oil exports, and on Western financial markets" (Kagarlitsky 2004: 271). While Russia continues to have nuclear weapons and continues to have a seat in the UN Security Council, as does China, it is no longer the superpower it was as the leading republic of the Soviet Union.

China

The Soviet Union over the course of its existence provided the political inspiration and space for revolutions to occur in China and various other countries. As Chase-Dunn observes,

> The idea that there was a real alternative to the end of history in the capitalist version of the European Enlightenment was kept alive by the existence of the Soviet Union, despite its grave imperfections.

> The Chinese, Cuban, Korean, Yugoslavian, Albanian and Vietnamese revolutions were able to learn from Russian mistakes of their own. The most obvious example of useful learning was Mao's turn to the peasantry. While the Bolsheviks had treated peasants as a conservative force (despite Lenin's analysis), thus putting the Party at odds with the majority of the Russian people, Mao embraced the peasantry as a revolutionary class. The later revolutions also benefited from the maneuverability that Soviet political/military power and the Cold War balance made possible. (Chase-Dunn 2010: 50)

Mao (1893–1976) viewed the Chinese Revolution as an anti-imperialist/anti-feudal project committed to creating a socialist society based on a federation of communes (Amin 2013: 14–18). In the process, he established a strong alliance with the peasant majority in China, which by and large supported him because he promised them that he would redistribute the land, a promise that Chiang Kai-Chek refused to make. Mao also regarded artisans, smaller traders, white-collar workers, and professional people as allies of the revolution and also viewed the "national bourgeoisie as a temporary ally" against a small oligarchy that consisted of "large landowners and wealthy businessmen and politicians acting as the agents of international capital" (Horvat 1982: 475). The Communist Party in China achieved massive support over the course of its history, with 58,000 members in 1927, 400,000 in 1937, 800,000 in 1940, 1.2 million in 1945, and 4.5 million in 1949, the year of the Chinese Revolution (Kolko 2006: 60). China under Mao developed a complicated love-hate relationship with Stalinism, which viewed Stalin as 70 percent good, 30 percent bad (Silber 1994: 44). For a long time, China included Stalin in its pantheon of "great Marxists," along with Marx, Engels, Lenin, and Mao (Silber 1994: 109). Under Mao, in contrast to the Soviet Union under Stalin, China pursued modernization by improving the socioeconomic status of the peasantry (Arrighi 2007: 374). According to Deutscher (1967: 86), whereas Stalinism constituted an "amalgam of Marxism with the savage barbarism of old Russia," Mao was an "amalgam of Leninism with Chinese patriarchalism and ancestral cults." Kagarlitsky (1995: 139) characterizes Mao as the Lenin and Stalin of the Chinese Revolution, even its Trotsky.

The first five years after the revolution included nationalization of most large capitalist firms, with compensation, but toleration and even support for smaller capitalist enterprises. The government redistributed agricultural land to the peasantry, but permitted some private ownership of it. The first Five-Year Plan of 1953–1957 emphasized the development of heavy industries, which relied heavily on assistance by Soviet technical experts, and the creation of peasant cooperatives. The Great Leap Forward beginning in 1958 relied heavily on steel production in backyard furnaces, resulting in the erection of imposing buildings and structures. It also entailed a massive collectivization campaign in which some 26,000 communes, each with population of around 100,000, replaced some 74,000 of the larger cooperatives (McLellan 1983: 125). In contrast to the widespread resistance to collectivization on the part of many peasants in the Soviet Union, Mao did not encounter a similar backlash as a result of implementing communization during the Great Leap Forward (Glassman 1995: 248).

At least 50 million Chinese died in the Great Famine of 1958–1962 (Lynch 2004: 171). When tensions developed between China and the USSR, Khruschev ordered the withdrawal of all Soviet technicians from China in 1960. In 1963–1964, China accused Yugoslavia and USSR of "revisionism" because of the former's adoption of "market socialism" and the latter's adoption of a "new program at its party's Twenty-second Congress" (Hodges 1981: 78).

In response to the shortcomings of the Great Leap Forward, Mao initiated the Great Proletarian Cultural Revolution in 1966, which was ostensibly designed to counteract the bureaucratization and the rise of the "capitalist roaders" within the Chinese Communist Party. He believed that the USSR had become a bureaucratic capitalist society—"a dictatorship of the big bourgeoisie, a dictatorship of the German fascist type," one that joined the imperialist camp along with the United States (quoted in Hodges 1981: 159). In Mao's view, both the United States and USSR "were ... the enemies of the underdeveloped countries in Asia, Africa, and Latin America but also of the industrialized nations in Western and Eastern Europe constituting a buffer zone between them" (Hodges 1981: 160). The Red Guards, consisting primarily of high school students, and the Revolutionary Rebels, consisting primarily of university students, served as the vanguard of the Cultural Revolution and targeted party elites, intellectuals, and professionals in the cities for their allegedly bourgeois tendencies, but clashes quickly emerged within the revolutionary vanguard itself as well as between students and workers and between students and the army. By September 1967, the army, with Mao's support, halted the excesses of the Cultural Revolution and Mao began to rebuild the Communist Party, even reinstating many party elites who had been purged. Anderson (2010: 68) maintains that the Cultural Revolution, in its effort to "shake up bureaucratic structures," did not entail "wholesale slaughter." Indeed, during the 1970s, former capitalists were given back their properties and bank accounts that had been confiscated during the Cultural Revolution. Chinese technocrats and managers were rehabilitated, allowing them to exert considerate input into the Chinese economy.

In 1972, Mao with the assistance of Chou En Lai orchestrated a rapprochement with the United States, which was facilitated by Henry Kissinger as Richard Nixon's secretary of state, in part apparently to outmaneuver the Soviet Union. After the death of Mao in 1976, a group called the "Gang of Four," which according to Lynch (2004: 188) was "more Maoist than Mao," briefly ruled China. The party elites, however, deposed the gang, which included Mao's last wife, Jiang Quing, and selected as their new leader Deng Xiaoping, who began to implement a program of modernization and economic reforms in 1979 under the guise of a "socialist market economy." With the implementation of the modernization program, Deng dissolved the communes and promoted small family agrarian production. However, since the beginning of the twenty-first century, the party leadership has called for "land consolidation to reach a sufficient scale to launch agricultural investment and more efficient management," thus paving the way of capitalist-style agrobusinesses (Xu 2013: 25).

China has embraced capitalist structures to the point that some experts argue that it now constitutes a state capitalist society in which, "though there is a high degree of public ownership, workers and peasants are exploited for the benefit of officials and managers" (Weil 1994: 17). Indeed, it was admitted into the World Trade Organization (WTO), with U.S. backing, in December 2001. Multinational corporations have profited handsomely due to the availability of a disciplined workforce paid low wages. China's massive industrial expansion, with annual economic growth rates of 9–10 percent at its zenith, has been supported by its heavy export of a wide array of products, ranging from automobiles to electronic goods to clothing. Coal, of which China has massive supplies, provides for 68.7 percent of China's total energy consumption, but also results in carbon dioxide emissions and sulphur dioxide emission, the latter of which causes acid rain. Modernization has contributed to China having sixteen of the twenty most polluted cities in the world (Gerth 2010: 185). Under Deng's policies, the costs of economic modernization and massive urbanization were tremendous, including environmental degradation, industrial and motor vehicle pollution, increasing greenhouse gas emissions, lack of adequate safety standards, growing socioeconomic inequality, and social conflicts. Increasing inequality in China contradicts revolutionary egalitarian ideals and has contributed to an increase in social struggles in both urban and rural areas, increasing from some 10,000 in 1993 to 87,000 in 2005 (Arrighi 2007: 377). China's demand for raw materials has resulted in importing iron ore and timber from Australia, Indonesia, and Brazil; oil from Sudan; and copper from Zambia (Bonini 2012: 63). In recent years China has undergone a decline in its annual growth rates to about 6–7 percent, still high when compared to developed capitalist countries, which has "contributed to a collapse of global commodity prices" (Li 2016: 1).

Amin (2013: 20) maintains that the characterization of China as state capitalist is vague and superficial so long as the specifics of the society are not better spelled out. He asserts that China is capitalist in that the party bureaucrats oversee a submissive and alienated workforce. China has embarked upon an ambitious project of national development in which it has extended its influence in numerous other countries, including ones in Africa. It is not clear to what extent the means of production in China has become privatized (Selenyi 2015: 47). The Chinese Communist Party has diminished its adherence to orthodox Marxist-Leninism and adopted Confucian and nationalistic ideals. At any rate, China has become an economic powerhouse. According to Rosa and Dietz (2010: 28), in 2003 it "consumed one-half of the world's cement production, one-third of its steel, nearly one-fourth of its copper, and nearly one-fifth of its aluminum." China has experienced a notable rising standard of living and acquisition of consumer goods and has a growing middle class. The more affluent sectors of China's population have become important players in the global culture of consumption. China accounts for around 28 percent of the world's meat consumption (Myers and Kent 2004: 20). China has overtaken both Japan and the United States in terms of the number of mobile phones and is the third-largest personal computer market in the world, although one must bear in mind that China's population exceeds that of the United States by four times and that of Japan more than tenfold. Nevertheless, growing inequality in China threatens to delegitimize a regime that still claims to adhere to socialist principles that most working-class people and peasants still accept (Amin 2013: 28).

In recent years, the Communist regime has embarked on an expansion of social benefits, including in housing, health care, and pensions as perhaps a means for countering growing social discontent (Amin 2013: 32). Conversely, Wang Hui, a member of the Chinese New Left, describes the result of a neoliberal agenda imposed by a strong authoritarian state as "consumerist nationalism" under which new political and economic elites engage in a pattern of ostentatious consumption at the expense of the working class and peasants who are exploited and remain on the margins of society.

German Democratic Republic

During the Cold War, a common perception in the West was the belief that the Soviet Union broke East Germany away from the rest of Germany as a "communist rump state" (Laquer 1895: 24). In reality, the partition of Germany occurred in the context of the Cold War, which quickly developed between the Western Allies and the USSR after World War II. Stalin wanted to create a buffer zone in Eastern Europe to discourage future invasions of Russia and the Soviet Union from Germany or other potentially Western states. While Soviet-sponsored regimes came to power as a result of this policy, Stalin apparently did not believe that Germany as a whole could be simply annexed into the Soviet bloc (Dennis 1988: 12). Until at least 1955, the Soviet Union proposed numerous plans that would have resulted in the formation of a reunited, neutral Germany, something along the lines of an "Austrian solution." Conversely, the United States did not place a priority on the political merger of the American, British, French, and Soviet zones but, with support from Conrad Adenauer, pushed for economic unity and a federal system for a new German state. On 7 June 1948 the Western Allies agreed that their respective occupation zones could draft a constitution for the creation of what became the Federal Republic of Germany (FRG). This development was followed by the Soviet blockade severing road, rail, and canal lines to West Berlin beginning on 24 June 1948, which was counteracted by the Berlin Airlift led by the United States. The German Democratic Republic (GDR) was formed on 7 October 1949 as a response to the creation of the Federal Republic of Germany on 24 September 1949.

The forty-year history of the GDR (1949–1989) was heavily influenced by the presence of its larger and economically more powerful neighbor to its immediate West. Under the Hallstein Doctrine formulated in the late 1950s, the FRG refused to enter into diplomatic relations with any state, except the USSR, which recognized the GDR. Eventually as a result of the Ostpolitik policy of Willy Brandt and his Social Democratic Republic, partial normalization of relations between the two countries commenced in 1972. Nevertheless, the FRG never granted full diplomatic recognition to the GDR and viewed all GDR citizens as FRG citizens. The Soviet Union granted permission for the formation of anti-fascist parties in the Soviet Zone on 10 June 1945. In April 1946 the German Communist Party (KPD) and the Social Democratic Party (SPD) merged to form the Socialist Unity Party of Germany (SED). In the course of its evolution into a Soviet-style Marxist-Leninist party, the SED purged its ranks of many former SPD members and communist advocates of a democratic road to socialism. The GDR nominally functioned as a multiparty state in it that permitted the existence of various other parties, including the Christian Democrats, the Liberal Democrats, and the Farmers' Party.

Soviet aid to East Germany was counteracted by war reparations exacted from the Soviet Union and the GDR during the period from 1945 to 1953 (McAdams 1985: 35–36). According to one source, reparations accounted for 33 percent of the Soviet Zone's GDP in 1947 (Childs 1983: 70). The first Five-Year Plan (1951–1955) stressed the production of massive reparations to the USSR, the construction of heavy industry, and the development of a military apparatus, much of which translated into austerity for the GDR masses (Childs 2001: 23). Ironically, the implementation of a more consumer-oriented economic policy after Stalin's death in March 1953 did not prevent the workers' rebellion of June 1953, which spread from East Berlin to many other GDR towns.

The glamor of the FRG economic miracle, the harsh economic conditions in the GDR, and the repressive policies of the GDR prompted approximately 3 million people, roughly one-sixth of East Germany's population, to emigrate from East Germany to West Germany between 1945 and 1961. Furthermore, West Germany retained the major portion of Germany's industry and natural resources, whereas East Germany lacked extensive natural resources and industrial facilities. While the Wall erected in 1961 violated the human rights of GDR citizens, it served to stabilize the GDR, at least partially, for the next twenty-eight years.

After 1961, the GDR embarked on a period of impressive economic growth. In 1963 the GDR implemented the New Economic System, under which state enterprises were allocated some decision-making responsibility and profit was identified as a key economic lever, emphasized placing more emphasis on quality goods than quantity of goods, and allowed some space for private enterprise. The 7th Party Congress in 1967 initiated the Economic System of Socialism, which attempted to incorporate systems theory and computer technology into the planning process. The SED regime announced its commitment to the enhancing the material and social needs of the GDR people. During the 1960s, the GDR experienced a "red economic miracle" during which it became the most affluent country within COMECON. World Bank figures listed the GDR in tenth place among industrial countries in 1980. The average monthly income of a worker increased from 558 Ostmarks in the 1960s to 1,140 Ostmarks in 1985 (Burrant 1988: 88–89). GDR agriculture generally fed the population at an adequate level, although fresh fruits and various vegetables were generally in short supply. Gorbachev and other Soviet officials often looked at the GDR economy for clues as to how to reform the Soviet economy. Central planning in the GDR may have operated better than in the USSR because of Germany's long historical experience with making bureaucracies rational and efficient. In the late 1960s, the GDR began to form *Kombinate*, groupings of related enterprises. By 1986, 132 Kombinate, with an average of 25,000 workers for each, existed at the national level and 93 existed at the regional level. Despite the GDR economic miracle, the command economy contained a number of glaring contradictions. Supply shortages, cautious managerial policies, and an 8.75-hour work day produced lower worker motivation. Due to alienation in the workplace and the larger society, the GDR workforce became increasingly apolitical as people turned their interests to the private sphere (Stand 2012: 68).

During the 1970s and the 1980s, the GDR increasingly imported sophisticated technology from advanced capitalist countries, particularly the FRG—a practice that further exacerbated the drive for hard currency through exporting the best products and the economic dependency of the GDR on the capitalist world system. Despite its limited energy resources, the GDR had been able to cover 65 percent of its energy requirements from domestic sources (Scharf 1984: 87). In 1985 brown coal supplied about 72 percent of the GDR's domestic energy needs (Gates 1988: 137). Unfortunately, an inferior technological infrastructure resulted in severe industrial pollution and greenhouse emissions, with the latter hardly being seen as problematic at the time, even on the part of the marginalized environmental movement in the GDR. Most of the GDR's imported raw materials and energy resources came from the USSR and Eastern Europe. The 1973 oil price explosion in the wake of the Arab-Israeli War and the global recession of the 1970s adversely impacted the GDR economy (Childs 2001: 25). As early as 1970, the GDR began to exhibit a trade deficit with the West and after 1975 with the USSR (Gates 1988: 133–134). The second international oil crisis in 1979 also adversely affected the GDR economy.

The increasing cost of domestic energy extraction and imported energy resources, the export of higher quality consumer goods to the West (in order to obtain hard currency to pay off debts to Western banks), and the heavy state subsidies for housing, basic food products, and social and health services, began to erode the relatively high material standard of living that GDR citizens had come to expect. Most GDR citizens were influenced by the hegemony of the capitalist culture of consumption, which they learned about vis-à-vis FRG media and through contacts with relatives in the FRG. The GDR state attempted to suppress environmentalists as well as reform-minded SED members who questioned the rationality of the culture of consumption. Unlike the FRG, the GDR lacked extensive access to cheap resources and the labor of southern Europe and the developing world that played a significant role in subsidizing the Western German culture of consumption. As Peter Marcuse aptly observes, "West Germany, Sweden, and the United States produce as efficiently as they do in part because they benefit from an international division of labor. If the capitalist standard of living were calculated on a comprehensive basis, it would have to include Haiti as well as Switzerland, Ethiopia as well as England, the underpaid and repressed millions of Korea, as well as the overpaid few of Wall Street" (Marcuse 1990: 51).

The Sixth SED Part Congress in April 1986 decided to further modernize the GDR economy by increasing exports of high-tech products, an effort that was to be facilitated with assistance from Stasi industrial espionage activities in the West and secret deals with Japanese manufacturers (Childs 2001: 26). Honecker backed the emphasis on microelectronics without possibly fully comprehending the costs involved. Childs reports: "In 1988 Scherer estimated that by the end of 1989 the GDR's indebtedness abroad would amount to 38.9 billion marks. He attempted to make it clear to Honecker that the GDR was on the verge of bankruptcy and presented reform proposals, including cutting back on costly, highly subsidised hi-tech enterprises. Honecker would not listen and even suspected conspiracy against him. In May 1989 Schuerer warned once again that the GDR was on verge of bankruptcy" (Childs 2001: 27).

Many GDR citizens, including rank-and-file SED members as well as SED members at low-level positions of leadership, welcomed the winds of change initiated by Gorbachev, not so much Perestroika but *glasnost*, in the late 1980s. However, SED elites resisted these changes and took various measures to suppress them. Once large numbers of GDR citizens, most of them in their twenties and thirties, discovered in summer 1988 an escape route to the West through Hungary, and, later, by seeking asylum in FRG embassies in Budapest, Prague, and Warsaw, the GDR began to quickly crumble. In an effort to stem the tide of mass exodus to the West and respond to mass demonstrations throughout the GDR, on 18 October 1989, the Politburo replaced the ailing Honecker with his heir apparent, Egon Krenz, as the first secretary of the SED. Mass demonstrations opposing Krenz's election forced the Politburo to resign on 7 and 8 November. Hans Modrow, a reform-oriented SED leader from Leipzig, became the first secretary and announced new travel and emigration regulations, free and secret elections, and a "contractual community" with the FRG. In the euphoria following the opening of the Berlin Wall on 9 November, it appeared for a brief time that the revolution might create a socialist-oriented democracy in the GDR. As Rossman (1990: 65) observes, the "explosive growth last fall of groups like 'New Forum,' 'Democracy Now,' and the revived Social Democratic Party, all of them anchored in the traditions of the Left, suggested that decaying Stalinism would be superseded by a reinvigorated democratic socialism, rather than a headlong rush into a D-mark capitalism." Ironically, East German trade unions remained largely removed from the events of the *Wende* (turn), either as critics of the existing regime or supporters of it (Stand 2012: 68).

Although the SED renamed itself the Party of Democratic Socialism (PDS), the expression "We are *the* people" became transformed into "We are *one* people." FRG Chancellor Helmut Kohl capitalized on the mood of the GDR masses and countered Modrow's suggestion of "contractual community" with a Ten-Point Plan for unification of the two Germanys. On 18 March 1990, the largest block of GDR citizens voted for a rapid unification plan supported by the Alliance of Germany, a multiparty coalition led by the Christian Democratic Union (CDU). In East Berlin, 35 percent of the electorate voted for the SPD, 30 percent for the PDS, and 6 percent for Bundnis 90 (an alliance consisting of the New Forum, Democracy Now, and the Peace and Rights Initiative). In the southern regions, however, "where about two-thirds of the GDR's industrial production originates, but where industrial and environmental neglect has also created the worst desolation, the Right won the critical advantage" (Minnerup 1990: 4). The euphoria of the reunification of the two Germanys in October 1990 quickly wore off as the reality that East Germans had become "second-class citizens" in the reunified Germany had begun to set in, a reality that has played itself out in various ways over the course of the past twenty-four years. The GDP of the new eastern German states diminished by over 60 percent in the first two years after unification and the collapse of many East German companies resulted in laying off of workers, contributing to unemployment rates in some regions rivaling those seen during the Great Depression (Zatlin 2007: 344). Sometime later, figures for 2008 show that, on the basis of employment income and investments alone, East Germans would have a disposable income of on average 68 percent that of their Western counterparts. With tax breaks and supplementary benefits such as pensions, child benefit payments, and so on, the disposable income of East Germans now reaches an average of just 78 percent of their West German counterparts. East Germans are clearly financially disadvantaged in comparison with West Germans (Hogwood 2012: 7).

While the GDR constitution formally provided for democratic processes in the polity, the workplace, and social life, the monopoly of power exercised by the SED elites precluded full expression of these ideals in the GDR. Whereas the term "socialism" had apparently become a dirty word in many other Eastern European societies, many GDR citizens continued to subscribe to it up until the last days of their country. Peter Marcuse, who spent a year as a Fulbright Scholar during 1989–1990 in the GDR, observes: "Our initial experience, after our arrival in mid-August (three months before the opening of the Berlin Wall), suggested a schizophrenic society; no one we spoke to defended the political structure, yet almost everyone professed a continued commitment to socialism and took the prevailing and social structure for granted, expecting, at best, small incremental changes here and there" (Marcuse 1991: 9).

A substantial nostalgia persists in eastern Germany for social life during the GDR era, in part evidenced by the fact that a 2007 poll indicates that about 20 percent of Germans "would like to see the Wall rebuilt," and the "desire to see the wall return is as high among former citizens of communist East Germany (the GDR) as it is among those from the west" (Connolly 2007: 20). The poll also reveals that reunified country is one where Wessi/Ossi tensions persist, suggesting that mentally there still are two Germanys.

Democratic People's Republic of Korea

Korea essentially functioned as a Japanese colony between 1910 and 1945, until Japan was defeated in World War II. Like the GDR, North Korea, officially called the Democratic People's Republic of Korea (DPRK), emerged out of the complex conditions of the Cold War (McWilliams and Piotrowski 2009). In the last days of World War II, the United States and the Soviet Union at the Potsdam Conference in July 1945 divided, theoretically only temporarily, the Korean Peninsula at the 38th parallel. Soviet occupation of northern Korea and U.S. occupation of southern Korea were to last until the defeat of Japan was achieved and a unified government could be established in Korea. In the meantime, Kim Il Sung and other Korean communists who had been living in exile in the USSR created a Soviet-type government in the north and began to carry out land reform. In the south the United States supported Syngman Rhee, who had lived in exile in the United States during the war and spearheaded an effort, with U.S. and UN support, in which his party won an UN-sanctioned election in May 1948, which led to the creation of the Republic of Korea (ROK), purportedly the government for the entire Korean Peninsula. Kim Il Sung and his fellow communists followed suit and proclaimed the establishment of the Democratic People's Party of Korea in September 1948.

When U.S. troops entered the South, they found a "working, socialist-oriented Korean government, the Korean People's Republic, maintaining order and production throughout the country through popularly run people's committees and mass organizations," but destroyed them in the south and imposed a militaristic government (Hart-Landsberg 1998: 12). One of the first battles between the DPRK and the ROK occurred on 4 May 1949, near Kaesong, a southern border city, a conflict that the ROK initiated (Hart-Landsberg 1998: 89). Ongoing tensions between South Korea and its U.S. supporters, and North Korea with its reluctant Soviet and Chinese supporters, provoked North Korea to invade the south on 25 June 1950. North Korea reportedly had received a report that South Korea planned to attack the northern part of the Ongjin Peninsula (Hart-Landsberg 1998: 118). China became involved in the bloody conflict after it looked like North Korea was on the verge of losing. The war ended in essentially a draw or truce that was signed on 23 July 1953, thus cementing permanently the division of the Korean Peninsula. Pyongyang essentially was obliterated as a result of the Korean War and an estimated 35 percent of all housing in North Korea was destroyed (Hunter 1999: 188).

Since that time, North Korea has developed into a largely isolated and demonized dynastic system that in some ways resembles former archaic states. Among postrevolutionary societies, North Korea exists almost in a league of its own, although at one time countries like Albania, Rumania, and Kampuchea also were often regarded as bizarre outliers, even within the world of postrevolutionary societies. According to Andrei Lankov (2013: 15), who lived as a Soviet exchange student in North Korea during the 1980s, what emerged into North Korea in the late 1950s was a "unique state that had almost no equivalents in 20th-century history—a truly fascinating topic for any cultural anthropologist, historian, or sociologist." North Korea initially adopted a development strategy very similar to China's strategy under Mao. In large part, North Korea was established as a Soviet client state in that its first leader, Kim Il Sung, was selected for this position by the Soviet military and assumed power in February 1946. Under his guidance, North Korea was organized around two principles, the first being *Juche* (self-reliance) and *Songon* (military first). Under Juche, which proclaimed Kim Il Sung as a great thinker and demanded absolute loyalty to him and the Korean Workers Party (KWP), it sought to embark on a development path different from those of either the USSR or China and between the early 1970s and the early 1990s it followed an "equidistance" policy between these two powers (Lankov 2013: 19). Kim Il Sung created a caste-like system consisting of three categories: "(1) the 'core' class included those who fought the anti-Japanese guerrilla war or war with South Korea and turned into high-ranking KWP cadres, (2) the 'wavering' class included the majority, and (3) the 'hostile' class included landowners, traditional elites, priests, doctors, merchants, and Japanese collaborators" (Woo 2014: 122). While initially Juche was defined as a creative form of Marxism-Leninism, reference to the latter disappeared in the mid 1980s.

During the Cultural Revolution, the Red Guard accused Kim Il Sung of being a neofeudal ruler who lived a life of luxury and decadence. Despite its strong ties to the Soviet Union, North Korea never opted to join COMECON. Given the facts that North Korea has undergone concerted efforts on the part of the United States to isolate and even overthrow it, in part by providing economic and military assistance to South Korea, one scholar argues that it "could scarcely be anything but suspicious and fearful" (McCormack 2008: 86).

North Korea organized its economy around centralized planning, heavy industry, and collective agriculture. Beginning in the 1960s, the North Korean government pursued the "'four modernizations of 'mechanization, electrification, irrigation, and chemicalization'," which resulted in a highly energy-intensive agricultural system (Haggard and Noland 2007: 26). During the 1960s and 1970s it implemented a Great Leap Outward policy under which it purchased plants and machinery from Western countries and Japan and "acquired the world's largest cement factory as well as a watch factory, paper mill, still mill, fertilizer and chemical plants, among many assets," an effort that incurred massive debts to the West and Japan (French 2014: 114). In 1976, North Korea ceased payments on its loans, having declared bankruptcy. At the beginning of the 1990s, both the Soviet Union and China demanded that North Korea use hard currency in its trade relations with them. The collapse of the Soviet Union had a devastating impact on North Korea's economy, including a large degree of deindustrialization. A standing army of 1.1 to 1.2 million has served to distort the North Korean economy, meaning, as is the case in many countries in the world, less funds for social and health services. Kim Il Sung died in 1994 and was replaced by his son Kim Jong-il, who adopted a policy of "military first politics" that privileged the Korean People's Army (KPA) over the Communist Party (Woo 2014).

The floods of 1995 and 1996, exacerbated by poorly designed terrace farming and poor forestry management, had a devastating impact on the North Korean economy and contributed to the Great North Korean Famine of 1996–1999, during which many lives were lost, although estimated figures vary wildly, with the most reliable figure being around 490,000 (Goodkind, West, and Johnson 2011). Meredith Woo-Cumings even suggested that the floods may have been the result of climate change, which interrupted a drying trend in northern China and North Korea starting in the 1950s: "These may or may not be linked to El Nino, because North Korea is located at much higher latitudes (38 degrees North to 45 N), far away from the main effects of El Nino whose influence is mostly tropical and subtropical. Rather they may be related to global warming and a natural phenomenon called the 'Arctic Oscillations,' which affects polar regions, including Eurasia and the North Pacific Ocean" (quoted in Cumings 2004: 180).

Conversely, the droughts of 1997 appear to have been impacted by the El Niño Southern Oscillation (ENSO) of 1997–1998. Unlike South Korea, which has much cultivable land, making it a veritable "rice bowl" due its extremely mountainous terrain, only about 20 percent of North Korea is cultivable. Furthermore, its cold climate translates into short growing seasons.

While in the past North Korea derived much of its electricity from hydropower, persistent drought has reduced reservoir levels so low that it became difficult to generate electricity, thus prompting the government to construct coal-powered plants (French 2014: 148). North Korea has oil and natural gas reserves, but has piped in these two fossil fuels for the most part from China because of inadequate extractive technologies. Heinberg (2004: 113) asserts that "North Korea is probably the most dramatic existing example of a modern industrial society falling into an energy-led collapse." Since the late 1990s North Korea developed an "identifiable two-track economy: the supply of materials and resources to the military and leadership would continue, while the civilian economy came increasingly to survive on humanitarian aid" (French 2014: 153).

Relations between North Korea and both the United States and South Korea underwent a brief thaw during 1994 starting with the Agreed Framework between the North Korea and the US. North Korea agreed to shut down its graphite-moderated nuclear reactors, which the United States suspected could be used to produce plutonium for nuclear weapons "in exchange for a package which included two light-water reactors, heavy fuel oil as interim compensation for electricity foregone, formal guarantee against nuclear attack and movement towards lifting of sanctions and normalisation of relations" (Beal 2005: 1). The United States could not follow through completely with its side of the agreement due to opposition from the Republicans who captured Congress in 1996. Kim Dae-jung, who was elected president of South Korea in 1997, entered into the "Sunshine Policy" with North Korea in June 2000, which temporarily eased tensions between the two countries. However, this brief period of rapprochement ended after George W. Bush assumed the U.S. presidency in January 2001 and labeled North Korea as part of an alleged tripartite Axis of Evil along with Iraq and Iran. During the Bush years, the United States openly discussed "developing new-generation nuclear weapons for use in Korea even while demanding that the DPRK abandon nuclear weapons" (Beal 2005: 52). After a failed nuclear test in 2006, North Korea succeeded in testing a nuclear weapon in May 2009, an apparent effort to deter potential aggressors.

Kim Jong-il died in December 2011 and was replaced by Kim Jong-un, the leader's youngest son. Kim Jong's eldest son, Kim Jong-nam, prior to his mysterious assassination in Malaysia in February 2017, had lived largely in semi-exile outside the country, particularly in China and Macao. Kim Jong-un, who had been educated in Switzerland and has been surrounded by advisors in order to compensate for his relative ignorance of the country that he leads, has proven to be the most public of the three Kims and has permitted more consumer items to flow into North Korea from the outside world, perhaps as a way of justifying his flamboyant lifestyle (Lankov 2013: 132–142). In essence, North Korea has evolved into a dynastic dictatorship under which its "leadership has no desire to replace nationalism with internationalism, much less see working-class socialist revolutions in South Korea or the West, events that would be highly destabilizing for the privileged state bureaucracy" (David-West 2013: 588).

North Korea embarked on a massive housing-building campaign that provided shelter to many North Koreans who became homeless during the Korean War of the early 1950s. Despite highly authoritarian and repressive rule under Kim Il Sung, Lankov maintains that most North Koreans lived fairly ordinary and quiet lives during the time that he resided in the country: "They thought about their families, they hoped to get a promotion, they wanted to educate their children, they were afraid of getting sick, they fell in love. They enjoyed romance, good food, and good books, and didn't mind a glass of liquor. The political and ideological was more prominent in their lives than in the lives of the average person elsewhere, but it still did not color most of their experiences" (Lankov 2013: 62).

Despite the existence of a relatively rudimentary health-care system, life expectancy in North Korea stood at seventy-two years, slightly lower than the South Korean figure at the time, but dropped to sixty-nine years after the devastating famine of the 1990s (Lankov 2013: 64). The WHO reported in 2008 that the infant mortality rate was 45 per 1,000 live births, slightly higher than China's infant mortality rate, but lower than those of many developing countries. Most North Korean hospitals are situated in run-down buildings with small crowded rooms. Conversely, the physician-patient ratio in North Korea is roughly on par with developed societies, 32.9 doctors per 10,000 people as opposed to 35.0 doctors per 10,000 people in France (Lankov 2013: 65).

Other than two assassination attempts by military officers on Kim Jong-il, the three Kims have managed to rule over North Korea with no serious political opposition from the masses, although some of them have defected to South Korea or other countries. In large part North Korea today constitutes a "dynastic power under national isolation, complete with ballistic missiles and mass poverty" (Therborn 2008: 11). While North Korea has engaged in more than its share of human rights violations and saber rattling, as Gavin McCormack (2008: 86) observes, in that it has "faced the threat of nuclear annihilations for more than half a century," it is not surprising that it has been "obsessed with its survival and has resorted to employing nuclear weapons as the *sine qua non* of national security."

Cuba

While it was one of the more prosperous Latin American countries, Cuba was ruled by a series of dictators during the period of 1902–1959, the last of them having been Fulgencio Batista. The Cuban Revolution of 1959 did not embark under Communist Party sponsorship as had been the case in China. Indeed, the insurgents who overthrew the Batista regime had significant support from the Cuban middle class, at least initially (Martinez-Fernandez 2014: 43). Fidel Castro, however, was highly influenced by Argentine Che Guevara, who did subscribe to Marxism-Leninism. According to Hodges, "In Cuba the turn toward socialism formally began with Fidel's May Day speech following the CIA-sponsored Bay of Pigs invasion in April 1961. In this speech he noted that the 1940 Cuban constitution had become outdated because of the socialization of major sectors of the economy.... Because the original populist-oriented 26 July Movement was ill prepared to administer the socialist reorganization of Cuba, Fidel made efforts to integrate it with the old communist party or Popular Socialist party (PSP)" (Hodges 1981: 162).

Castro announced in December 1961 that he was a Marxist-Leninist, despite his public admission that he had read very little of Marx or Lenin.

In part as a response to the United States having installed missile bases in Turkey, the Soviet Union installed missile bases in its new ally in the Western Hemisphere, namely, Cuba. As part of the negotiations surrounding the withdrawal of Soviet missile bases in Cuba resulting from the Cuban Missile Crisis of 1962, the United States agreed not to invade Cuba but refused to lift its economic blockade on the country. As a result of the blockade, the United States forced Cuba to purchase oil from the USSR and to nationalize its oil refineries (Kagarlitsky 1995: 154). Out of this process came the establishment of a new political party, the United Party of the Socialist Revolution, in February 1963, which became the Communist Party of Cuba in 1965. Cuba joined COMECON in 1970, became a full-fledged member of it in 1972, and began to coordinate its Five-Year Plans with those of the USSR by 1974. In a speech, Che Guevara argued that voluntary labor is an important ingredient for developing worker consciousness and paving the way for a new society, particularly if carried out in tasks outside one's ordinary routines, such as cutting sugar cane (Hodges 1981: 165–166). Cuba became highly economically dependent on the Soviet Union for finance and exported sugar to the Soviet Union, for which it received a preferential price of 42 cents per pound at the beginning of the 1990s, compared to the world market price of 9 cents (Morris 2014: 15). Cuban imports took up 40 percent of the country's GDP, including 50 percent of its food supply and 90 percent of its oil supply. While the revolutionary government initially had planned to develop a diversified agricultural economy and to industrialize, dependence on the Soviet Union forced Cuba to revert back to high levels of sugar production. The Cuban leaders convened the Organisation of Latin American Solidarity and Tri-Continental Conferences, at which they mildly criticized the Soviet Union for imposing its policies on Third World Communist Parties, but to no avail. In 1964 Castro attempted to mediate the tensions between the USSR and China, and "dispatched Guevara on diplomatic missions to both countries," also to no avail (Martinez-Fernandez 2014: 97–98).

Although Castro supported the Soviet invasion of Czechoslovakia in 1968, Cuba sometimes annoyed the USSR with its willingness to take a more radical stance by aiding revolutionary efforts outside of its borders, including in Latin America, Africa, the Middle East, and Southeast Asia. Furthermore, Cuba engaged in various foreign aid programs, including ones providing free education to international students and establishing medical missions abroad (Martinez-Fernandez 2014: 93).

Like many other postrevolutionary societies, Cuba began to liberalize its economy by the early 1980s. Les Holmes observes, "This was seen, for instance, in the legalization of private farmers' markets at the beginning of the decade. But at precisely the time when so many communist states in Europe (including the USSR) were intensifying the moves toward market economics, Castro announced that his government was concerned at the negative effects private markets were having on socialist morality in Cuba and shut them down (1986). As Gorbachev adopted ever more measures that Castro interpreted as endangering socialism, so the Cuba leader distanced himself and his country from the USSR" (Holmes 1997: 125).

With the loss of Soviet support, Cuba found itself with a fragile economy that various American businesspeople, including many of Cuban extraction, have been looking to take over for over four decades. Cuba's economic performance placed it thirteenth among the twenty-seven ex-COMECON countries for which the World Bank has complete data, but went into a deep recession in the early 1990s, like other "transition economies" (Morris 2014). Cuba needed about a decade to restore real per capita national income to its 1990 level and by 2013 it had risen around another 40 percent.

Cuba has managed to provide its citizens with free social services, including health care, transportation and telephones, low-rent housing, and school scholarships. It has one of the highest life-expectancy rates of ex–Soviet bloc countries and has one of the highest life-expectancy rates in Latin America. While obviously Cuba has produced some remarkable achievements since its revolution in 1959, it exhibits authoritarian tendencies because "power remains in the hands of a small and generally unelected elite" (Hardt and Negri 2009: 81). Kagarlitsky (1995:155) maintains that the "Cuban Revolution has shown once again the result of attempts to build socialism in a single backward country" and under external hostility. Conversely, Samuel Farber (2011), a Marxian historian, maintains that Fidel Castro and Cuban leadership in general did not need to be as authoritarian as they were and continue to be under the present leadership of Raul Castro, despite the U.S. economic embargo. At any rate, a common characteristic of virtually all postrevolutionary or socialist-oriented societies, including the Soviet Union, China, the GDR, North Korea, and Cuba, has been a psychology of besiegement that developed due to authentic external threats but may have been in many instances overexaggerated, thus resulting in authoritarian patterns that actually undermined the transition to socialist democracy.

Interpretations of the Nature of Postrevolutionary Societies

The historical reality that successful socialist-inspired revolutions occurred in semi-peripheral and peripheral capitalist countries rather than core or developed capitalist countries has contributed to a considerable amount of debate among Marxist scholars on the nature of postrevolutionary societies. Characterizations of societies, such as the former Soviet Union and other Eastern European countries, China, North Korea, Vietnam, Cuba, Mozambique, and Ethiopia, tend to fall into one of the following categories: (1) actually existing or state socialism, (2) transitional forms between capitalism and socialism, (3) state capitalism, and (4) new class societies or social formations (Linden 2007).

Actually Existing or State Socialism

Some scholars contend that postrevolutionary societies are basically socialist in nature, despite their ongoing contradictions (Sherman 1972; Sanderson 2010: 108–113; Lebowitz 2012). Syzmanski argues that "state socialism, where the initiative as well as the day-to-day operational decision making is in the hands of state officials, is just as much a socialist society as decentralized socialism, where initiative and day-to-day decision making are in the hands of the producing classes themselves. Whether or not a society is in fact socialist ... must be determined ... on the basis of whether or not decisions are made in the interests of the producing classes" (Syzmanski 1979: 24–25).

Syzmanski developed an elaborate typology of socialist systems, which is illustrated in Figure 2.1 below. He argues that socialism may evolve from one form into another. For example, whereas, in his assessment, the Soviet Union was a charismatic state socialist form between the 1920s and the early 1950s, it became a technocratic socialist one following the death of Stalin. Following Szymanski, the communal socialism of China during the Cultural Revolution evolved into bureaucratic state socialism following the demise of the "Gang of Four." He also argued that bureaucratic state socialism can regress into state capitalism as state and party officials acquire a number of privileges, and market socialism can develop over time into market capitalism.

Transition between Capitalism and Socialism

Probably most Western Marxists do not share Szymanski's belief that workers exerted considerable input, either directly or indirectly, in economic and political decision making in the former Soviet Union and most other postrevolutionary societies. Perhaps the most common alternative interpretation of these societies is that they constitute transitional forms between capitalism and socialism. Trotsky viewed the USSR as a hybrid social formation situated between capitalism and socialism, socialist in the sense that it lacked a capitalist class that extracted surplus value from the working class and that its economy was based on the production of

Figure 2.1. Types of "Socialist" Systems

A. State Socialism
   1. Charismatic socialism: predominant role of leaders
   2. Bureaucratic socialism: predominant role of bureaucracy
   3. State market socialism: predominant role of markets
   4. Technocratic socialism: predominant role of technocrats
B. Decentralized Socialism
   1. Decentralized market socialism: predominant role of markets with decision making by producers' collectives
   2. Communal socialism: predominant role of collective decision making without reliance on markets

---

Source: Adapted from Szymanski (1979: 130).

goods for use rather than exchange but capitalist in the sense that the party elites continued to exploit the working class. He viewed the Soviet Thermidor, a period of retreat from more radical goals and strategies, as an increasingly severe regressive development in all areas, but not one that resulted in a reinstatement of capitalism (Mandel 1992: 46). Trotsky believed that the USSR constituted an essentially new socioeconomic formation that despite its deformations had to be defended by both the Soviet and international working class against any attempt at capitalist restoration. He regarded the Soviet bureaucratic dictatorship as a historically transitional stage in which capitalism could be restored if it was not replaced by a soviet democracy orchestrated from below (Mandel 1992: 47). In *Revolution Betrayed*, he called on communists to mount a revolution against the Soviet bureaucracy as a necessary step toward achieving socialism. Trotsky favored a "transitional program" at the opening of the Fourth International in 1938, which called for a "sliding scale of wages and hours, workers' control of industry, opening the books of the employers, militant picket lines, workers' self-defence guards and labour-based militias, factory councils, soviets, the expropriation (without compensation) of industry and the banks, and, as a crowning demand toward which all other transitional demands point, a workers' government" (Smith and Dumont 2011: 127). He saw Stalinism as a consequence of "heritage of oppression, misery, and ignorance," of "cultural level of the country, the social composition of the population, the pressure of barbaric world imperialism" (quoted in Lane 1981: 86).

In part, following Trotsky's characterization of the Soviet Union as a "degenerated workers' state," Mandel maintains that it manifested aspects of capitalism and socialism. On the socialist side, "State ownership of all important industrial, transportation, and financial (i.e., of the means of production and circulation), combined with legal (constitutional) suppression of the right to their private appropriation, centralized economic planning and state monopoly of foreign trade, imply the absence of generalized production and the rule and the rule of value (and thus value) in the U.S.S.R." (Mandel 1981: 35).

Conversely, the survival of partial commodity production due to several factors, including the pressure of the world market and the underdevelopment of the forces of production, prevented the achievement of socialism in the Soviet Union. Instead, a privileged bureaucratic stratum of the working class came to control, but not own, the means of production. While Mandel admits that the nature of the Chinese workers' state differs from that of the Soviet Union and Eastern European countries, he maintains that the fundamental character of a society in transition from capitalism to socialism, halted in that usurpation of power by a privileged bureaucracy that can be removed only through an anti-bureaucratic political revolution, is also essentially applicable to China (Mandel 1981: 159).

Wallerstein (1979) also developed a version of the transition model. In his view, "socialist" countries were largely semi-peripheral units in the capitalist world economy, although the Soviet Union was prior to its demise emerging into a core power. He asserts that the "October Revolution changed the fate of world politics and undermined, just by having occurred, a major pillar of the political stability of the world capitalist world system" (Wallerstein 1979: 234). Initially, Wallerstein believed that the transition from a capitalist world system to a socialist world government would be a protracted (and by no means an inevitable) one, which could take another 100 to 150 years. During this interim period, "Several state machineries (and others will) have come under the control of socialist parties which are both seeking to bring about this worldwide transition and seeking to prefigure within their state boundaries some aspects of a socialist mode of production (such as collective ownership of the means of production, the institution of non-material incentives for production), etc." (Wallerstein 1979: 56).

Wallerstein (1998) argued at a later point that capitalism has only perhaps fifty or so years left and that humanity is heading toward a great historical transition in which it in all likelihood will be replaced by some type of global socialist system. In an even more recent projection, he is more cautious and notes that while the capitalist world system is in a terminal structural crisis that will culminate in a turbulent transition within the next twenty-five to fifty years, whether the outcome will be a more equitable social system or a more inequitable one is uncertain (Wallerstein 2007: 382).

State Capitalism

The notion that the Soviet Union prior to its collapse was a state capitalist society has had a number of adherents (Cliff 1974; Resnick and Wolff 2002). Charles Bettelheim (1976, 1978) argues that programs such as the New Economic Policy restored capitalism in the Soviet Union during the 1920s and 1930s. He argues: "What was supposed to give rise to an increasingly socialist relations has instead produced relations that are essentially capitalist, so that behind the screen of 'economic plans,' it is the laws of capitalist accumulation, and so of profit, that decide how the means of production are utilized" (Bettelheim 1976: 44).

A "bureaucratic state bourgeoisie" consisting of party and state bosses came to control, but not own, the means of production. State enterprises in Soviet-type societies, Bettelheim argues, reproduced capitalist relations of production in two ways: (1) by separating workers from control and ownership of the means of production and (2) through a system of commodity exchange between state enterprises. In addition to collectively appropriating surplus value or profit for the workers, the state bourgeoisie imposed its power by means of a highly centralized, authoritarian party apparatus. Christopher Chase-Dunn (1982: 34) argues that socialist movements that have assumed state power in various semi-peripheral and peripheral countries have not implemented socialism per se and "remain parts of the larger capitalist world-economy."

Social democrats and communists both advocated and implemented a greater role for government involvement in the economy (Wolff 2012: 108). Whereas social democrats want the state to regulate and partially socialize capitalist enterprises, communists want the state to "more extensively and intensively regulate and monitor but also to more thoroughly socialize the means of production" (Wolff 2012: 108). In essence, both want different forms of *state capitalism*, a system that essentially prevented social democratic and postrevolutionary societies from becoming more democratic and egalitarian and essentially prevented workers' democracy, which resulted in alienation. While in postrevolutionary societies state bureaucrats do not own the means of production per se, they in essence appropriate surplus values from the enterprises they manage.

Aside from what China may have constituted prior to the post-Deng era, which commenced in 1979, numerous scholars of different political stripes have contended that it has evolved into a form of state capitalism. Reportedly state-owned companies have experienced a steady decline in their share of assets, profits, and sales in the Chinese economy. According to Rothkopf, "While the state still plays a larger role in the Chinese economy than in any other major economy by far, companies that are not majority owned by the state account for two-thirds of output. These companies are also responsible for the lion's share of profits in the economy, and according to one South Korean estimate, they contribute almost three-quarters of Chinese GDP" (Rothkopf 2012: 352–353).

Many Chinese workers have come to resemble the working class in conventional capitalist developed and developing societies in that they strike for higher wages, better working conditions and social benefits, and even "greater workers' self-organization and independence" (Hart-Landsberg and Burkett 2005: 35). Furthermore, they have campaigned for trade unions independent of the state.

New Class Societies

Various scholars have referred to postrevolutionary societies as new class societies or social formations, neither capitalist nor socialist per se. For

Rudolf Bahro (1978), "actually existing socialism" in the Soviet Union and Eastern European countries, including the German Democratic Republic, served as a noncapitalist road to industrialization. He argues that these societies resembled the Asiatic mode of production in that they were based on state control rather than on capitalist ownership and control. Paul Sweezy (1980: 147) arrives at a somewhat similar conclusion: "[T]he most important difference between capitalism and post-revolutionary society is that this overwhelming dominance of capital has been broken and replaced by the direct rule of a new working class which derives its power and privileges not from ownership and/or control of capital but from the unmediated control of the state and its multiform apparatuses of coercion." The USSR ceased to have a property-owning class but was governed by a "new ruling class" (Sweezy 1980: 130). Sweezy (1980: 138) asserts that a proletarian revolution can result in a new type of society that is neither capitalist nor socialist. Thus the transition between capitalism and socialism in the USSR was at least temporarily delayed by bureaucratic deformation, thus making it a new class society in its own right. He maintains that in Soviet-type societies the state functioned as the "master of the economy," in contrast to capitalist societies where the state is the "servant of the economy" (Sweezy 1980: 142). In the USSR and other Soviet-type postrevolutionary societies, a new ruling class consisted of a "group of disparate individuals summoned to occupy the command posts" of a one-party state (Sweezy 1980: 146).

Samir Amin (1985) contends that postrevolutionary societies constitute class societies based on a state mode of production. Thus he characterizes societies such as USSR, Yugoslavia, Albania, Hungary, China, Vietnam, Cuba, etc., as "statist societies" that resulted from popular revolutions in backward countries or both peripheral or semi-peripheral zones of the capitalist world system (Amin 1985: 21). These societies faced a dual task of rapidly developing their productive forces and carrying out various other endeavors achieved elsewhere under capitalist development. Amin (2008a:21 characterizes postrevolutionary regimes in Asia and Africa as by and large having been "national-populist" and sometimes were "inspired by forms of organization in the experiences of 'really existing socialism,'" which came to embody single-party, authoritarian, command economic systems. Amin (1985: 53) posited the following theses about socialist-oriented revolutions:

- They constitute "moments of constitution of social and political forces capable of bringing forward national strategies of modernization and development."
- Statist societies should "pass through a moment of 'separation' or isolation from world capitalist system."

- They lose an authentic claim to exhibiting a "socialist" nature.
- Statist societies eventually aspire to be reincorporated in the capitalist world system.

Branko Horvat (1982: 21) utilizes the term *estatism* in referring to postrevolutionary societies: "A society will be called *estatist* if its ruling stratra profess basic tenets of traditional socialist ideology, such as emancipation of private productive property and the emancipation of exploited classes, but revise the socialist approach in regard to one crucially aspect: the role of the state." Instead of the traditional socialist view of the state as a repressive apparatus, in an *estatist* society, "[a]ll economic and political power is concentrated in the hands of the ruling political organization, which openly claims the monopoly of political power" (Horvat 1982: 21–22). In a somewhat similar vein, Moshe Lewin (2005: 379–382) argues that the USSR "should be placed in the same category as traditional regimes where ownership of high patrimonial estate equalled state power," which inherited aspects of the previous Tsarist regime and essentially functioned as a "development state in its initial stages."

In appraising various theories as to what was or is the nature of postrevolutionary societies, while I definitely do not view them as having been authentically socialist, I have viewed them as transitions between capitalism and socialism that need to undergo democratic revolutions that would make them authentically socialist; I am now not so sure whether this characterization of them is accurate. Obviously many postrevolutionary societies have been fully absorbed into the capitalist world system and have become capitalist. This would be true of Russia, Ukraine, the Baltic Republics, the former GDR, Poland, the Czech Republic, various Central Asian republics, and formerly Leninist-oriented African states. Perhaps China has become a state capitalist society but perhaps it still constitutes some sort of transitional society situated between capitalism and socialism, given that many Chinese at some level or another continue to be committed to socialist ideals. I am most likely to see Cuba as a transitional society because of its continuing commitment to socialist ideals, but obviously socialism cannot be created in one country.

Positive and Negative Features of Postrevolutionary Societies

Postrevolutionary societies created human resources, which were provided with a relatively high level of subsistence, and social and health services that were essential for economic development in underdeveloped or developing societies. Conversely, this program of development all too often entailed great human suffering, pain, and death in certain countries, including the Soviet Union, China, and North Korea, especially in the collectivization of agriculture and political repression. It is important to point out that the impersonal market forces of capitalist development have produced an even greater extent of tragedy over the long run and on a global scale, a point that critics of postrevolutionary societies often overlook. Silber captures a sense of some of the achievements of transformation of the Soviet Union from a backward agrarian country in 1917 to an industrial powerhouse by the late 1930s in the following observation: "[T]he new system brought Soviet citizens a higher and more egalitarian standard of living, along with guarantees of employment, free healthcare and public education without precedent elsewhere in the world. These gains were particularly noteworthy among the non-Russian nationalities, enabling the communists to forge what appeared to be a harmonious union between the scores of different peoples and nations of the old czarist empire" (Silber 1994: 116).

Postrevolutionary societies generally exhibited less economic inequality than occurred in capitalist society, less public ownership of the means of production, and less priority placed on human services, and they did not pursue capital penetration of other countries (Parenti 1997: 50). They were generally far ahead of capitalist societies at roughly the same level of economic development on criteria such as employment, education, health, social welfare, and land reform (Sweezy 1980: 147). In commenting on the purported economic failure of postrevolutionary societies, Saral Sarkar, an Indian who ended up living in the Federal Republic of Germany, observes that economically they were only relative failures. He maintains: "A Third World person like me, who could observe the standard in the GDR, was not all convinced that its economy was in a catastrophic state ... Even the foreign indebtedness of these 'socialist' countries was low compared to that of most Third World countries, and the threat of insolvency could have been removed if they had given up their ambition of catching up with the West" (Sarkar 1999: 25).

Indeed, as noted, I became intimately acquainted with socio-political life in the GDR as a result of being a Fulbright Scholar at Humboldt University in East Berlin (Baer 1998). In essence, I witnessed firsthand the GDR in what proved to be close to its last days, but during a time when virtually no one anticipated its collapse. Indeed, Erich Honecker, the head of the GDR state, gave a speech in October 1988 in which he said that the Wall would stand for another fifty years if necessary. While many of my informants generally recognized that advanced capitalist countries, including the FRG, exhibit serious contradictions, such as poverty and high levels of unemployment, they often felt that their state did not provide them with an accurate portrayal of life in the West. Conversely, many of my cultural consultants believed that advanced capitalist societies tended to be more democratic, provided greater freedom of expression, were economically more efficient, and provided greater access to material amenities. I often had the impression that many East Germans held an idealized view of the West—one that was conditioned by their viewing Western films and television programs and the credibility gap they perceived in terms of what the GDR state told them about life in the West. When I pointed out that the relatively high material standard of living for the majority in the advanced capitalist countries, such as the FRG and the United States, is in part related to the impoverishment and underdevelopment of the Third World, many East Germans agreed with me in theory, but had difficulty recognizing this reality. Most of them had not seen abject poverty firsthand either in the developed or developing worlds. Indeed, the GDR had one of the best child-care systems in the world (Eagleton 2011: 14). While indeed a legitimation crisis existed in the GDR, many East Germans adhered to the ideals of socialism but lamented that socialism in full practice did not exist in their society. Many East Germans since the unification have fond memories, an *Ostnotalagie*, about their lives in the GDR. Michael Brie (2007: 46) reports that "[s]tudies commissioned by Deutsche Shell AG provide evidence that the majority of young East Germans have positive memories of their life in the GDR."

Economy and Workplace

The command economy and central planning in the USSR facilitated rapid industrialization during the 1930s and 1940s, which during the latter period allowed the country to fend off the invasion by Germany. Table 2.5 below depicts economic growth in the capitalist core and postrevolutionary societies between 1961 and 1988.

Table 2.5 indicates that whereas the USSR and postrevolutionary societies overall exhibited comparable growth rates to the capitalist core countries during the 1960s, by the 1980s the USSR and the postrevolutionary societies exhibited lower growth rates than the capitalist core countries but China exhibited a high growth rate under the auspices of a rapid program of industrialization and modernization.

The centrally planned economy in the Soviet Union resulted in various inefficiencies and problems. In attempting to meet production targets, the focus was on meeting the target, with little attention being given to quality. Overall the long run, the Soviet economy proved to be inadequate in providing a wide range of consumer goods and services. Many goods

Table 2.5. Economic Growth in the Capitalist Core and Postrevolutionary Societies, 1961–1968.

	Average annual real GNP growth rate		
	1961–1970	1971–1980	1981–1988
Capitalist core	4.95	3.70	2.62
Postrevolutionary societies*	4.70	3.45	2.27
USSR	4.90	2.85	2.08
China	4.05	5.70	8.68

*USSR, Eastern Europe, Yugoslavia, China, North Korea
Source: Adapted from Boswell and Chase-Dunn (2000: 101).

were overproduced and many other goods remained in inventory because of distribution problems. Furthermore, the command system resulted in the hoarding of supplies than contributed to shortages of goods for sale.

The focus on industrial production contributed to a neglect of efficient agricultural production, resulting in serious food shortages at time. During the period of 1976–1980, the Soviet Union experienced "three unusually bad weather years which adversely affected the harvests," resulting in less for urbanites and contributing to lower work motivation and productivity (Zimbalist and Sherman 2015: 193).

Postrevolutionary societies generally provided security of employment, a relatively leisurely pace of work, and availability of alternative jobs. Official unions protected rights of workers, despite the fact that leaders were chosen by party elites. However, workplaces in postrevolutionary societies were characterized by an inefficient work organization, a slow work pace, toleration of disruptions to work regime, and a general disregard for the quality of work. Lebowitz (2012: 108) characterizes the Soviet economy as "dysfunctional" in terms of "the waste, the stockpiling of labour and resources, the poor quality products, extreme alienation and low productivity of workers, the lack of correct information to engage in planning, the departmentalism, the plan evasions, and the inability to control enterprise managers." Furthermore, between the 1950s and 1980s, it experienced declining economic growth and failure to keep up with innovations in computer technology. Overall, postrevolutionary societies were "less efficient, less flexible, and more resistant to innovation" than were advanced capitalist economies (Silber 1994: 144)

The USSR had an informal economy, sometimes referred to as a "second economy" or underground economy, that entailed some production and the selling and purchasing of various goods and services by individuals functioning outside of the formal economy. The authorities tolerated this sector because the formal economy was "notoriously deficient" in providing for certain goods and services (Sweezy 1980: 143).

Although the SED leadership sought to impose tax increases and cuts in social services and consumer items in 1952 in order to invest in heavy industry and defense, the reduction in workers' living standards resulted in worker rebellions in East Berlin and across 272 cities and towns throughout the GDR. According to Vu (2014: 444), "Soviet tanks brought the country under control, but SED leaders were forced to make quick concessions to defuse the tension: the new norms were rescinded and the government thereafter consistently spent more on consumer goods and welfare." The GDR was generally acknowledged as having been the most technologically and economically developed society in the Soviet bloc, with Czechoslovakia not far behind in terms of these dimensions. While there has been a tendency to compare the GDR with the FRG in terms of economic performance, Stokes (2000: 8) maintains that in the late 1950s and the early 1960s, the GDR resembled the U.K. and Japan in that all three were "highly industrialized, with considerable scientific and technological capabilities" and had largely government-directed centralized economies. Carl Zeiss Works, a manufacturer of optical lenses and equipment in Jena, maintained a "very high level of technological capability through the GDR's existence and was a major foreign-exchange earner for the often cash-strapped country" (Stokes 2000: 17). While alienation certainly existed in GDR workplaces, some factories functioned as virtually self-contained worlds. A former supervisor of the Chemiefaserwerk (CFW), which produced artificial silk, in Premnitz reported: "When I came to the CFW in 1964, this was still a self-sufficient world. There were 7500 employees. The factory maintained a nursery (garden centre), a butchery, a piggery, a hospital, two dentists, a laundry, childcare facilities, cultural centres, sports facilities, a large accommodation block for apprentices, a school for vocational training, a library, a bath house, five canteens and so on.... The CFW had flats [for its employees] and financed almost all sporting and social initiatives for the town of Premnitz" (quoted in Madarasz 2006: 138).

Gail S. Henderson and Myron S. Cohen (1984) conducted an ethnography of the Second Attached Hospital complex, which includes staff dormitories and various auxiliary buildings situated on the outskirts of Wuhan in China; I will describe it in the "ethnographic present." The medical college is adjacent to the hospital's grounds. About two-thirds of the employees at the hospital belong to its attached *danwei*, or work unit. The *danwei* functions as a quasi-urban village that not only provides housing and other services but serves as the center of its members' economic, political, and social life. The *danwei* is the "vehicle through which staff may communicate with higher-level authorities" (Henderson and Cohen 1984: 7). About a third of the approximately 830 hospital workers live outside the complex and some residents of the complex work outside of it. The hospital complex includes day-care centers, schools, and businesses. An estimated 75 to 80 percent of the hospital and medical staff are married to others on staff. In contrast to residential patterns in capitalist countries, physicians often live next door to cooks or maintenance workers. The hospital is responsible for dispatching health workers for a fifteen-county area. A special ward provides medical treatment for high-level cadres. Although the hospital emphasizes biomedical treatment procedures, patients may request admission to the combined Western and traditional Chinese medicine ward. Although work units are hierarchical units whose staff are assigned and whose leaders are appointed, some provisions have been made for feedback from *danwei* workers. Although the input sought by supervisory personnel from their subordinates hardly qualifies as proletarian democracy, it is not any less rigid than authority structures in most workplaces in capitalist societies.

Social Stratification and Social Benefits

Overall, postrevolutionary societies have exhibited less social inequality in terms of income and wealth than capitalist societies, whether in the developed or developing worlds. Hodges reports: "In 1921 there was little difference between the wages of skilled and unskilled workers. But with the introduction of the New Economic Policy (NEP) in 1922 wage scales were revised upward" (Hodges 1981: 119).

The wage differential in the USSR rose from 8:1 in 1924 to 30:1 in 1926 (Hodges 1981: 81). After 1934 the USSR ceased compiling statistics indicating the maximum and minimum incomes for specific jobs (Hodges 1981: 120). Reportedly, in "1966 the difference in pay between the highest-paid bureaucrats and the lowest-paid workers in the Soviet Union was approximately 30:1—roughly what it had been in 1926" (Hodges 1981: 86). Sherman (1995: 201) maintains that from 1928 on, the Soviet Union essentially had a class system with two major classes: (1) the political ruling class, which "controlled the economy, extracted the surplus, determined the use of the surplus" and the working class, which labored for "money income, doing industrial, agricultural, or professional work." The working class represented a wide range of income and prestige, "since it included workers from highly paid artistic and sports stars to very poor paid menial workers" (Sherman 1995: 205). Elsewhere, Sherman along with Wood identifies the four main groups in the Soviet ruling class: the top echelons of the Communist Party, including members of the Politburo and the Central Committee; the top echelons of the government bureaucracy, such as the Council of Ministers; the top military officers; and the top economic managers and planners (Sherman and Wood 1989: 344). In comparing the Soviet and U.S. power elites, Sherman and Wood maintain: "Whereas the very wealthy and powerful of the American ruling class number up to a million or so, the similar group in the Soviet Union is much smaller, less than a hundred thousand. For an economic equalitarian standpoint, this is quite good in the sense that the Soviet leaders take a very much smaller amount of the total national income for their own use than do the American rulers.... On the other hand, it means that political power is much more concentrated than in the United States" (Sherman and Wood 1989: 345).

The Chinese Revolution eventuated in a marked improvement of the material standard of living of the Chinese people. Whereas in 1950 the average Chinese person lived at a bare subsistence level and had life expectancy of about forty years, by 2008 Chinese incomes had risen to 80 percent of the global average and life expectancy had risen to over seventy years (Harrison 2014: 394–395).

Revolutionary Cuba implemented food rations in the mid 1960s, which ensured that the poorest people were guaranteed a relatively adequate diet commensurate with people at higher levels of income. The Cuban state also implemented "free and public services, including health care, education, child care, bus transportation, sports events, and various utility" and established polyclinics in keeping with the Soviet model of health care beginning in 1965 (Martinez-Fernandez 2014: 104). In large part due to the influence of Che, social inequality in Cuba tended to be lower than it has been in most postrevolutionary societies: "As late as 1970 the minimum wage was 60 pesos per month for agricultural workers and 85 pesos for urban and industrial workers. With a few exceptions, the spread was slightly less than 6:1, and the total spread was less than 8:1" (Hodges 1981: 171).

Despite the rhetoric of social equality, Soviet bloc societies exhibited a social hierarchy, which included the party elites (known in the Soviet Union as the *nomenklatura*), the intelligentsia and technical specialists, the workers, and the peasants (Filtzer 2014: 509). Filtzer reports: "Members of the elite had sumptuous flats, large country cottages, chauffeured cars, access to Western clothes, furniture, and foods, and the right to read Western newspapers and watch Western films. If they or their families felt ill, they would be treated in special hospitals with the best (often Western) medical equipment and well-trained doctors and surgeons" (Filtzer 2014: 506).

Industrial managers, popular entertainers, and the higher status authors and artists also enjoyed perks not available to the masses. However, until the 1980s, the gap between the party elites and workers in China was much less than had been the case in the Soviet bloc countries.

Overall the status of women improved, often remarkably, in postrevolutionary societies, even though patriarchy remained well-entrenched, evidenced by the fact that women often were still expected to perform domestic tasks alongside their employed jobs. In the Soviet bloc countries, women's employment tended to be concentrated in low-wage industrial and clerical positions and agriculture. By the 1970s, women accounted for about half of university students in most Soviet bloc countries and they made up about a quarter of parliamentary representatives (Smith 2014: 31). The status of women improved under collectivization in China, but they did not earn the same work points as men. Women were virtually absent at the highest levels of political leadership as evidenced by the fact that the USSR only ever had one female voting member of its Politburo and no woman was ever the member of the GDR Politburo. Harsch observes: "Very few women became a state minister and the ones who did served as minister of culture or education or consumer affairs, relatively marginal ministries. An exception was Hilde Benjamin, minister of justice in East Germany (GDR) from 1953 to 1967. Under her tenure, the administration of justice, including many judgeships, became notably feminized" (Harsch 2014: 492).

Postrevolutionary societies had or have women's mass organizations; however, they have had had minimal influence in policy making. Beginning in the 1960s, women assumed positions in health care, social welfare, government administration, retail trade/sales, and hospitality in large numbers. Women have gained considerable visibility, for example, in the highly touted Cuban health-care system.

Postrevolutionary societies brought "social security, rudimentary welfare, improved health care and education to people who had lacked these things under the old order" and were better off in these areas than their counterparts in "vast areas of the world, such as Latin America, Africa, parts of Asia and the Middle East" (Smith 2014: 32). China has witnessed a profound improvement in life expectancy at birth, from 36.3 years in 1960, to 66.8 years in 1980, and 70.3 years in 2000 (Li 2008: 34).

Environmental Problems

The Soviet Union beginning in the 1930s underwent a tremendous amount of growth in part due to relatively easy access to the rich natural resources of a vast country, indeed the largest country that has ever existed in the world, one that accounted for about one-sixth of the earth's land surface, stretched over eleven time zones, and exhibited considerable climate variability. Unfortunately, the Soviet Union and other postrevolutionary societies have had, by and large, a poor environmental record. While the point can be contested, this tragic reality prompted eco-Marxist Reiner Grundmann (1991b: 6) to assert that "it seems that socialist countries present an even worst ecological record than capitalist countries."

The fast-paced drive for industrialization, in part rooted in the threat posed by developed capitalist countries, particularly the United States, contributed to serious environmental damage. The first Five-Year Plan (1928–1932) entailed a commitment to rapid economic growth to propel the Soviet Union quickly out of the historical economic underdevelopment of Russia. In the words of Stalin himself, "We are fifty to one hundred years behind the most advanced countries. We must close this gap in the span of ten years. Either we do that or they will sweep us away" (quoted in Weiner 1988: 168).

Ecologists working in government planning agencies questioned the sustainability of such a course of action but obviously were in no position to stand in the way of Stalin's and the party elite's notions of progress.

The managerial objective of producing maximum output at minimal cost resulted in high levels of air, water, and soil pollution and a lack of safety precautions in industrial and nuclear power plants. Feshman and Friendly (1992: 40) maintain that the "plan and its fulfilment became engines of destruction geared to consume, not conserve, the natural wealth and human strength of the Soviet Union." Early ambitious development projects included the construction of the White Sea–Baltic Sea Canal in the early 1930s and the Great Stalin Plan for the Transformation of Nature, initiated in the late 1940s, which sought to divert some of the river flow of northern rivers southward to Central Asia and southern Russia (Ziegler 1987: 26). Under Khrushchev, a glaring example of environmental devastation was soil depletion due to agricultural practices in the Virgin Lands experiment in Siberia. While agricultural production dramatically increased, "by the end of the decade inadequate replenishment of the soil had transformed much of the Virgin Lands into a giant dustbowl" (Holmes 1997: 226). Soviet efforts beginning in the 1960s to increase irrigated cotton production also resulted in serious environmental degradation, particularly on the Aral Sea, which derived its water sources from the Amu Darya and Syr Darya rivers. In the Soviet Union and Eastern European countries, agriculture contributed to "environmental damage through misuse of irrigation systems and pesticides, and as a result of gas and effluent output from larger livestock rearing complexes" (Gorz and Kurek 2001: 200). The Soviet Union and Eastern European countries built numerous hydroelectric dams and steel mills, the latter in particular with outmoded technology, resulting in serious environmental degradation (McNeill and Engelke 2014: 195).

The Soviet Union exhibited the worst instances of radioactive contamination, the most spectacular being that of Chernobyl nuclear power plant, and Czechoslovakia and Poland had the highest levels of industrial pollution in Europe and perhaps the world (Commoner 1990: 219–220). Most Soviet cities were characterized by high levels of air and water pollution in the last days of the Soviet Union (Oldfield and Shaw 2001: 175). Table 2.6 below depicts some of the serious environmental problems that existed in various regions in the Soviet Union in what proved to be its last days.

In the last decades of the USSR, an environmental social paradigm posed a mild challenge to the dominant social paradigm by calling for elimination of waste and the use of science and technology to manage

Table 2.6. Regions of the Soviet Union Suffering Serious Environmental Problems.

Region	Location	Environmental problems
Aral Sea	Central Asia	Severe desiccation due to reduced water inflow
Areas affected by Chernobyl fallout	Belarus, Russia, Ukraine	Radioactive contamination
Donbass	Ukraine	High levels of air and water pollution due to industrial activity
Moldova	Moldova	Large areas with land and soil degradation due to agricultural activity
Black Sea	Azerbaijan, Georgia, Kazakhstan, Russia, Turkmenistan, Ukraine	Air, water, and sea pollution
Middle Volga	Russia	Air and water pollution, depletion of water, land degradation from mining, soil erosion from wind
Urals industrial zone	Russia	Land degradation from mining, air and water pollution, forest degradation
Baikal	Russia	Air and water pollution, depletion of fish, forest degradation

Source: Adapted from Oldfeld and Shaw (2001: 177–178).

environmental problems, but it did not question the faith in ongoing economic growth (Ziegler 1987: 42–43).

In the case of various Eastern European countries, although many of them had passed legislation for environmental protection and conservation, in reality these measures were not strictly enforced (Dingsdale and Loczy 2001: 189). Due to its intensive industrial projects, including chemical production in the Leipzig and Halle areas, the GDR reportedly in the early 1980s had the "most severely polluted water in East Europe," even more polluted than water in Czechoslovakia and Poland (Holmes 1997: 226). Many environmental specialists declared that by the late 1980s, most of the GDR and large portions of Poland and Czechoslovakia constituted the "most heavily polluted region on earth" (Holmes 1997: 227). Environmental problems in the Soviet Union and Eastern European countries resulted in the emergence of environmental groups, some of them officially sanctioned and encouraged by *glasnost* in the Soviet Union.

Yih maintains that such instances of environmental devastation are or were rooted in the conditions under which postrevolutionary societies developed, which included:

> relative underdevelopment, external aggression, and, especially for the small, dependent economies of the Third World, a disadvantaged position in the international market. The corresponding pressures to satisfy the material needs of the populations, ensure military defense, and continue producing and exporting cash crops and raw materials for foreign exchange, have led to an emphasis by socialist policymakers on the accumulation by the state, the uncritical adoption of many features of capitalist development, and a largely abysmal record vis-à-vis the environment (although there are exceptions, of course). (Yih 1990: 22)

A commitment to full employment inadvertently reduced the drive to use fuel-intensive technologies as a way of replacing labor-intensive jobs. Also, Soviet citizens were not encouraged to participate in a culture of consumption characteristic of capitalism that encourages people through advertising to purchase more and more consumer items ranging from large automobiles to a wide array of electronic devices and appliances.

Furthermore, the weak development of democratic institutions in postrevolutionary societies and bureaucratic suppression of information about the environmental impact of agricultural and industrial practices had until recently inhibited the emergence of an independent environmental movement (O'Connor 1989: 99). Although *glasnost* permitted the emergence of a small Green movement in the Soviet Union, the official policy of Perestroika, with its emphasis on production, and the serious disruption of the Soviet economy in what proved to be the last days of this postrevolutionary society, served as impediments to the implementation of environmental protection regulations. The Soviet Union and other postrevolutionary societies were caught up in a feverish competition with the developed capitalist countries in which the environmental impacts of industrialization and arms production were overlooked. On the positive side, however, Fidel Castro argued in a speech on 13 March 1968 that an underdeveloped society, such as Cuba, cannot tolerate waste (Hodges 1981: 167).

Hou and Xu (2012) delineate three eras of ecological crisis in China since the 1949 Revolution. The first occurred during the Great Leap Forward, when steel production resulted in extensive deforestation and increased agricultural production, which resulted in farmers engaging in deep plowing and close planting, practices that destroyed soil fertility. The second occurred during the Cultural Revolution during which intensive crop production intended to eradicate food scarcity contributed to neglecting forestry, animal husbandry, fisheries, and other agricultural activities. The third was part and parcel of the drive for modernization, which resulted in wastes from the manufacturing sector being dumped into the rivers, an increase in desertification, and extensive air pollution in most cities, with three hundred of them having pollution levels below WHO standards and high levels of greenhouse gas emissions. In the late 1950s, out of the tens of thousands of dams launched during the Great Leap Forward of the late 1950s, almost three thousand of them had collapsed by 1980 due to poor construction (Geall 2013: 54). In terms of environmental devastation, China under modernization has probably become worse than the Soviet Union and the Eastern European countries ever were. I spent three weeks in July 2007 as an academic staff person with the geography field school at the University of Melbourne in parts of eastern China, including Beijing and Shanghai. The only place where we saw clear skies was the steppes of Inner Mongolia. Pieke reports: "Air, water, and soil in many areas are now so polluted that they have become virtually unusable, and the environment has become a major cause of local political contention. ... In some places, the damage is irreparable, or, perversely, pollution has become so much part of life that the local population either is ignorant of its harmful effects or has become so dependent on payoffs from polluting factories that they actually resist change" (Pieke 2014: 126).

A major strain on China's environment historically has been rapid population growth, in large part stimulated by poverty. It had a population of about 535 million in 1949, 840 million in 1970, and over a billion in the early 1980s (McWilliams and Piotrowski 2009: 351). Mao, according to McNeill and Engelke (2014: 157), had "refused to be alarmed by China's rapidly growing population in part because accepted Soviet orthodoxy on the matter." Despite the implementation of the "One Child Family Policy" in 1979, China's population today stands around 1.4 billion people, in part due to exemptions allowed under the policy as well as violations of it.

China's annual greenhouse gas emissions grew by 80 percent between 1990 and 2007 and surpassed those of the United States in 2007. As the "factory of the world," China derives 70 percent of its energy from coal, most of it obtained domestically but some of it imported from other countries, including increasingly Australia. While there has been some shift to renewable energy sources in China in recent years, it hardly has been enough to stave off the emissions resulting from a political-economic system committed to maintain high rates of economic growth. While China still lags behind developed countries in both per capita resource consumption and per capita greenhouse gas emissions, as John Gulick (2011: 27) observes, its program of economic development "could very well unleash into the atmosphere the extra increment of carbon dioxide that catalyzes runaway global warming, a catastrophe that would not only put paid to the chimera of Chinese hegemony, but would also devastate China's hundreds of millions of rural poor as severely as any other human collectivity in the world-system."

To give China its due, while it continues to be committed to high economic growth, it has begun to invest in renewable energy sources to a greater extent than the United States, Australia, and other G20 countries (McMichael 2012: 279). Various progressive scholars are calling for the development of a "socialist-ecological civilization" in China that stresses people's well-being, social benefits, and ecological sustainability over continual economic growth (Huan 2010b: 200–202).

Social Democratic/Welfare Capitalist Societies

Social democracy or welfare capitalism is often viewed as a "third way" or an intermediate path that combines the best elements of capitalism and socialism. Ironically, Marxists in Russia, including Lenin and Trotsky, referred to themselves as "social democrats" and their party as the Russian Social Democratic Workers Party. In the April Theses that Lenin (1964: 24) wrote on his return to St. Petersburg in 1917 he proposed that "Instead of 'Social Democracy', whose official leaders *throughout* the world have betrayed and deserted to the bourgeoisie ... we must call ourselves the *Communist Party*."

By the eve of World War I, the German Social Democratic Party was the largest social democratic party in the world and highly influential among working-class Germans. Virtually all socialist or social democratic parties, with the exception of the U.S. Socialist Party under the leadership of Eugene Debs, supported nationalist agendas in their respective countries, including Germany, despite having earlier declared that World War I constituted an "imperialist war" in which various imperial powers were contending with each other to further their national ambitions, particularly in Africa and Asia. Left-wing social democrats broke with their respective national parties and formed the basis of a new Communist International.

Social democrats ideologically have been committed to parliamentary democracy and co-existing with both more radical and conservative parties. In Germany social democracy flourished to some extent during the Weimar Republic, disappeared during the Nazi period of 1933–1945, and resurfaced in the Federal Republic of Germany but was subsumed within the Socialist Unity Party in the German Democratic Republic. British social democracy has by and large operated under the guise of Fabian socialism and with the support of the Labour Party. After World War II, social democracy again flourished in various Western European countries, particularly Germany, the Low Countries, the Scandinavian countries, France, Austria, and Italy but also eventually Spain after the overthrow of the fascist regime under Franco in the late 1970s. While Western European social democratic parties contained a left-wing stream that sought rapprochement with Soviet bloc countries, as was the case with SPD German Chancellor Willy Brandt, most social democrats assumed an anti-communist stance and sided with the United States and Britain in an "Atlanticist" consensus.

Social democracy seeks to make capitalism more socially just and democratic and advocates a mixed economy with predominance of private ownership but some public ownership of the means of production. Social democratic parties have varied on the issue of public ownership of the means of production. Padgett and Paterson observe:

> A commitment to nationalisation of the basic means of production was, until the 1950s, a central feature of party programmes in parties such as the British Labour Party as well as those such as the German Democratic Party (SPD) which had been more influenced by Marxist ideas. Subsequently, the impact of the Cold War, allied to the pervasive character of postwar posterity, moved most of the parties to weaken their commitment to nationalisation. This change was most dramatically exemplified by the Bad Godesberg programme of the German party (1959) which served to redefine social democracy. (Padgett and Paterson 1991: 1–2)

Social democracy became associated with Keynesian economics, which advocates state regulation of the privatized economy and welfare capitalism. For Sanderson (1995: 374), social democracy, particularly as it has functioned over several decades in Sweden, "means a capitalist system that has a large number of built-in protections in terms of minimum standards of income, health care, education, and other aspects of the modern welfare state, but at the same a great deal of economic planning." In Scandinavia, social democrats view the capitalism as an entity that must be regulated, asserting that "the market is an excellent servant, but a poor master" (quoted in Brandal, Bratberg, and Thorsen 2013: 9). In various countries, including Britain, France, Germany, and Australia, social democratic parties, including the Labour Party in Britain, the Socialist Party in France, the Social Democratic Party in Germany, and the Australian Labor Party, embraced aspects of neoliberalism, including deregulation and privatization, while in power beginning in the 1980s (Moschonas 2002: 220–245). In the case of the SPD, over the course of "its seven years of power, the Schroeder government committed almost every sin one could imagine from a supposed left-wing government: leading Germany into the war against Yugoslavia (its first foreign engagement since 1945); cracking down on social security, cutting unemployment and welfare payments under the notorious 'Hartz Laws'; partially privatizing the pension system (further deregulating financial markets); lowering taxes on the high incomes and business" (Spehr 2012: 161–162).

Social democratic parties have lost much of their working-class character and have by and large become "people's parties" with a broad constituency, including professional people. As Moschonas (2002: 306) observes, "contemporary social democracy seeks to accommodate in its broad church highly diverse social groups, with opposed interests and values: capitalists and workers, 'excluded' and 'excluders', smart, snobbish districts and council estates; but also left-wing intellectuals with a 'universalist' culture, the traditional middle classes, and popular strata prey to xenophobia and 'law-and-order' appeals." In contrast to northern Europe where the ties between social democratic parties and the working class have historically been strong, this has not been the case in Mediterranean Europe. Indeed, in Italy many working-class people have had a strong link with the Communist Party. In Australia the Australian Labor Party (ALP), which drew heavily on Fabian socialism and once called itself a "democratic socialist party," has had strong ties with trade unions, but these have become nominal as professional politicians have come to dominate the party leadership.

Social democracy in Denmark, Sweden, and Norway takes the guise of the "Nordic model," which has permitted partial socialization and redistribution of income, support of property rights, and generous provision social, educational, and health services. To the left of the three social democratic parties, namely, the Social Democrats of Denmark, the Social Democratic Labour Party of Sweden, and the Norwegian Labour Party, are the Socialist People's Party in Denmark, the Socialist Left Party in Norway, and the Left Party in Sweden (Brandal, Bratberg, and Thorsen 2013: 3). The Nordic social democratic model exhibits a strong social welfare system consisting of the following traits:

- Financing of the social welfare state largely from general revenues
- An extensive system of social services, particularly for health care and education, but also day care for preschoolers and care of elderly people
- A robust family policy that allows women to enter the labor force on equal terms by providing day care for children
- Job protection policies ranging from low (Denmark) to high (Sweden)
- Corparatist industrial-labor relations with a system of centralized collective bargaining
- A stated commitment to full employment (Meyer and Hinchman 2007: 137–138).

In contrast to social democratic societies such as the Scandinavian societies and Germany, the social welfare regime is much weaker in "liberal, Anglo-Saxon" societies such as the United States, the U.K., and Australia. Gorbachev during his stint as the Soviet head of state envisioned the Soviet Union evolving into a social democratic society along the lines of the Scandinavian countries.

Social democrats, compared to Green parties around the world, including in Europe, tend to be weak on environmental issues. While increasingly expressing concern about the environment, social democrats are generally committed to the growth paradigm and speak of "sustainable development" and "green growth." In Germany, the Social Democrats entered into a coalition government with the Greens during the stint that Gerhard Schroeder served as chancellor. Brandal et al. report: "In the Swedish general elections in 2010, the social democrats failed to regain power due to their inability to win a majority with the Left Party and the Greens. As of 2013 Norway and Denmark are governed by centre-left coalitions, in the former case as a socialist-socialist-agrarian coalition, and in the latter as socialist-social democrat-radical one. Only in Sweden, however, do we find a viable green party. In contrast, in Norwegian and Danish politics, the green dimension has been absorbed first and foremost by socialists and radicals" (Brandal, Bratberg, and Thorsen 2013: 140).

While the Socialist International has "consistently pressed for a more equitable distribution of global resources," many social democrats in advanced capitalist countries materially benefit from the unequal exchange between developed and developing countries within the context of the capitalist world system (Padgett and Paterson 1991: 222). Western European countries where social democracy has more or less thrived since World War II, aside from Norway, have relied on relatively inexpensive external resources from the Middle East, Russia, Africa, and Latin America, have at times, along with the United States, supported dictatorial regimes, such as Saudi Arabia, and have sought to undermine potentially progressive ones, with Iran in the early 1950s perhaps being the clearest example. During its heyday, social democrats tended to function as "sterile administers who serve capitalism even when they at times slightly slowed down the rate of immiseration and despair" (Suvin 2016: 108). Under neoliberalism, social democracy in various European countries has gone into serious decline. For example, the German SPD only garnered 142 out of 622 seats (23 percent) in the Bundestag in the 2009, its lowest electoral count since 1949. Conversely, various far-right-wing parties have grown in popularity in many European countries, including France, the Netherlands, and Denmark.

Conclusion

Postrevolutionary societies in large part had functioned or continue to function as "development states" in that they have generally propelled peripheral or economically underdeveloped societies economically forward, although this often proved and continues to be an awkward process. Indeed, in the aftermath of World War II, the Soviet economic system served as a model of development for Eastern European countries, and later China, North Korea, Vietnam, Cuba, and various African countries. Conversely, various postrevolutionary societies, including China, Yugoslavia, Rumania, and Albania, distanced themselves politically from the Soviet Union at times, for a variety of reasons.

The collapse of the Soviet Union and its bloc in Eastern Europe did not mean the end of postrevolutionary or socialist-oriented societies, although it may have contributed to their demise in various countries in Africa, such as Ethiopia, Mozambique, and Angola. Conversely, they continue to function in various manifestations into the first decades of the twenty-first century in China, Vietnam, Cuba, and North Korea. Regardless of the degree to which these regimes subscribe to Leninism, it emerged about a century ago in a very different time and is not a perspective that can be simply be transplanted onto our present globalized world. The collapse of the Soviet bloc proved to be a shock to many Western leftists who had hoped that somehow the Soviet Union and its satellites could reform themselves into democratic socialist societies. Unfortunately, as Tariq Ali (2009: 75) asserts, every socialist-oriented revolution, with the partial exception of Cuba, has resulted in some type of "monolithic state." In a similar vein, Wright (2010: 107) observes that the "empirical cases we have of ruptures with capitalism have resulted in authoritarian state bureaucratic forms of economic organization rather than anything approaching a democratic-egalitarian alternative to capitalism." Wars appear to have played a crucial role in the emergence of many Leninist states and served as a significant factor in contributing to their authoritarianism. According to Kolko (2006: 53), "Virtually all Leninist parties have burgeoned and become important in astonishingly brief periods because of crises linked to wars and their immediate aftermath." This certainly proved to be the case for the Soviet Union, the various East European postrevolutionary societies, North Korea, Vietnam, and Cambodia.

In the aftermath of the collapse of the Soviet bloc, many leftists turned to other theoretical perspectives, including poststructuralism and postmodernism, and came to presume that capitalism was the end of history or the only game in town, one that hopefully could be humanized through a variety of incremental reforms. However, as Terry Eagleton observes,

> It may be that some kind of dictatorial government was well-nigh inevitable in the atrocious conditions of the early Soviet Union; but this did not have to mean Stalinism, or anything like it. Taken overall, Maoism and Stalinism were botched, bloody experiments which made the very idea of socialism stink in the nostrils of many of those elsewhere in the world who had most to benefit from it. But what about capitalism? As I write, unemployment in the West is already in the millions and is mounting steadily higher, and the capitalist economies have been prevented from imploding only by the appropriation of trillions of dollars from their hard-pressed citizens. (Eagleton 2011: 15)

The history of postrevolutionary or socialist-oriented societies in the twentieth century was very mixed, to the point that many believe that socialism is historically a dead project. However, when one considers not only issues of social inequality but also the limits to growth and looming climate crises as we enter more and more into the twenty-first century, capitalism, even a reformed and supposedly more environmentally friendly capitalism, may spell the end of much of humanity. The contradictions associated with capitalism strongly suggest that it is not a viable political economic and social system for billions of people and is environmentally unsustainable. This begs the question as to whether the concept of socialism can be rejuvenated in the twenty-first century to ensure social parity, democratic processes, and environmental sustainability for humanity. Can we create a democratic eco-socialism and a socialism for the twenty-first century? This is an issue that I explore in chapter 5.

CHAPTER 3

Technoliberal and Countercultural Visions of the Future

Despite the common belief that capitalism is the "end of history" and that "there is no alternative" (TINA), it has become increasingly apparently that late capitalism or neoliberalism exhibits numerous contradictions and shortcomings. The growing realization of the seriousness of the global ecological crisis and anthropogenic climate change has prompted the development of numerous visions of the future, most of them relatively mainstream and others more radical. Certain mainstream visions even question some of the basic parameters of global capitalism without fully calling for transcending it (Heinberg 2011). Historian W. Warren Wagar (1991) delineates three paradigms of the future: (1) the technoliberal paradigm, (2) the countercultural paradigm, and (3) the radical paradigm. He asserts that the technoliberal could be termed "the 'liberal' paradigm, or perhaps even the 'conservative' or 'neo-conservative' or 'capitalist' paradigm" but aside from how it is termed, it manifests "abiding faith in the power of technology and managerial technique to solve problems and help preserve liberty" (Wagar 1991: 36). The countercultural paradigm calls for cultural transformation but although attacking certain core values of modern civilization, some (if not many) counterculturalists do not fully reject the "economic and political world-system of capitalism" (Wagar 1991: 41). Finally, the radical paradigm rejects the existing economic and political world system and calls for its transcendence with a more socially just and equitable world system. In this chapter, I provide a partial coverage of various technoliberal and countercultural visions of the future, while in chapter 5 I focus on radical visions of the future.

Technoliberal Visions of the Future

Wagar identifies two groups of technoliberal future visionaries, the more conservative ones being the "marketeers" and the more progressive ones being the "welfarers" (Wagar 1991: 36). Herbert Kahn, the founder of the Hudson Institute, had been the leading marketeer in futures research until he died in 1983. Given that Wagar classifies more technoliberal futurists as welfarers, I focus on some of the research groups and scholars who fall into this category, namely, the Global Scenarios Group; sociologists Ulrich Beck and Stephen K. Sanderson; Lester R. Brown and Pat Murphy as proponents of Plans B and C, respectively; various proponents of postgrowth or slow-growth models; proponents of the climate emergency mobilization model; and Frans Berkhout's vision of Anthropocene futures.

The Global Scenarios Group Scenarios

Paul Raskin, an American physicist who became a scenario builder associated with the Stockholm Environment Institute, is the founder of the Global Scenario Group. Allen Hammond (1998), an associate of the group, delineated three possible future scenarios: (1) Market World, (2) Fortress World, and (3) Transformed World. In its later work, the Global Scenario Group refined its scheme to include three possible future scenarios for humanity with respect to the ecological crisis, with two subscenarios in each of the broader scenarios that are depicted in Figure 3.1 below.

The Conventional Worlds scenarios entail two futures, one in which neoliberalism continues to serve as the prevailing order and another that recognizes the need for some policy reforms under which governments and international bodies regulate the market to some degree in order to reduce poverty and ensure "sustainable development," as first defined by the Brundtland Commission's report in 1987. Barbarization consists of two interrelated subscenarios in which conflict and crises lead to social

Figure 3.1. Possible Future Scenarios

Conventional Worlds
    Market forces: Adam Smith
    Policy reform: John Maynard Keynes and commitment to "sustainable development"

Barbarization
    Breakdown: Thomas Malthus
    Fortress world: Thomas Hobbes

Great Transitions
    Eco-communalism: William Morris, Gandhi, and E.F. Schumacher
    New sustainability paradigm

Source: Adapted from Raskin et al. (2002).

chaos, resulting in the collapse of institutions, a situation that results in the creation of a Fortress World consisting of authoritarian institutions under which the "world divides into a kind of global apartheid with the elite in interconnected, protected enclaves and an impoverished majority outside," a replication on a large scale of contemporary gated communities around the world (Raskin et al. 2002: 15).

The Global Scenario Group report quickly glosses over the eco-communalism scenario, describing it as a "vision of bioregionalism, localism, face-to-face democracy and economic autarky" (Raskin et al. 2002: 15). The report regards the New Sustainability Paradigm as the more attainable Great Transition scenario under which corporations become more humanized, resulting in a commitment to greater social parity and environmental sustainability (which includes heavy reliance on solar energy). People will reside in cohesive communities in which they are situated relatively close to work, shopping centers, and recreational facilities.

John Bellamy Foster (2005) laments that the Global Scenario Group does not discuss its eco-communalism scenario in greater detail, which theoretically would require an ecological revolution based on democratic and eco-socialist principles that are discussed in the next chapter. He views the New Sustainability Paradigm as flawed because ultimately private corporations are driven by profit making and only pay lip service to meeting social and ecological needs.

Sociological Perspectives on the Future

The prominent German sociologist Ulrich Beck in his various books about *risk society* tends to eschew any specific reference to global capitalism, but repeatedly refers to it as *industrial modernity* or *Western modernity* (Beck 2007). Nevertheless he recognizes that "[I]n the light of climate change, the apparently independent and autonomous system of industrial modernization has begun a process of self-dissolution and self-transformation. This radical turn marks the current phase in which modernization is become reflexive, which means: we have to open up to global dialogues and conflicts about redefining modernity.... It has to include multiple extra-European voices, experiences and expectations concerning the future of modernity" (Beck 2010: 264).

Beck calls for a form of *cosmopolitanism* that transcends national interests and has the potential to create a green modernity. In keeping with this emphasis on *reflexive modernization*, among at least those who take the findings of climate science seriously, ecological modernization has become a virtually hegemonic stance that asserts that environmental sustainability and effective climate change can be implemented by adopting more efficient, environmentally friendly and low carbon–emitting energy sources and manufacturing processes. Ecological modernization is part and parcel of what goes under the rubrics of *green capitalism* or even *climate capitalism*. Newell and Patterson assert: "So the challenge of climate change means, in effect, either abandoning capitalism, or seeking a way to find a way for it to grow while gradually replacing coal, oil, and gas. Assuming the former is unlikely in the short term, the questions to be asked are, what can growth be based on? What are the energy sources to power a decarbonised economy? ... What kind of climate capitalism do we want? Can it be made to serve desirable social, as well as environmental, ends?" (Newell and Patterson 2010: 9).

In essence, they seek to decouple economic expansion from greenhouse gas emissions and environmental degradation.

Sociologist Stephen K. Sanderson (2015) delineates five macro-trends over the course of the next fifty years, which I list below along with various micro-trends:

- Technology will continue to advance, and to advance with increasing rapidity.
  - Nanotechnology will probably constitute the next major technological breakthrough.
  - Renewable energy sources will in large part replace oil and gas.
  - Automobiles will be replaced by self-piloting cars and worldwide tunnels will transport people over longer distances, effectively eliminating airplanes.
  - Major developments in biotechnology will occur that will undoubtedly lengthen life-span.
- Substantial democratization will occur through the semi-periphery and parts of the periphery.
  - Some regions, particularly sub-Saharan Africa, will undergo a new authoritarianism to contain social unrest.
- There will be increasing ethnic conflict on a global scale.
  - Immigration from developing countries to developed countries may contribute to ethnic conflict in the latter.
- Globalization will continue and will intensify to mind-numbing levels.
  - The global economy will be dominated by a few gigantic corporations, which will integrate production and finance.
  - An increasing number of organizations will be necessary for international political governance.
- A world-annihilating war is a very real possibility.
- Numerous efforts will be made to establish a world state.
  - Even if a world state does not eventuate, worldwide political bodies will play an increasing role in social life.

More recently, Sanderson modified and expanded his listing of future scenarios to nine:

- Economic development will continue but very unevenly around the world.
- Most sub-Saharan African countries will continue to disintegrate and some may even implode.
- Substantial democratization will take place through much of the "less-developed world."
- Ethnic conflict will increase around the world.
- The remaining five or six "Leninist" societies, namely, China, North Korea, Vietnam, Cuba, Laos, and possibly Ethiopia, which is still dominated by the Ethiopian People's Revolutionary Democratic Front, will collapse, a scenario under which "industrialization and rapid capitalist development occurring in China will eventually undermine its Communist regime" and the two Koreas will unite.
- The center of the capitalist world system will shift from the United States and Western Europe to East Asia, particularly a reunited China that might evolve into the new world hegemon.
- Decades-long efforts to create a world state will continue.
- Economic, political, and sociocultural globalization with continue at dizzying levels.
- Technological innovations will continue an increasing rate (Sanderson 2015: 190–198).

While interesting and insightful, Sanderson's two future scenarios surprisingly fail to address issues such as social disparities within and between nation-states or the ecological crisis and climate change. To be fair to him, in 2005 the vast majority of social scientists were not giving much, if any, attention to the impact of climate change on the future of humanity, despite the fact that the Intergovernmental Panel on Climate Change had been created in 1988. However, since then the situation has very much changed with a growing number of social scientists giving attention to numerous dimensions of climate change. Over the past decade or so, a growing number of social scientists, particularly anthropologists, sociologists, political scientists, and human geographers, have become interested in various dimensions of climate change, particularly how it impacts human societies, how it is driven by anthropogenic activities, and how human societies are responding to it (Roberts and Parks 2007; Giddens 2009; Lever-Tracy 2010; Dryzek et al. 2011; Baer and Singer 2014). In light of this fact, Sanderson's (2015) failure to touch on both the ecological crisis and climate change in any meaningful way strikes me as a glaring oversight and a puzzling one at that.

Plans B and C

Lester R. Brown (2009: 23–24) has devised several versions of *Plan B*, a scheme essentially designed to save civilization, which has four components: "cutting net carbon dioxide emissions 80 percent by 2020, stabilizing population at 8 billion or lower, eradicating poverty, and restoring the earth's natural systems, including its soils, aquifers, forest, grasslands, and fisheries." He proposes an energy-efficient revolution, which would include stabilizing the climate by shifting from fossil fuels to renewable energy sources, developing sustainable cities, and ensuring that everyone has adequate food. However, his scheme does not challenge the treadmill of production and consumption and the need to constantly grow, which are integral parts of global capitalism. Furthermore, while Brown acknowledges the need to eradicate poverty, his scheme does not even suggest the need for a drastic redistribution of wealth and a shift to pronounced social parity.

Pat Murphy has drawn up a *Plan C* that includes the following components:

- A drastic reduction in the consumption of fossil fuel energy and fossil fuel–derived products
- A shift from a growing economy to a contracting economy
- An emphasis on small communities
- The consumption of less food, dietary changes, reduction in meat consumption, purchase of local organic food, preservation and storage of food, and creation of gardens and/or henhouses
- A shift to energy-efficient cars and sharing rides
- The erection of smaller homes (Murphy 2008: 111–125).

For the most part, the suggestions delineated in *Plan C* might be more appropriate for most people in developed countries and the more affluent sectors of developing societies than the poor in developing societies. Murphy's call for a shift from a growing economy to a contracting economy is laudable but ultimately incompatible with global capitalism and thus would require a transcendence of it, something that he does not explicitly suggest.

Postgrowth or Slow-Growth Models

Harkening back to the work of Herman Daly, who called for a steady-state or zero-growth economy some time ago, a growing number of ecological economists and environmentalists have been calling for a postgrowth or at least a slow-growth or "green growth" economy, although still within the parameters of global capitalism. Daly himself recently proposed the following policies for achieving a steady state economy:

- Developing "cap-auction-trade systems" for basic resources, under which renewable resources are not being depleted faster than they can be regenerated
  - Not permitting nonrenewable resources to be depleted without being able to replace them with substitutes
  - Not permitting wastes to be dumped into the environment faster than they can be absorbed by it
- Tax shifting
- Limiting social inequality
- Overhauling the banking sector
- Managing trade for the public good
- Increasing leisure time
- Stabilizing the population
- Restructuring national accounts
- Promoting just global governance (Daly 2015: 22).

James Gustave Speth (2008), a longtime environmental insider who had worked with the Carter and Clinton administrations and served as the administrator of the UN Development Programme between 1993 and 1997, calls for a drastic transformation of global capitalism that would entail transitioning to a postgrowth society in which people would live with enough, not always more. He maintains that large-scale environmental and socioeconomic change will need to go beyond incremental reforms, including in the case of the United States, degrowth in certain sectors but growth in selected areas, including jobs and better incomes for poor and working Americans, availability of good health care and education, investments in public infrastructure, development of climate and environmentally friendly green technologies, and assistance for "sustainable, people-centered development" for the half of humanity living in poverty (Speth 2012: 3–4).

More recently, he has delineated the four guideposts to steer the United States toward a postgrowth political economy:

- Guidepost 1 entails basic justice, real global security, environmental sustainability, authentic popular sovereignty, and economic democracy. This would include more social equality, along Nordic and Japanese lines; no large-scale poverty and income insecurity; excellent health-care and educational systems; the virtual eradication of racial and ethnic disparities; rare military interventions and minimal arms sales; and stringent public control of corporations.

- Guidepost 2 entails new values, motivations, and thought patterns in which people would view themselves as a part of nature and that would reward people who foster community and solidarity, and prioritize family and personal relationships over materialism and consumerism.
- Guidepost 3 seeks to keep many options and choices open for future generations and would entail prevention of severe climate change, depletion of natural resources, and a corporate-dominated plutocracy.
- Guide 4 entails various "virtues of necessity," such as protection of the climate and ecosystems, adjustment to increased energy and resource scarcities, relocalization, and new business models that are more socially just and environmentally sustainable and result in more durable goods (Speth 2012: 71–92).

Speth is essentially a radical green social democrat who, while recognizing the flaws of U.S. capitalism, has not fully come to terms with completely transcending it. In this regard, however, he is in good company not only in the United States but around the world. Conversely, Speth, who is based at the University of Vermont's Law School, is co-founder of the Next System Project, which asserts:

> It is time to explore genuine alternatives and new models—"the next system." It is time to debate what it will take to move our country to a very different place, one where outcomes that are truly sustainable, equitable, and democratic are commonplace.
>
> Let's begin a real conversation—locally, nationally, and at all levels in between—on how to respond to the profound challenge of our time in history.
>
> We need to think through and then build a new political economy that takes us beyond the current system that is failing all around us. Systemic problems require systemic solutions.
>
> We must think boldly about what is required to deal with the systemic difficulties facing the United States.
>
> An extraordinary amount of experimentation is taking place in communities across the United States—and around the world. These sophisticated and thoughtful proposals for transformative change suggest that it is possible to build a new and better America (Next System Project n.d.).

Richard Heinberg (2011: 277) also makes suggestions for achieving a postgrowth world that includes Transition Towns and Common Security Clubs, which he admits are at the present time not well-known, and a national network of locally based Community Economic Laboratories, which would consist of a wide range of organizations (such as food co-ops, community gardens, and health centers) and businesses "dedicated to providing the armature around which a new economy can be woven." Contrary to ethnographic evidence on foraging and horticultural village societies, he believes that humans have an "innate propensity to maximize population and consumption" and "difficulty making sacrifices in present in order to reduce future costs" (Heinberg 2015: 111). While Heinberg acknowledges that a small percentage of humanity may be willing to live more simply so that future generation have enough resources, he suspects that the great majority of people are not so altruistic. Heinberg (2015: 108) anticipates that humanity in the near future will be forced to co-exist somehow in an "energy-constrained world," but that the path to such a world will inevitably be filled with conflict.

Norwegian Jorgen Randers (2010) presents a global forecast for 2052 that delineates various scenarios in several areas. For population and consumption, he predicts that population will peak: economic growth will continue, but on a more reduced scale; the costs of adapting to climate change and dealing with the disasters from it will increase drastically; and states around the world will increasingly intervene in addressing social and environmental crises. In terms of energy consumption and CO_2 emissions, Randers predicts continuing improvements in energy efficiency, increases in energy consumption but with some leveling off, a peaking of CO_2 emissions by around 2030, and serious problems resulting from an average global temperature exceeding 2°C. He also predicts a trend toward dematerialization due to a slower growth in productivity and the development of mega-cities but accompanied by better health and militaries combating new threats. Finally, Randers envisions a new zeitgeist, which will entail increased localization, less fixation on economic growth, and a reformed capitalist system that includes a strong role for government intervention. In chapter 10 of his book, he projects five regional futures, namely, ones for the United States; China, which he terms the "new hegemon"; the OECD minus the United States; and the BRIC countries (consisting of Brazil, Russia, India, and China); and ten "emerging economies" (Indonesia, Mexico, Vietnam, Turkey, Iran, Thailand, Ukraine, Argentina, Venezuela, and Saudi Arabia). Randers (2010: 323–351) even goes so far as to provide readers with twenty "pieces of personal advice" as to how they should prepare for the future, such as focusing on "satisfaction rather than income," "don't teach your children to love the wilderness," "live in a place that is not overly exposed to climate change," "do more than your fair share—to avoid a bad conscience in the future," and "in politics, accept that equal access to limited resources will trump free speech."

Paul Gilding (2011: 184) argues that the elephant in the room is not climate change per se, which he views as merely a symptom of a larger problem, but, namely, the "delusion that we can have infinite quantitative growth, that we can keep having more and more stuff, on a finite planet." He proposes various steps necessary to ride through the Great Disruption that looms ahead for humanity and that will end economic growth:

- A large-scale economic shift that would greatly reduce greenhouse gas emissions within twenty years;
- The adoption of low-risk and reversible geoengineering actions;
- The removal of around six gigatons of CO_2 from the atmosphere for around 100 years and the longtime storage of this CO_2 in underground basins, and in biomass;
- The reduction of deforestation and other logging by 50 percent;
- The shutdown of some one thousand coal power plants within five years;
- The retrofitting of one thousand coal power plants with carbon capture and sequestration (CCS);
- The erection of a wind turbine or solar plant in every town;
- The creation of huge wind and solar farms in suitable locations;
- And various other changes.

Although Gilding indicates that there is a need for a drastic economic shift, he appears to leave it up to the supposedly more progressive actors in the corporate sector to nudge governments to take action on climate change. Citing examples such as the U.S. Climate Action Partnership, which includes various companies, and the Corporate Leaders Group on Climate Change, he asks, "[W]hat if these companies behaved as if the economy and their future prosperity and survival were at risk? What if they organized a global market coalition of pension funds, companies, and consumers that was so powerful, it overwhelmed the opposition from other corporations resisting change and forced government to act?" (Gilding 2011: 238).

In reality, the belief that corporate leaders, with perhaps a few notable exceptions, will do anything that interferes with their primary commitment to profit making and economic expansion seems politically naïve. Conversely, they are more likely to dawn the mantles of "corporate social responsibility" and "sustainable development" as strategies for warding off public criticism.

Various parties have been calling for a Green New Deal (United Nations Environment Programme 2009, Simms 2009). In one version of the Green New Deal, Tim Jackson (2009: 7), economics commissioner of the Sustainable Development Commission in the U.K., argues that the Global Financial Crisis of 2008 offered humanity a "unique opportunity to address financial and ecological sustainability together" by questioning the "underlying vision of prosperity built on continual growth." Unfortunately, corporate and political elites by and large have not taken such advice on board. In keeping with the Jevons paradox, Jackson acknowledges that improvements in energy and carbon intensity tend to be offset by increased economic growth. He calls on governments to invest in public infrastructure, reduce social inequality, redistribute existing jobs and reduce work hours, reverse the culture of consumption, implement resource/emissions caps, and shift to alternative energy sources that will help stabilize CO_2 emissions. Jackson (2009: 103) argues that a "macroeconomy predicated on continual expansion of debt-ridden materialistic consumption is unsustainable ecologically, problematic socially, and unstable economically." As a green Keynesian economist, Jackson appears to assume that global capitalism can function as a steady-state or nongrowth economic system, when history repeatedly has told us that, by its very nature, it must grow or die out. As Magdoff and Foster (2011: 56) argue, the "notion of a capitalist no-growth utopia violates the basic motive force of capitalism." Companies that do not grow inevitably collapse or are taken over by more productive firms.

The idea of a green economy or green growth, which has received support from the World Bank, the IMF, and the OECD, is based on the notion that somehow economic growth can be "decoupled" from pollution, environmental degradation, greenhouse gas emissions, and resource consumption. However, as Peter Ferguson (2015: 22) observes, "in a growing economy, absolute decoupling is likely to remain elusive because gains in resource efficiency are almost always absorbed by increases in resource management," an illustration of the enigmatic Jevons paradox or rebound effect. Despite his recognition that the notion of green economy, like the notion of sustainable development, "largely constitutes business as usual," he perplexingly asserts "it retains the latent potential to transform itself from a weak to a strong articulation, and thus towards a post-growth society." However, in reality, if ever implemented in any serious way, the green economy would essentially constitute yet another reformist reform that serves to maintain global capitalism.

The Climate Emergency Mobilization Paradigm

Jorgen Randers and Paul Gilding (2010) maintain that a climate emergency plan is likely to emerge prior to 2020 when global society finally more fully recognizes the threat of climate change to humanity. They assert that their action plan can keep global warming below 1°C above pre-industrial levels. The one-degree war plan draws inspiration from the mobilization that Britain orchestrated under the leadership of Winston Churchill as well as the United States during World War II against the military threat of Nazi Germany. Randers and Gilding envision three phases in their plan:

1. Climate war (years 1–5) would launch global society to reduce greenhouse gas emissions within five years.
2. Climate neutrality (years 5–20) would "lock in the 50 per cent emergency reductions, and move the world to net zero climate emissions by year 20" (Randers and Gilding 2010: 175).
3. Climate recovery years (years 20–100) would entail stabilization of the global climate system and the creation of a sustainable global economy.

Their climate war model is extremely detailed, but for purposes of illustration I list several of their recommendations:

1. Reduce deforestation and other logging by 50 percent;
2. Shut down 1,000 coal power plants within five years;
3. Ration electricity;
4. Construct a wind turbine or solar plant in every town;
5. Create huge wind and solar farms in desert areas;
6. Ration use of highly polluting cars to reduce transport emissions by 50 percent;
7. Gradually reduce the world's aircraft by 50 percent;
8. Launch "shop less, live more" campaigns (Randers and Gilding 2010: 179–181).

Randers and Gilding propose the creation of a climate war command by countries participating in the plan, which would draw on advice from the IMF and the IPCC as well as various multinational military commands, presumably such as NATO. This command structure would introduce a global tax of U.S. $100 per ton of CO_2 emissions, which would have two aims: "to fund the war effort (i.e. the development and implementation of the various actions described above) and to alleviate the resulting hardship—primarily among the poor (globally speaking)" (Randers and Gilding 2010: 183). The climate emergency plan does not call for a radical restructuring of global capitalism, in part indicated by the prediction that eventually "wealth levels would gradually move back toward current levels, though distribution would be more even" (Randers and Gilding 2010: 186).

While in part agreeing with the spirit of various climate emergency plans, Delina and Diesendorf (2013) express a number of concerns about them. They maintain that the ten-year transition scenarios are utopian and unachievable but maintain that "under circumstances of strong government interventions, 25-30-year transitions are possible for a number of developed countries, and so is a 40-year global transition, provided we allow for international trade in renewable energy" (Delina and Diesendorf 2013: 377). Delina and Diesendorf (2013: 378) also express concern that climate emergency plans could prove to be authoritarian in situations where they do not obtain popular support, with "no guarantee that a state of normal democracy would return" after their implementation. Given that there is a tendency for multinational corporations in most countries to make or break governments and politicians, it seems that at least for the foreseeable future, few governments are strong enough, although China may be an exception, to implement a climate emergency plan. Finally, while the threat of the Axis powers became increasingly apparent in both the West and the East during the late 1930s and early 1940s, most economic and political elites, let alone ordinary people, do not at the present time appear to perceive the existence of a climate emergency since heat waves, droughts, bushfires, hurricanes, floods, etc., in their minds come and go. Eventually, however, this perception is likely to alter as climatic crises intensify around the globe.

On Anthropocene Futures

Dutch atmospheric chemist Paul Crutzen (Crutzen and Stoermer 2000) has suggested that the geological age in which humans are living should be renamed the Anthropocene (Age of People). He argues that for the past 150 years, more so than natural forces, it is human activity that has had the most impact on shaping the biogeological environments of Earth and its climate. Crutzen maintains that the Anthropocene began in the latter part of the eighteenth century, when analyses of air trapped in polar ice showed the beginning of growing global concentrations of carbon dioxide and methane. This date also coincides with the James Watt's design of the steam engine in 1784. However, William F. Ruddiman (2005) has pushed the beginnings of the Anthropocene further back in his recognition that CO_2 emissions began to slowly increase as humans started to clear the land in their shift from foraging to farming about eight thousand years ago in places such as China, India, and Europe and that methane emissions began to increase around five thousand years ago as various populations started to irrigate for rice production and raise livestock. However, in his tripartite typology of the Anthropocene, geologist

Andrew Glickson (2013: 91) pushes the Anthropocene even further back with the *Early Anthropocene* starting roughly 2 million years ago with the discovery of fire, the *Middle Anthropocene* starting with the development of "extensive grain farming" in the Near East, and finally the *Late Anthropocene* commencing with the "onset of combustion of fossil fuels," starting out with coal. Anthropologist Leslie C. White (2008: 118) referred to coal as the "king, or the father, of the Fuel Revolution," which served as part and parcel of "modern capitalist culture," a fact that suggests that the Late Anthropocene might be more appropriately termed the *Capitolocene* (Moore 2016).

Geographer Frans Berkhout (2014: 156) maintains that social scientists can play a major role in predicting and forecasting of the future, including Anthropocene futures, despite their reluctance to venture into this endeavor. Indeed, reflecting on the first Grand Challenge Symposia organized by the Smithsonian Institution in 2012, anthropologist Shirley Fiske (2012) emphasized that "the meaning of the Anthropocene is ethical and moral—how do we want the future to look and what can we do with the knowledge we have?" Aside from human intentions, it is clear that, barring a global catastrophe (e.g., a major meteor impact, all-out nuclear war, an unstoppable infectious disease pandemic, or worst-case global warming), humanity will likely remain a major environmental force for the foreseeable future. Despite his recognition of the limitations to all future scenarios, Berkhout (2014: 156–158) delineates some possible Anthropocene futures:

- The possibility of peak oil can be expected to result in a drive for non-oil energy sources, something that has already happened with fracking of shale gas and the gradual turn to renewable energy sources, particularly solar and wind energy.
- The adverse impact of climate change on agricultural production can be expected to "generate the search for new, more diversified but intensified global food production systems."
- A diminishing availability of phosphorous for agriculture may stimulate low-phosphorous forms of agricultural production.
- New planetary risks will be very unlikely unevenly distributed, creating "winner" and "loser" scenarios around the world, at least in the short run.
- In other words, "[s]hort of a real cataclysm, it is likely that 'good' and 'bad' Anthropocenes will continue to exist side-by-side."

In essence, much of the discussion of winners and losers in assessing the impact of climate change on human societies closely follows a neoclassical microeconomic or neoliberal discourse. For example, Ward asserts: "Among the winners and losers will be locales that today are too cold to be desirable for year-round dwelling. Through geographic accident, most such places are in the North Hemisphere. The biggest victors will be Canada, Alaska, Greenland, Russia, and Scandinavia, and in the Southern Hemisphere, Argentina most of all. Perhaps future world power will not be relocated to the countries in the Southern Hemisphere, as is often predicted, but stay concentrated, if not redistributed, in the Northern Hemisphere" (Ward 2010: 196).

Al Gore on the Future

Despite the fact that I have reviewed various mainstream future scenarios, perhaps yet one more futurist might be worth considering, namely, Al Gore, a former U.S. vice-president and a politician who many have argued was robbed of the U.S. presidency. Al Gore's film and accompanying book *An Inconvenient Truth* did much to propel climate change into public consciousness around the world in 2006 (Gore 2006). While to his credit, Gore popularized the findings of climate science, his solutions to the climate crisis are framed very much within the parameters of green capitalism and ecological modernization by advocating emissions trading schemes, ecological modernization, tree plantations, and techno-fixes as sufficient climate change mitigation strategies. In *Earth in the Balance*, the first edition of which appeared in 1992, Gore (2007: 307–337) proposes implementation of a Global Marshall Plan, which would entail the following elements: (1) stabilization of the world population; (2) the development and sharing of "appropriate technologies"; and (3) the development of "new global economics." He argues that the definition of GNP or GDP should be changed to include environmental costs and benefits, or treated as what mainstream environmental economists term *externalities*. In his book *Our Choice*, Gore (2009) lays out his views on ecological modernization, which include an overall endorsement of energy efficiency, retrofitted buildings, hybrid cars, greater reliance on public transport, and renewable sources of energy (solar, wind, and geothermal) as climate change mitigation strategies.

In the wake of his film, Gore created the Climate Project and the Alliance for Climate Project, which merged in June 2011 into the Climate Reality Project, which in turn has trained several thousand Climate Reality Leaders around the world. Indeed, on 25–27 June 2014, along with some five hundred people, I underwent training for the Climate Reality Project Corps in Melbourne (Baer 2015b). While Gore had been primarily the target of criticism from climate change skeptics, progressive environmentalists have critiqued him on various points, not only the limitations of his green capitalistic solutions but also the tremendous amount of energy and greenhouse gas emissions that he generates at his Tennessee farm and in his travels by private jet around the world, as well as his failure to promote vegetarianism. To his credit, however, Gore had advocated civil disobedience at sites where coal-fired power plants are being constructed.

In his most recent book, Gore (2013) takes a stab at futurology. He maintains that in confronting the future, humanity must make some crucial choices in confronting the global economy, namely, creating a planetary electronic communications system; a new balance of global political, economic, and military forces; rapid sustainable growth both demographically and economically; the development of new biological, biochemical, genetic, and material sciences technologies; and the "emergence of a radically new relationship between the aggregate power of human civilization and the Earth's ecological systems," including the climatic system (Gore 2013: xiv–xv). On the positive side, Gore contends that there is room for optimism about the future: "For the present, war seems to be declining. Global poverty is declining. Some fearsome diseases are being conquered and others are being held at bay. Lifespans are lengthening. Standards of living and average incomes—at least on a global basis—are improving. Knowledge and literacy are spreading. The tools and technologies we are developing—including Internet-based communication—are growing in power and efficacy. Our general understanding of our world, indeed, our universe (or multiverse!) has been growing exponentially" (Gore 2013: xxx).

Although some of these points can be contested or need to be qualified, Gore does recognize that not all is well in the world and that global capitalism is a system with many contradictions, including disproportionate corporate input into political decision making, the concentration of wealth among a relatively small number of people, growing social inequality, and adverse impacts on the environment. He is a strong advocate for "Sustainable Capitalism," in which the narrow economic interests of "Earth, Inc." or a "new hyper-connected, tightly integrated, highly interactive, and technologically revolutionized economy" (Gore 2013: 4) would be overridden by the broad, enlightened interests of the "Global Mind," which encompasses "new business models, social organizations, and patterns of behavior" (Gore 2013: 45) that promise to solve many of humanity's seemingly intractable problems. At the same time, Gore (2013: 143) is aware that Earth, Inc. has the capacity to use the Global Mind to its own advantages as evidenced by its "enhanced ability to manufacture consent for political decisions that serve their interests rather than the public interest—and provides corporations with an enhanced ability to manufacture wants in order to increase consumption of commodities and manufactured products," developments that he recognizes are environmentally unsustainable. Ironically, although he chronicles many of the major contradictions of the capitalist world system in numerous areas, including economic and ecological crises, Gore seems to have an abiding faith that a reformed and sustainable capitalist world order, with the United States at its helm, will save the day in the future by "fixing the prevailing flaws and distortions in capitalism and self-governance. It means controlling the corrosive corruption of money in politics, breaking the suffocating rule of interests, and restoring the healthy functioning of collective decision making in representative democracy to promote the public interest. It means reforming markets and making capitalism sustainable by aligning incentives with our long-term interest. It means, for example, taxing carbon pollution and reducing taxes on work—raising revenue from what we burn, not we earn" (Gore 2013: 368–369).

While he is correct in asserting that humanity has reached a fork in the road, one path that will lead to climate destruction, depletion of natural resources, and the possible end of civilization and the other path to a better future, Gore's assertion that this will be achieved within the parameters of a reformed and supposedly greener capitalism seems highly dubious.

Ultimately, green capitalism generally fails to address the treadmill of production and consumption that contributes to the depletion of natural resources, environmental degradation, and climate change. It tends to either be oblivious to social justice issues or at best to downplay them or pay them lip service. It is important to note that some components of ecological modernization (a perspective that falls under green capitalism) —such as renewable sources of energy (solar, wind, and geothermal), improved efficiency and building construction and design, and a massive shift from private vehicles to an energy-efficient public transport system—have the potential to serve as important mitigation strategies. However, as anthropologist Alf Hornborg persuasively argues, "What ecological modernization has achieved is a neutralization of the formerly widespread intuition that industrial capitalism is at odds with global ecology.... The discursive shift since the 1970s has been geared to disengaging concerns about environment and development from the criticism of industrial capitalism as such. But the central question about capitalism should be the same now as it was in the days of Marx: Is the growth of capital of benefit to everybody, or only to a few at the expense of others" (Hornborg 2001: 25–26).

Thus, ultimately, technological innovations that on the surface appear to be more environmentally sustainable and energy efficient must be part and parcel of a shift to a steady-state or zero-growth global economy if they are to circumvent the Jevons paradox or the *rebound effect* in which consumption tends to increase as products because more efficient, a phenomenon that is consistent with the growth paradigm associated with global capitalism.

Countercultural Visions of the Future

While countercultural futurists advocate for the replacement of fossil fuels with renewable energy sources, they emphasize simpler or labor-intensive technologies, decentralized authority, and the emergence of locally based cooperatives and communities. According to Wagar (1991: 43), "[c]ounterculturalists rarely make a wholesale assault on capitalism, as such, but it is difficult to imagine how capitalism could survive, much less thrive, in a world tailored to their specifications." Given limitations of space, I focus on future scenarios delineated by sustainability advocate Graeme Taylor (2008) and Australian permaculturalist David Holmgren (2009).

Graeme Taylor on Sustainable Futures

Graeme Taylor (2008) coordinates BEST Futures, a project that seeks to develop theories and tools that support sustainable solutions in a world where the global economy is depleting its resources and contributing to climate change. In his rather thought-provoking book *Evolution's Edge*, he delineates three possible future scenarios depicted below:

- Scenario 1: Business as usual
    - Failure on the part of the majority of political and business leaders due to ignorance, vested interest in the present system, and inertia to act quickly enough to preserve major ecosystems.
    - Ongoing economic growth and depletion of the planet's "natural capital."
    - Accelerating destruction of major ecosystems.
    - Accelerating rate of global warming and climate change.
- Scenario 2: Adjusting the existing system
    - Adopting supposedly expedient solutions, such as producing biofuels from food crops, building nuclear power plants, and resorting to carbon and capture sequestration strategies at coal-fired power plants.
    - Other possible piecemeal and incremental solutions, some of which may be sustainable, such as halting the construction of new

freeways and airports and investing in rail transportation, but are insufficient for preventing the global economy from continuing on an environmentally destructive path.
- Scenario 3: Transformational change
  - The adoption of a world view that recognizes that economic systems depend on their environments.
  - A shift from a consumer society to a conserver society. (Taylor 2008: 203–206)

In reality, despite the fact that Taylor does not explicitly adopt a democratic eco-socialist perspective, which is discussed in detail in the following chapter, many of his ideas for creating a more sustainable future are compatible with it. These include the following points:

- "Our collective survival depends on human economies operating within the carrying capacity of the Earth's ecosystems."
- "Human economies will only be sustainable when they meet essential human and biophysical needs for health and wholeness. Cultural and genetic diversity is necessary for ecological and social health, wholeness, and resilience."
- "Power and resources will have to be redistributed in order to meet essential human and biophysical needs."
- "A sustainable global system is not possible without peace and accountable governance." (Taylor 2008: 230)

The Permaculture Paradigm as a Vision for the Future

David Holmgren (2009: 60), an Australian who along with Bill Mollison developed the permaculture concept, delineates the following four "energy-descent and climate change scenarios."
- Brown tech
  - Slow oil decline
  - Fast climate change
- Green tech
  - Slow oil decline
  - Slow climate change
- Earth steward
  - Fast oil decline
  - Slow climate change
- Lifeboats
  - Fast oil decline
  - Fast climate change

The brown-tech scenario assumes that the production of oil will peak between 2005 and 2010 and that a corporatist or fascist political system that entails a merger of corporate and state power adopts assertive policies to address the energy peak and climate change. The new governing regime "gives priority to getting more energy out of lower-grade nonrenewable resources (e.g., tar sands, coal, and uranium) and biofuels from industrial agriculture and industry" (Holmgren 2009: 62). Continued reliance on fossil fuels exacerbates climate change, which results in a reduction of food production, which in turn forces fascist regimes to construct or replace urban infrastructure impacted by storms and sea-level rise. This scenario entails resource wars, displacement from rural areas to crowded urban ones, the introduction of stringent population control measures in some countries, and a social divide "between the falling numbers of 'haves' dependent on a job in the 'system' and the relatively lawless, loose but perhaps communitarian 'have-nots' with their highly flexible and nomadic subcultures living from the wastes of the 'system' and the wilds of nature" (Holmgren 2009: 66–67). Eventually the brown-tech scenario devolves into the lifeboats scenario.

The green-tech scenario entails a shift to renewable energy sources and a "resurgence of rural and regional economies," with organic farming, ecological management, and resource allocation becoming normative practices (Holmgren 2009: 69). A social divide emerges with wealthy farmers and agrobusinesses employing "both high technology and cheap labor from migrant workers" in the more fertile and less populated regions. Conversely, in regions "with poorer and steeper land and more diversified land ownership, smaller polyculture systems designed using permaculture principles spread wealth more evenly through local communities" (Holmgren 2009: 70). State and city governments implement a shift to more compact cities and towns with strong public transportation systems. This scenario results in substantial reductions in greenhouse gas emissions, thus minimizing adverse climate change impacts for at least several decades. A purportedly "sustainability" elite imposes restrictions on the drivers of consumer capitalism, such as advertising, and seeks to foster a shift toward a postmaterialist society that blends together elements from feminism, environmentalism, and indigenous and traditional cultural values.

Under the earth-steward scenario, the decline in the production of fossil fuels prompted by their declining availability adversely impacts the global financial system, "resulting in severe economic depression and perhaps some further short, sharp resources" (Holmgren 2009: 75). The ensuing collapse of the "global consumer economy" and the energy-intensive large-scale farming system slows down and then reverses global warming but is accompanied by the "abandonment of even highly productive land," food shortages, rationing, black markets, and food and energy riots (Holmgren 2009: 76). Due to the collapse of large businesses and the difficulty of maintaining urban infrastructures, many people migrate to smaller towns, villages, and farms with more robust local economies. The better-off farmers emerge into feudal-like lords who provide exurbanites with food and housing in exchange for their labor. Furthermore, "[a]n explosion of home businesses based on building and equipment retrofit, maintenance, and salvage starts to build a diversified economy" (Holmgren 2009: 79). New bioregional governments appear in some places, implement land reform, and cancel debts, which have become meaningless due to the collapse of banks and other financial institutions, thus permitting people to remain on their properties.

The lifeboats scenario constitutes the most dystopian and dire of Holmgren's four future scenarios, particularly given that "supplies of high-quality fossil fuels decline rapidly, the economy fails, and human contributions to global warming collapses, but lag effects and positive feedbacks in the climate system continue to drive an acceleration of global warming" (Holmgren 2009: 82). The lifeboats scenario blends the worst elements of the brown-tech and earth-steward scenarios and entails warfare, including nuclear warfare, at the local, national, and global levels, which, combined with waves of hunger and epidemics, results in the "halving of global population in a few decades" (Holmgren 2009: 82). While cities by and large are abandoned, they provide places where suburban and rural dwellers can salvage materials, especially metal ones. Civilization essentially has collapsed with people being scattered about in foraging and small farming communities where they seek refuge from local warlords and pirates. On the positive side, however, people in the lifeboats scenario retain knowledge of their past cultures, which when "combined with a moderately habitable environment allow new civilizations to emerge" (Holmgren 2009: 85).

According to Holmgren (2009: 81), the earth-steward scenario in some ways "might be considered as the archetypal one of the energy-descent future and the one in which permaculture principles and strategies are most powerfully applied." In contrast to the lifeboats scenario, which is based on patriarchy and a warrior cult, the earth-steward scenario embraces feminism and earth spirituality, but it is not clear how these would be compatible with neofeudalistic structures in which landlords would most likely be males. Ironically, while asserting that Cuba under the leadership of Fidel Castro had embodied "many the elements of the brown-tech world," Holmgren (2009: 92) maintains that in the aftermath of the Special Period of the 1990s Cuba developed a bioregional scale of governance "more akin to that proposed for the green-tech scenario" given that it has shifted to urban agriculture, sustainable farming techniques, a diet more oriented toward vegetables, fruits, and nuts than meat and dairy products, and heavy use of bicycles as a form of transportation. He suggests that all four scenarios may actually emerge simultaneously or as "nested scenarios" in which they range from the household or local level to the national level to the global level. Thus, national governments and large corporations are likely to adopt policies and practices characteristic of the brown-tech scenario, city and state or bioregional governments those characteristic of the green-tech scenario, small businesses and local governments those characteristic of the earth-steward scenario, and households and local communities those characteristic of the lifeboats scenario. Holmgren argues:

> I think the nested concept is one of the most insightful and empowering ways to think about these scenarios because it helps us to understand the apparent contradictions between different perspectives and motivations of different groups in society and even contradiction within our own thoughts and behaviors. For example, it is common for people to have private thoughts about the lifeboats or perhaps earth-steward futures, while most of people's public behavior as workers and consumers reinforces brown tech or perhaps green tech. The private thoughts are often internally critiqued as antisocial or at least naïve, while the public actions are often internally critiqued as driven by powerful outside forces. (Holmgren 2009: 103–104)

Conclusion

My listing of mainstream future and countercultural scenarios by no means is exhaustive, but does illustrate that more or less mainstream and countercultural thinkers are grappling with many of the possible consequences if humanity continues to pursue "business-as-usual" and ways to move toward more equitable and environmentally sustainable alternatives. In reality, there is often a certain overlap between mainstream and countercultural perspectives on the future. Bearing this thought in mind, I end this chapter with an overview of a recent future scenario depicted by Paul Raskin in a recent update of his thinking on the Great Transition. In *Journey to Earthland,* he maintains that humanity is in the process of transitioning from the Modern Era, which has lasted nearly one thousand years, to the Planetary Phase of Civilization, during which humanity is confronting an "unprecedented moment of uncertainty and opportunity" (Raskin 2016: 19). Relying on science fiction and using 2084, the centennial of George Orwell's grim depiction of the state of humanity, as a timeline, Raskin presents a somewhat optimistic depiction of the stages in Great Transition between 2001 and 2084.

- Takeoff of the planetary phase (1980–2001)
    - Global capitalism achieved hegemony
    - Massive protests occurred at intergovernmental meetings
    - Global marketing and entertainment industries promoted mass consumption among the affluent, resulting in aspirational yearning among the have-nots and thwarted expectations among the young, angry, and poor masses
- A rolling crisis (2001–2023)
    - War, violence, displacement, pandemic, recession, and environmental disruption became widespread
    - Critiques of global capitalism and neoliberalism grew more systemic and radical
    - Collective resistance gained momentum
    - A global citizens' movement (GCM) convened its inaugural Intercontinental Congress in 2021 and laid the groundwork for the Earthland Parliamentary Assembly (EPA)
- General emergency (2023–2028)
    - Devastating impacts resulted from anthropogenic climate change
    - The GCM played a critical role by prodding confused governments to act on sustainability and climate goals that had languished since the UN adopted them in 2015
- The reform era (2028–2048)
    - The old order began to reassert itself
    - The UN established a New Global Deal, which pushed for resilience economies intending to force markets to operate within more humane social norms and environmental limits
    - The alliance of policy reforms became untenable
- The emergence of the Commonwealth of Earthland (2048–present)
    - The EPA adopted world constitution of 2048
    - A Revolutionary turn toward planetary civilization takes off in full swing consisting of the pillars of enhancement of quality of life, human solidarity, and eco-centrism (Raskin 2016: 72–75).

While by 2084 the World Assembly "sits at the pinnacle of the formal political structure," it is made up of "both regional representatives and at-large members selected by popular vote in world-wide elections" (Raskin 2016: 87). Conversely, Earthland exhibits three different regional forms of democracy. *Agoria* is the most conventional or mainstream region in that private corporations continue to operate, however within a "comprehensive regulatory framework designed to align business behavior with social goals" (Raskin 2016: 89). It also is the most urban region. *Ecodemia* gives priority to economic democracy consisting of numerous worker- and community-owned enterprises. Finally, *Arcadia* is the most anarchistic region with a strong emphasis on local community decision making and simpler lifestyles than in the other two regions. In terms of social justice, "Earthland has become more equitable and tolerant than any country of the past, the fruit of the long campaign to mend deep fissures of class privilege, male domination, and bigotry of all shades" (Raskin 2016: 103). While caps on total personal assets and limits on inheritance have ended the era of the superrich and redistributive tax structures and a guaranteed minimum standard of living have eradicated abject poverty, social inequality persists in that "in a typical region, the highest earning 10 percent have incomes three to five times greater than the poorest 10 percent (national ratios a century ago were six to twenty)" (Raskin 2016: 103).

In terms of environmental sustainability objectives, Earthland's population has peaked and stabilized at just less than eight billion people. It has set a target of reducing atmospheric carbon concentrations to 350 ppm in the near future and "climate visionaries recently launched 280.org, a one-hundred-year to return concentrations to pre-industrial levels" (Raskin 2016: 107). Earthland has left behind the fossil fuel era and adopted renewable energy sources and an array of technological innovations, including nanotechnology and biofabrication, which have resulted in lighter, more durable, and longer-lasting products. Earthland has by no means achieved utopia but in various ways has achieved real utopian objectives.

The project of restoring the richness, resilience, and stability of the biosphere remains a vast collective cultural and political enterprise. People monitor sustainability indicators as closely as sports results or weather forecasts and nearly everyone is actively engaged through community initiatives or GAIA's (Global Assembly for Integrated Action) global campaign. At last, humanity understands the moral and biophysical imperative to care for the ecosphere, a hard-learned lesson that, future generations may be assured, shall not be forgotten. In our time, the wounded earth is healing; someday, the bitter scars from the past will fade away like yesterday's nightmare (Raskin 2016: 107).

Raskin's Earthland in 2084 appears to be a robust green social democratic world with regional variation, but he leaves it up to readers to envision exactly where his three regions are located. Perhaps something on the order of Earthland might constitute a transitional stage between the existing capitalist world system and a democratic eco-socialist world system.

CHAPTER 4

Efforts to Reconceptualize Socialism

Despite all the baggage associated with the term *socialism* and the desire of various leftist thinkers to substitute other terms for it, it is important for socialists or Marxian scholars to grapple with the ideals of socialism and social experiments that have been labeled socialist, both at the national and local levels. In the aftermath of the collapse of the Soviet bloc, many progressive people, even ones involved in social movements and Green parties, argue that humanity has tried socialism and this attempt has essentially failed. In light of this, perhaps the best we can do is to reform or humanize capitalism. However, radical historian Staughton Lynd reminds us that: "Capitalism took centuries to come into existence and it did so from a myriad of false starts, bastardized ventures, and outright failures. The good society that will come after capitalism should be expected to experience similar birth pangs" (Lynd and Grubacic 2008: 83).

In other words, socialism remains very much a vision, one with which various individuals and groups continue to grapple, often by seeking to frame it in new guises. As Stilwell (1992: 211) argues, "liberated from Stalinist legacy, it now makes sense to start asking what a progressive socialism involves." The question is, from where can we seek socialism to re-emerge like a phoenix in today's world at a time when the power of global capitalism appears to be extremely hegemonic in both its influence on politicians as well as ordinary people around the world, some obviously more than others? Boswell and Chase-Dunn maintain: "Semiperipheral areas are the terrain upon which the strongest efforts to establish socialism have been made in the past, and this is likely to be true of the future as well, even if names for it change. These states often have sufficient resources to be able to stave off core attempts at overthrow and to provide some protection to socialist institutions (such as national health care) if the political conditions for their emergence should rise" (Boswell and Chase-Dunn 2000: 216).

In his delineation of five linked crises, namely, a crisis of hegemony and global governance, a crisis of inequality and democracy, a crisis in the relationship between humans and the natural environment, a crisis in the global capitalist system, and a crisis in the New Global Left. Chase-Dunn (2013: 179) suggests that capitalism may undergo a transformation from the "rule of finance capital and the military complex" to one that shifts to a "global green Keynesianism" in which technocrats and civil society play a pivotal role. However, green Keynesianism or some version of "green capitalism," despite its commitment to ecological modernization and emissions trading schemes, is still committed to ongoing economic growth, which does not address the reality of limited resources and does not seriously address issues of social justice or equity.

In the nineteenth century, various revolutionaries and reformers sought to develop alternatives to an increasingly globalizing capitalist world system. Efforts at the national level to create such an alternative started out with the Bolshevik Revolution in Russia and included subsequent revolutions in other countries, including China in 1949, Vietnam in 1954, Cuba in 1959, and Nicaragua in 1979. Unfortunately, as Wright (2010: 106) observes, "these attempts at ruptural transformation ... have never been able to sustain an extended process of democratic experimentalist institution-building" for a variety of complex reasons, some of which I have explored in chapter 2. Scholars have spilled much ink trying to determine whether these societies constituted examples of state socialism, actually existing socialism, degenerated workers' states or postrevolutionary societies that constituted transitional societies between capitalism and socialism that required some kind of democratic revolution, state capitalism, or new class societies and why many of these societies, particularly the Soviet Union but also China, became fully incorporated into the capitalist world system. Suffice it to say that their failure to achieve authentically democratic socialist societies was related to both internal forces specific to each of these societies and external forces that created a hostile environment. Historically, given the economic and military external pressures that revolutionary regimes experienced from advanced capitalist countries, they tended to attempt to consolidate their power and build robust institutions by turning to authoritarian measures. In addition to this, revolutionary regimes generally came to power in economically underdeveloped countries, which certainly was the case in Russia and China as well as elsewhere around the world.

While the powers-that-be around the world are seeking to address climate change within the parameters of global capitalism, as Simms (2009: 184) observes, "global warming probably means the death of capitalism as the dominant organising framework for the global economy." Thus, it is imperative to think outside the box and construct an alternative to global capitalism as the ultimate climate mitigation strategy, even though it will not be achieved anytime soon, if indeed ever. As humanity enters an era of dangerous climate change accompanied by tumultuous environmental and social consequences, it will have to consider alternatives that hopefully will circumvent dystopian scenarios on the order delineated earlier.

Marxian-Inspired Future Scenarios

Perhaps because Marx was cautious about creating a detailed blueprint for what socialism and communism as visions for the future might look like, Marxists or socialists seem to have been reluctant to envision future scenarios, utopian or dystopian. Fredric Jameson (2005: 199), the world-renowned literary critic, quipped that is easier to imagine the end of the world than the end of capitalism. Wagar, a staunch proponent of future studies and world systems theory, laments that neo-Marxian scholars in general and world systems theorists specifically have tended to eschew future studies, despite the fact that they aspire to the eventual transcendence of global capitalism with a socialist world system. He argues that the "traditional aversion of leftist thinkers to futures inquiry (an overreaction to nineteenth-century utopian socialism) does not remove the fact that Marxism or historical/dialectical materialism in its many varieties is a powerful tool for seeking to understand the global future." Nevertheless, various leftist scholars, such Ernst Bloch, Herbert Marcuse, Michael Harrington, and Barry Commoner, touched in some of their writings on what a postcapitalist world might look like.

Virtually all social scientists and historians writing in the early 1990s about possible future scenarios did not touch on climate change. Wagar (1991: 62) was ahead of his time on this issue. Citing the burning of fossil fuels and the wholesale destruction of both temperate and tropical forests, with the release of methane due to rice and cattle production and dairy farming, and the release of chlorofluorocarbons due to their use by both manufacturers and consumers, he asserts that the evidence indicates that "human beings in the next century will be living on a hotter planet."

Nevertheless, drawing on the frequent Marxian projection of the future of humanity being one between "socialism" or "barbarism," some time ago world systems theorist Walter L. Goldfrank (1987) delineated four possible future scenarios, namely, Barbarism I, Barbarism II, Socialism I, and Socialism II. Barbarism I would entail a nuclear holocaust that is most likely to occur sometime between 2015 and 2050 and has a 15 percent chance of occurring. Barbarism II or "global fascism" would entail a world order along the lines of Nazi Germany or Stalinist USSR in which there would be state expropriation of private property and rigid control of the labor force, and has a 30 percent chance of occurring. Socialism I would consist of a welfare state based on something like Swedish social democracy and has a 50 percent chance of occurring. In this scenario, while capitalism would be preserved, a populist world party would gradually erode the privileges of the transnational capitalist class and redistribute its profits, with the offspring of the capitalists becoming technocrats and being part of a gradual socialization of the global economy. Socialism II would be a democratic socialist world state and has a 20 percent chance of occurring, with planned industrial and agricultural production but with space for private production of many consumer goods. The capitalist class will have lost its power base and most enterprises would be controlled by workers and communities.

Eco-socialist Martin Ryle (1988: 7–8) depicts three possible dystopian future scenarios that might be created in order to create a "sustainable society":

- "an authoritarian capitalist or post-capitalist society, with rigid and marked hierarchies of wealth and power, in which those at the top enjoyed ecologically profligate lifestyles amidst 'unspoiled' surroundings, protected by armed police from the mass of the people, who would endure an impoverished and 'sustainable' material standard of living in dangerously polluted habitats."
- "'barrack socialism' [that would occur in 'actually existing socialist' regimes] in which an ecologically well-informed bureaucratic elite directed the economy in accordance with environmental and resource constraints, but in which the population participated as more or less reluctant helots."
- "a kind of anarchy—not necessarily a pacific and libertarian kind."

Some aspects of these three scenarios are depicted in a science fiction account titled *The Short History of the Future* in which historian W. Warren Wagar (1991), a world systems theory proponent, envisions three possible future scenarios: (1) the broadening of capitalist hegemony and representative democracy into the twenty-first century, (2) socialism, and (3) a worldwide web of decentralized and eco-utopian countercultural communities. The first scenario depicts a dystopian world that begins to collapse with a nuclear holocaust in 2044, followed by two contrasting, partially utopian scenarios that emerge in the twenty-first and twenty-second centuries (Wagar 1992). Despite various technological innovations, which give the global economy brief spurts of growth, capitalist production and consumption patterns continue to result in greenhouse gas emissions that contribute to ongoing climate change. Thus, by 2040, the atmosphere contains 555 ppm of CO_2 and there are even more alarming increases in methane and chlorofluorocarbons resulting from the burning of fossil and biomass fuels, fertilizer use, and the decay of organic matter in rice paddies. The average global temperature increases 7.56°F (4.2°C) between the 1980s and 2040, resulting in ongoing melting of the polar ice caps and glaciers, a rise in sea level, heavy flooding in some regions, and the decline of food production. These developments result in the Catastrophe of 2044, which is marked by the outbreak of worldwide nuclear warfare and a sudden shift to global cooling. The global war of 2044–2046 results in the devastation of most of the cities of North America, Europe, Japan, China, and the Indian subcontinent and rampant epidemics. In the aftermath of a nuclear holocaust, the power center of the world shifts to countries south of the twenty-fifth parallel. The world party, with its technocratic socialist-oriented agenda, comes to power in much of the world and forms the Commonwealth in 2062, with Melbourne as its titular capital. Despite significant achievements, both in terms of social parity and environmental sustainability, the World Party gradually begins to lose credibility. In its stead, humanity by 2157 is divided into 41,525 autonomous mystical communities of varying size, each with its own distinctive governance, small technologically based economy, and social structure.

Reconceptualization of Socialism

The collapse of communist regimes created a crisis for people on the left around the world. Many leftists had hoped that somehow these societies, which were characterized in a variety of ways, would undergo changes that would transform them into democratic and ecologically sensitive socialist societies. Various radicals advocated shedding the concept of socialism altogether and replacing it with other terms, such as *radical democracy, economic democracy, global democracy,* and even *Earth democracy.*

Efforts to replace the term *socialism* with new ones are understandable given the fate of postrevolutionary or socialist-oriented societies. However, I believe that leftists need to come to terms with the achievements and flaws of these societies as I have attempted to do in chapter 2 and to revisit the ideals of socialism and to critically assess the social experiments that have been labeled socialist, at both the national and local levels. As Samir Amin maintains, "[T]he expression of the demands of counterculture is fraught with difficulty—because socialist culture is not there in front of our eyes. It is part of a future to be invented, a project of civilization, open to the creativity of the imagination" (Amin 2009: 2).

In essence, socialism remains a vision, one that various individuals and groups seek to frame in ever-evolving new guises. Ultimately, of course, the issue is not the label but the social reality to be created and the principles of social and environmental relationships that make up the new reality.

Democratic Socialism

Numerous Marxists have asserted that socialism is inherently more democratic than capitalist societies could ever be and thus democracy is an inherent component of socialism. Miliband (1994: 51) delineates three core propositions that define socialism: (1) democracy, (2) egalitarianism, and (3) socialization or public ownership of a predominant portion of the economy. Although some areas of a socialist society would require centralized planning and coordination, democratic socialism recognizes the need for widespread decentralized economic, political, and social structures that would permit the greatest amount of popular decision making possible. According to Miliband (1994: 74), "Socialist democracy would encourage the revolution of as much responsibility as possible to citizen associations at the grass roots, with effective participation in the running of educational institutions, health facilities, housing associations and other bodies which have a direct bearing on the lives of people concerned." He envisions three distinct economic sectors: (1) a predominant and varied public sector, (2) a sizable cooperative sector, and (3) a sizeable private sector consisting primarily of small and medium companies that would play a significant role in providing various goods, services, and amenities (Miliband 1994: 110).

Boggs (1995: x) maintains that future strategies for change will need to be "more anti-bureaucratic, pluralistic, ecological, and feminist than anything experienced within the vast history of Marxian socialism." Socialist democracy would not be identical with total state ownership and centralized planning but could include collective, cooperative, and even individual property. Leo Panitch (2001: 223) delineates ten dimensions essential to incorporating a "utopian sensibility with concern for capacity-building," namely, overcoming alienation, attenuating the division of labor, transforming consumption patterns, envisioning alternative lifestyles, socializing markets, ecological planning, increasing global equality, democratic communication, expanding democratic processes, and creating a sense of community. Tariq Ali (2009: 88) argues that twenty-first-century socialism should include political pluralism, freedom of speech, access to the media, the right to form trade unions, and cultural liberty.

In part inspired by attempts to introduce market reforms in various postrevolutionary countries, such as Yugoslavia after 1950, Hungary where a New Economic Mechanism was introduced in 1968, China starting in 1978 with decollectivatization of agriculture, Poland between 1981 and 1989, and the USSR with Perestroika under Gorbachev starting in 1985, John Roemer (1994) and others have made advocated *market* socialism as a means of permitting postrevolutionary societies to address problems of economic inefficiency by resorting to market mechanisms.

Roemer (1994: 27) maintains that socialists have fetishized the notion of public ownership or socialization and should make room for market forces on the grounds that "any complex society must use markets in order to produce and distribute goods that people need for self-realization and welfare." Conversely, while recognizing that planning still would have a place under market socialism, he observes that large capitalist companies essentially function as centrally planned structures and that the public sector has increased drastically in capitalist societies. Roemer (1994: 46–53) delineates three possible forms of market socialism: (1) the worker-managed firm, (2) the traditional management form, which allows for a more equal income distribution, and (3) a form that would create quasi-public banks, progressive taxation, and legislation that would limit the power of the rich over the economy and politics, thus paving the way for equalization of incomes and wealth. He contends that "socialists should count themselves victorious in short run if they can design systems that bring about degree of income inequality and level of public services that exist in Nordic social democracies" (Roemer 1994: 54).

Roemer (1994: 118) views market socialism as a "short-term proposal" on the road to a more authentic form of socialism. He admits that the collapse of the Soviet Union constituted a setback for socialism, but argues that Soviet systems ultimately failed not because they lacked egalitarian goals but because of their resistance to markets, which contributed to a loss of incentives and competition. Roemer argues that advanced capitalist societies are not good candidates for market socialism because the vast majority of people have been pacified with social democratic concessions, although it is important to note that these concessions have been greatly eroded in most of these societies over the course of the past two decades. Roemer (1994: 129) maintains that the most likely candidates for market socialism are "authoritarian developing countries in which the rate of growth and pattern of distribution of wealth have been insufficient to improve conditions of large working and peasant class" and where "extreme material inequality has engendered leftist parties with a popular base." Although he does not want to add to the literature of what he terms "transitology," he maintains that the "potential of any socialist movement depends upon its ability to provide blueprints," especially given that the Soviet Union as the first effort to create a socialist society failed (Roemer 1994: 130). Unfortunately, Roemer's scheme does not call for immediate redistribution of wealth or income. Furthermore, it does not enhance workers' democracy. Saral Sarkar (1999: 193) maintains that most advocates of market socialism either ignore or gloss over issues of environmental sustainability, noting: "In most of their writings, I have found only unserious passing references to ecology and the need to include environmental protection as a consideration in plans and policies."

Erik Olin Wright makes a distinction between what has conventionally gone under the name of socialism or what he terms *statism* and an authentic form of socialism. The former refers to an "economic structure within which the means of production are owned by the state and the allocation and use of resources for different social purposes is accomplished through the exercise of state power," whereas socialism is an "economic structure within which the means of production are socially owned and the allocation and use of resources for different social purposes is accomplished through the exercise of what can be termed 'social power'" (Wright 2010: 120–121). Wright delineates seven pathways that potentially could contribute to social empowerment:

- Statist socialism
- Social democratic statist economic regulation
- Associational democracy
- Social capitalism
- Cooperative market economy
- Social economy
- Participatory socialism: statist socialism with empowered participation (Wright 2010: 131–149)

"Statist socialism" served as the backbone of traditional Marxist ideals of revolutionary socialism in which the party represented the working class and soviets theoretically "would directly involve workers' associations in the exercise of power in both the state and the production" (Wright 2010: 131). Unfortunately, history tells us that these ideals were never achieved for very complex reasons that I have attempted to delineate in part in chapter 2. "Social democratic statist economic regulation" includes a variety of interventions, including "pollution control, workplace health and safety rules, product safety standards, skill credentialing in labor markets, minimum wages, and other labor market regulations" that in and of themselves are progressive measures but are not sufficient to bring about socialism (Wright 2010: 134). Indeed, under capitalist regimes, while labor unions, environmental groups, and other social movements may have played an important role in bringing about such reforms, the capitalist class and its political allies may in time, particularly in the guise of neoliberalism, withdraw at least some of them. "Associational democracy encompasses a wide range of institutional devices through which collective associations in civil society directly participate in various kinds of governance, characteristically along with state agencies and business associations" (Wright 2010: 136). As long as such associations are not manipulated by corporate and governmental elites, they constitute a partial pathway to social empowerment. "Social capitalism" refers to the involvement of labor unions, consumer groups, and various social movements (such as anti-sweatshop, labor standards, and fair trade movements) in exerting input over the "allocation, organization, and use of various sorts," such as in the case of retirement funds and health and safety monitoring (Wright 2010: 137). Whereas a worker-owned cooperative company constitutes a type of social capitalism, a "cooperative market economy is one in which individual cooperative firms join together in larger associations of cooperatives—what might be termed a cooperative of cooperatives—which collectively provide finance, training, problem-solving services, and other kinds of support for each other" (Wright 2010: 139–140). The "social economy" consists of various voluntary associations and NGOs, which organize non-profit-oriented economic activity, which is "distinct from capitalist market production, state organized production, and household production," for purposes of fulfilling various human needs, such as housing as carried out by Habit for Humanity. Finally, under "participatory socialism: statist socialism with empowered participation," while the state is "directly involved in the organization and production of the economic activity," workers and community members are involved in economic decision making (Wright 2010: 143).

Each of the seven pathways incorporates some degree of economic democracy. Wright maintains: "Taken individually, movement along one or another of these pathways might not pose much of a challenge to capitalism, but substantial movement along them taken together would constitute a fundamental transformation of capitalism's class relations and the structures of power and privilege rooted in them. Capitalism might still remain a component of the hybrid configuration of power relations governing economic activity, but it would be a subordinated capitalism heavily constrained within limits set by the deepened democratization of both state and economy" (Wright 2010: 144–145).

Socialist democracy would not be synonymous with total state ownership and centralized planning but would entail "several forms of property—collective, cooperative and small private or individual property" and even some small businesses (Lorimer 1997: 22).

In recent years various leftists have been revisiting the concept not only of socialism but also of communism (Ali 2009; Badiou 2010; Bosteels 2014). Renowned French philosopher Alan Badiou, a former member of a small Maoist group, maintains that the communism has to be reinvented, despite the travesties that went under the guise of socialism or communism. She states: "We're at a crossroads, in a period of great ferment that's strongly reminiscent of the late 1840s. In this general context, as often happens, a very big step backward is needed in order to make a new leap forward. This means going back to the original communism and salvaging the basic characteristics of the Idea itself so as to adapt it appropriately to the modern world. Now more than ever, we can, we must, and we will reactivate the communist hypothesis" (Badiou and Gauchet 2016: 48).

Similarly, Jodi Dean (2012: 8) argues that communism has "again become a discourse and vocabulary for expression of universal, egalitarian and revolutionary ideals." He argues that an alternative goal other than capitalism is needed in "light of the planetary climate disaster and ever intensifying global class war as states redistribute wealth to the rich in the name of austerity" (Dean 2012: 1). Despite Dean's acknowledgment of the destructive impacts of climate change, most scholars who have re-embraced the notion of communism have not discussed the global ecological crisis and how communism, let alone socialism, might serve to solve it (Douzina and Zizek 2010). Unfortunately, Badiou and Zizek have dismissively asserted that "ecology has become the new opium for the masses" (quoted in Foster and Clark 2016: 9).

Eco-Socialism

In the past, Marxian political economy has tended to give at best passing consideration to environmental factors. Historically, however, there have been exceptions to this tendency, which harkens back to "powerful ecological currents in the Bolshevik Party," including Lenin's interest in conservation and ecological insights expressed by Bukharin in his book *Philosophical Ababeque* (Wall 2010a: 77). Nature reserves were created in the Soviet Union in the first decade of the Bolshevik Revolution (Wallis 2008: 110). McLaughlin reports:

> The growth of nature preserves from 1925 to 1929 in the USSR is notable. In 1925, there were 1,041,045 hectares (=4,019 square miles) of state nature preserves. In only four years this area had almost quadrupled to 3,934,428 hectares (=15,191 square miles). However, these preserves were managed by different agencies with different ends. Some nature reserves were simply game preserves where commercially valuable species could propagate. Others were viewed as scientific preserves ("etalon") with the aim of studying the workings of nature to bring the economic practices of society in line with the carrying capacity of ecosystems (McLaughlin 1990: 82).

Socialist-oriented local authorities resisted the auto-mobilization of city life in post–World War Europe. Various Marxist theorists, including Herbert Marcuse, Eric Fromm, E.P. Thompson and Raymond Williams in the U.K., Andre Gorz and Jean-Paul Deleage in France, Barry Commoner in the United States, Wolfgang Harich in the German Democratic Republic, and Rudolf Bahro after his emigration from the GDR to West Germany, have served as precursors to present-day eco-socialism (Wall 2010a: 82–89; Loewy 2015: xi).

Over the past three decades or so, various leftists have become more sensitive to the environmental travesties that have occurred not only in both developed and developing capitalist societies but also in postrevolutionary societies. A growing number of neo-Marxian scholars as well as other radical scholars have been attempting to integrate ecological considerations into their analyses of various types of social formations and societies. Such endeavors have been referred to as the "political economy of ecology" (O'Connor 1989); "eco-Marxism" (Agger 1979; Dealeage 1989; Benton 1989; Grundmann 1991a,b); "Green socialism" (McLaughlin 1990); "eco-socialism" (Ryle 1988; Burkett 2014; Foster 2000, 2009; Kovel 2007); radical ecology (Merchant 1996); and "socialist ecology," and "social ecology" or "eco-anarchism" (Price 2012). Foster maintains: "Although contributions to ecological thought within the Marxist tradition have existed since the beginning—going back to Marx himself—eco-socialism, as a distinct tradition of inquiry, arose primarily in the late 1980s and early '90s under the hegemony of green theory (and in the context of the crisis of Marxism following the downfall of Soviet-type societies). The general approach adopted was one of grafting Marxian conceptions onto already existing green theory—or, in some cases, grafting green theory onto Marxism" (Foster 2014b: 57).

I touch on eco-anarchism later in this chapter, particularly in terms of its potential relevance to Marxian approaches to *eco-socialism*. In reality, as Qingzhi Huan (2010a: 3) observes, eco-socialism is "many things to many people." Much of the interest in political ecology or eco-socialism stems from Green politics in Europe, particularly Germany, as well as the more radical sectors of the Northern American and Australasian environmental movements and has been inspired by the work of neo-Marxian scholars (Commoner 1972, Gorz 1980, and Bahro 1982). Members of the leftist tendency or the *fundis* in Germany referred to themselves as eco-socialists, although others viewed themselves as eco-anarchists (Loewy 2015: xii).

As an advocate of "ecological Marxism," James O'Connor (1988) argues that capitalism essentially undermines the ecological conditions of its own survival. For him, whereas the "first contradiction of capitalism" is the reality that capitalists exploit workers, the "second contradiction of capitalism" is the fact that capitalist development is incompatible with environmental sustainability. O'Connor (1998: 166) maintains that the "examples of capitalist accumulation impairing or destroying capital's own conditions, hence threatening its own profits and capacity to produce and accumulate more capital, are many and varied."

Eco-socialism seeks to come to grips with the growth paradigm inherent in capitalism and to which postrevolutionary societies also subscribed. As Silber argues, "Today the notion of unlimited growth in production has been called into question by aggravated problems of industrial pollution, toxic waste, nuclear perils, ecological degradation, looming environmental catastrophe and finite resources.... Rather than an idyllic world based on super-abundance, humanity's future is at least as likely to be shaped by the need to ensure justice and equity in a society characterized by limits" (Silber 1994: 260–261).

More recently, Pepper (2010: 35) observes that "eco-socialists have largely accepted the ecocentric arguments of some radical environmentalists, that there are indeed limits to economic and population growth imposed by the earth's carrying capacity."

Leftist scholars appear to differ on whether economic growth would be possible under eco-socialist parameters. Jones (2011: 57) argues that a "postcapitalist, ecological socialist mode of production would have more potential for sustainable growth because ecological considerations would be built into planning decisions from the outset." Pepper (1994: 234) argues that an ecologically sound socialist society will not develop until most people are prepared to create and maintain it, which means that eco-socialism must be concerned about jobs for working-class people. He warns that the "red-green" project faces the danger of too easily dismissing the working class and its potential contribution to the revolutionary process.

As environmental sociologist John Bellamy Foster so aptly argues,

> It is important to recognize that there is now an *ecology* as well as a political economy of revolutionary change. The emergence in our time of sustainable human development, in various revolutionary interstices within the global periphery, could mark the beginning of a universal revolt against both world alienation and human self-estrangement. Such a revolt, if consistent, could have only one objective: the creation of a society of associated producers rationally regulating their metabolic relation to nature and doing so not only in accordance with their own needs but also those of future generations and life as a whole. Today, the transition to socialism and the transition to an ecological society are one (Foster 2009: 277).

Albert seeks to address critiques of earlier versions of Parecon, the notion of participatory economics, that claimed that it was not sensitive enough to environmental factors. In his recent rendition of Parecon, he argues that it is environmentally sustainable in that "there is not pressure to accumulate" and that participatory planning will allow its citizens not only to "make wise choices for their own interests, but for their children and grandchildren as well, regarding not only direct production and consumption, but also the myriad ripple effects of economic activity on the environment" (Albert 2014: 71–73). Albert (2014: 149) also calls for a more nuanced and multifocused Marxist practice that would incorporate feminist, multiculturalist, and anarchist conceptions of society and history.

Schwartzman (2009: 11) argues that the fulfilment of eco-socialism requires a "robust theory that goes beyond familiar binary aspects of 'red' and 'green' practice." In conceptualizing eco-socialism, he maintains that Marxian political economy will have to draw heavily on the physical and informational sciences, partly climate science, ecology, biogeochemistry, and thermodynamics, in order to inform "technologies of renewable energy, green production, and agroecologies" (Schwartzman 2009: 11).

Drawing on his seventeen contradictions of capitalism, Harvey (2014: 294–296) delineates seventeen mandates to "frame and hopefully animate political praxis." While most of these touch on traditional socialist concerns, mandates fifteen and sixteen touch directly or indirectly on concerns conversant with eco-socialism. Mandate fifteen states: "The economy converges on zero growth (though with room for uneven geographical developments) in a world in which the greatest possible development of both individual and collective human capacities and powers and the perpetual search for novelty prevail as social norms to displace the mania for perpetual compound growth" (Harvey 2014: 296).

This mandate implicitly recognizes that many people, particularly in the developing world, will need to undergo growth but that the affluent in both the developed and developing worlds will need to undergo degrowth in order for the global economy to approach zero growth. In mandate sixteen, Harvey states: "The appropriation and production of natural forces for human needs should proceed apace but with the maximum regard for the protection of ecosystems, maximum attention paid to the recycling of nutrients, energy and physical matter to the sites from when they came, and to an overwhelming sense of re-enchantment with the beauty of the natural world, of which we are a part and to which we can and do contribute through our works" (Harvey 2014: 296–297).

Ecological Marxism or eco-socialism has made some headway among Marxist scholars in China (Wang 2012: 12). In terms of Western Marxist theory, Chinese scholars have been attracted by the ecological crisis theory of William Liess and Ben Aggar, O'Connor's notion of the "two contradictions of capitalism," Kovel's theory of an eco-socialist revolution, and the work of Foster and Burkett on Marx's ecology. Wang, He, and Fan (2014: 49) believe that ecological Marxism "can help remind the CCP of its ecological responsibility, and because environmental issues involve the vital interests of the people, it can remind the Party of its social responsibility as well." Qingzhi Huan (2016) calls for the creation of "socialist eco-civilization," which combines socialism with its emphasis on social justice and ecological sustainability, in China. Although the 17th Congress of the Communist Party of China endorsed the notion of "ecological civilization," which would entail developing "an energy and resource efficient, environmentally friendly structure of industries, pattern of growth, and mode of consumption" (quoted in Wang et al. 2014: 49) and the 18th Congress incorporated this concept into its constitution in November 2012, it remains to be seen whether the party elites will follow through with such ambitious goals. As Wang et al. (2014: 51) admit, the "mainstream in China still tends to rely on the modernistic, technologically determinist, developmentalist way of thinking to address the problems facing China."

Socialist Eco-Feminism

Eco-feminism emerged out of the feminist, peace, and environmental movements in the late 1970s and 1980s. French writer Francoise d'Eaubonne coined the term *ecofeminism* in 1974, calling on women to lead a movement to save the planet (Merchant 1996: 5). Eco-feminism seeks to not only eradicate injustice against women and the environment, but all forms of social injustice. The meltdown of Three Mile Island in Pennsylvania resulted in women convening an eco-feminist conference—"Women and Life on Earth: A Conference on Eco-Feminism in the Eighties" in March 1980 (Shiva, Mies, and Salleh 2014: 13–14). The Chernobyl disaster in Ukraine, the construction of nuclear power plants in Germany, and the disastrous release of toxic fumes from a Union Carbide pesticide plant in Bhopal, India, serve as events promoting eco-feminism in both developed and developing countries. As Merchant (1996: 7) observes, "[m]any ecofeminists advocate some form of an environmental ethic that deals with the twin oppressions of the domination of women and nature through an ethic of care and nurture that arises out of women's culturally constructed experiences."

Eco-feminism assumes various genres, including liberal eco-feminism, cultural eco-feminism, social eco-feminism, and socialist eco-feminism (Merchant 2005: 200–211). Liberal ecofeminists in particular seek to liberate women while preserving and nurturing the environment. Spiritual eco-feminists focus on the female principle as cardinal to Mother Earth who nurtures and sustains all human beings. For instance, Australian spiritual eco-feminist Freya Mathews (1991) draws inspiration from her claim that indigenous people view the earth as the great Mother and womb of life, which they view as sacred. Radical eco-feminists challenge the North/South divide associated with capitalist patriarchy, in which "nature is subordinated to man; woman to man; consumption; and the local to the global, and so on" (Shiva, Mies, and Salleh 2014: 5). Radical eco-feminism maintains that patriarchal societies are built on five interrelated pillars: sexism, racism, classism, capitalism, and environmental degradation. In differentiating a radical eco-feminism from other genres of eco-feminism, Shiva, Mies, and Salleh argue:

> To "catch up" with men in their society, as many women still see as the main goal of the feminist movement, particularly those who promote a policy of equalization, implies a demand for a greater, or equal share of what, in the existing paradigm, men take from nature. This, indeed, has to a large extent happened in Western society: modern chemistry, household technology, and pharmacy were proclaimed as women's saviours, because they would 'emancipate' them for household drudgery. Today we realize that much environmental pollution and destruction is causally lined to modern household technology. Therefore, can the concept of emancipation be compatible with preserving the earth as our life base? (Shiva, Mies, and Salleh 2014: 7).

Socialist ecofeminism makes the category of reproduction as opposed to production central to the notion of achieving a socially just and environmentally sustainable world system. Ariel Salleh, an Australian sociologist, has acted as a longtime proponent of a socialist eco-feminism. She argues that eco-feminism reaches for an "Earth democracy" that "reframes environment and peace, gender, socialist, and postcolonial concerns beyond the single-issue approach fostered by bourgeois right and its institutions" (Salleh 1997: x). Salleh laments that eco-feminism has been attacked by various eco-philosophies, including eco-socialism. In her view, Earth democracy must be inclusive of not only working-class women but also subsistence farmers and indigenous hunters and gatherers (Salleh 1997: 9). Salleh (1997: 12–13) argues that ecological crisis stems from a "Eurocentric capitalist partriarchal culture built on the domination of nature, and domination of Women 'as nature'." Nature must benefit all human beings, not just a few privileged ones as is the case under the parameters of capitalism. While Salleh (1997: 92) asserts that eco-feminists "do not differentiate women by stratifications of class, race, age, and so forth, since the nature-woman-labour nexus as a fundamental contradiction defies these conceptual boundaries," it appears that many eco-feminists have embraced a class analysis or explicit critique of capitalism. Indeed. Salleh (1997: 103) maintains that "feminism in the North needs to become more sensitive to ethnic and class difference." She argues that global justice and sustainability demand that the "North will have to review its high-tech consumption in favour of more species-egalitarian models by which the South provisions itself." However, it is important to note that due to corporate globalization, the upper echelons of the South have embraced the agenda of global capitalism.

Carolyn Merchant succinctly delineates the basic questions that a socialist eco-feminism must pose:

1. What is at stake for women and for nature when production in traditional societies is disrupted by colonial and capitalist development?
2. What is at stake for women and for nature when traditional methods and norms of biological reproduction are disrupted by interventionist technologies (such as chemical methods of birth control, sterilization, amniocentesis, rented wombs, and baby markets) and by chemical and nuclear pollutants in soils, waters, and air (pesticides, herbicides, toxic chemicals, and nuclear radiation)?
3. What would an ecofeminist social transformation look like?
4. What forms might socialist societies take that would be healthy for all women and men and for nature? (Merchant 2005: 208).

Renowned eco-feminist Vandana Shiva has formulated a perspective that she terms *Earth Democracy*, entailing ten principles:

1. All species, people, and cultures have intrinsic worth.
2. The earth community is a democracy of all life.
3. Diversity in nature and culture must be defended.
4. All beings have a natural right to sustenance.

5. Earth democracy is based on living commons and economic democracy.
6. Living economies are built on local economies.
7. Earth democracy is a living democracy.
8. Earth democracy is based on living cultures.
9. Living cultures are life nourishing.
10. Earth democracy globalizes peace, care, and compassion (Shiva 2005).

She maintains: "Earth Democracy enables us to envision and create living democracies. Living democracy enables democratic participation in all matters of life and death—the food we eat or do not have access to: the water we drink or are denied due to privatization or pollution; the air we breathe or are poisoned by. Living democracies are based on the intrinsic worth of all species, all people, all cultures: a just and equal sharing of this earth's vital resources; and sharing the decisions about the use of the earth's resources" (Shiva 2005: 6).

Leigh Brownhill (2010: 96) contends that Shiva's conception of Earth democracy "sounds and looks a lot like ecosocialism, though it may not be called ecosocialism by those who are living it."

Efforts to Organize Eco-Socialism

At the organizational level, eco-socialism is situated "amongst green parties, social movements, socialist groups and indigenous networks" (Wall 2010a: 2). Green Left is an eco-socialist network within the Green Party of the U.K. The Ecosocialist International Network (EIN) was created in October 2007 in a suburb of Paris at a meeting that brought together fifty activists, including ones from Australia, Canada, and Brazil (Wall 2010a: 72). EIN held its second meeting at the World Social Forum in Belem in January 2009, with strong representation by "indigenous, left and green activists mainly from Latin America" (Wall 2010a: 72). It drew its inspiration in large part from the Ecosocialist Manifesto written by Joel Kovel and Michael Loewy. Below are some of the positions that they take in their document:

- "In our view, the crises of ecology and those of societal breakdown are profoundly interrelated and should be seen as different manifestations of the same structural forms."
- "We believe that the present capitalist system cannot regulate, much less overcome, the crises it has set going. It cannot solve the ecological crisis because to do so requires setting limits upon ambula-

tion—an acceptable option for a system predicated upon the rule: Grow or Die!"
- "If we say that capital is radically unsustainable and breaks down into the barbarism outlined above, then we are also saying that we need to build a 'socialism' capable of overcoming the crises capital has set going. And if socialisms past have failed to do so, then it is our obligation, if we choose against submitting to a barbarous end, to struggle for one that succeeds."
- "Ecosocialism will be international and universal, or it will be nothing. The crises of our time can and must be seen as revolutionary opportunities, which is our obligation to affirm and bring into existence" (Kovel and Loewy 2001).

Wall (2010a: 90–121) provides a useful overview of eco-socialism in various regions and countries, including Africa, Asia, Europe, Australia, North America, and Latin America. In Australia, the Socialist Alliance, a political party that runs candidates in selected federal, state, and city council electorates, has a strong eco-socialist orientation as is manifested in many of the articles that it publishes in *Green Left Weekly*. Socialist Alliance promotes eco-socialist analyses and climate justice approaches to climate change. It sponsored a "Climate Change/Social Change" conference in Sydney on 11–13 April 2008 and another "Climate Change/Social Change" conference in Melbourne on 30 September 30–3 October 2011, with John Bellamy Foster being a speaker at both conferences. Solidarity, a smaller Australian socialist group, also has a strong eco-socialist orientation. The Australian Greens have eco-socialist members but for the most part they do not constitute a strong identifiable presence.

To a large degree, at the organizational level, eco-socialism remains an internet phenomenon. Ecosocialist Horizons states on its website that it "seeks to advance ecosocialism as a world-view and as a movement capable of offering real answers to the crises caused by capitalism." Its three principal aims are:

- "Providing news and analysis of ecosocialist concern through a multi-media website and other publications."
- "Educating our members to produce creative work and to organize events and actions."
- "Organizing convergences to advance diverse struggles towards an ecosocialist horizon." (http://ecosocialisthorizons.com/)

Eco-socialists who have published articles on the Ecosocialist Horizons website include Joel Kovel, Saral Sarkar, and Daniel Tanuro.

Climate & Capitalism is an online eco-socialist journal edited by Canadian Ian Angus (http://climateandcapitalism.com). In February 2012, it formed a working relationship with *Monthly Review: An Independent Socialist Magazine* in which both organs maintain editorial independence but share technical resources and collaborate politically. *Monthly Review*, due largely to the influence of its editor John Bellamy Foster, exhibits a strong eco-socialist stance as does the journal *Capitalism, Nature, Socialism*.

What Can Eco-Socialists Learn from Eco-Anarchists?

Socialists and anarchists share an anti-capitalist stance and view capitalism as the cause of numerous social and environmental problems. Radical historian Staughton Lynd proposes a synthesis of Marxism and anarchism. He argues: "[D]uring the past century and a half neither Marxism nor anarchism has been able to carry out the transformative task alone. Marxism has produced a series of fearsome dictatorships. Anarchism has offered a number of glorious anticipations, all of them short-lived and many of them drowned in blood" (Lynd and Grubacic 2008: 12).

In terms of environmental issues, Anderson (2006: 256) maintains that "[a]narchism has a long history of environmentalism, from early anarchist thinkers such as Peter Kropotkin to the influential social ecologist Murray Bookchin (1991) who links the exploitation of nature to the exploitation of human beings," in much the same vein as eco-socialists. Pepper (1993: 207) maintains that while socialists tend to conceive of nature as a social construction and are anthropocentric in their stance toward it, anarchists tend "to see nature as external to society but [believe] the latter should conform to nature's laws and regard natures as a template" and tend to adopt to a stance toward nature that is neither anthropocentric nor biocentric. Greek eco-anarchist Takis Fotopoulos maintains that "modern hierarchical society," which for him includes both the capitalist market economy and "socialist" statism, is highly oriented toward economic growth, which has glaring environmental contractions:

- "First, it is simply not physically possible for the wasteful consumption standards, which are today enjoyed by the '40 percent societies' in the North and the elites in the South to be universalized and enjoyed by the world population."
- "Second, a universalized growth economy is not environmentally sustainable, at the present state of technological knowledge and cost of 'environmentally friendly' technologies" (Fotopoulos 1997: 72).

Ted Trainer is an Australian eco-anarchist whom I visited briefly at the University of New South Wales where he was an honorary research fellow in late 2009 and with whom I have corresponded since then periodically. While I do not agree with Trainer on all issues, including the role of the state and social movements in bringing about a socially just and environmental society, his thinking has influenced me in various ways. He is a scholar whose work eco-socialists should consider seriously. While Trainer has written several books, I summarize some of the principal points in his most recent book titled *The Transition to a Sustainable and Just World* in which he explores the following themes about capitalist society in general:

- It is enormously unsustainable
- It has a highly unjust global economy
- It fosters over-consumption
- "The defining feature of the coming era—scarcity"
- It cannot fix itself
- It requires massive and radical changes
- The alternative entails a "Simpler Way" (Trainer 2010: 2–12).

Stated in this way, without going into further specifics, these arguments are consistent with eco-socialism.

Trainer delineates the following five core principles of the Simpler Way:

- Material living standards on the whole must diminish
- There is a need for small-scale, highly self-sufficient local economies and communities
- Local communities based on cooperative and participatory principles control their own internal affairs, largely independent of international and global social structures
- A radically different economic system needs to be developed, one that is socially controlled, oriented toward meeting needs as distinct from maximizing profits and not driven by market forces and a growth paradigm
- These changes require a radical shift in values and world view, especially away from competition, greed, and acquisitiveness (Trainer 2010: 6–13).

Trainer's first principle applies primarily to people who are relatively affluent in developed and developing societies and obviously the super-rich. However, it clearly does not apply to people who are abjectly poor in developing countries, particularly ones in sub-Saharan Africa, or the indigent poor and homeless in developed countries. Many of these people need to undergo "appropriate development," which for Trainer (2010: 129–130) "should be about improving all aspects of a society, including the quality of food and water and health services, the opportunities for leisure and cultural activity, the level of debate and discussion, the processes for government and administration, the moral standards, the geographic and aesthetic conditions in which people live, citizenship and social responsibility, openness and accountability, social cohesion, equity, concern for those less fortunate, the quality of life, security, the conditions of the poorest, and especially ecological sustainability."

Appropriate development focuses on "developing what is needed" and may actually entail reducing the GDP, at least presumably in the developed countries and the materially better-off developing countries (Trainer 2010: 130). It would focus on the local economy and would minimize involvement in the national and global economies (Trainer 2010: 136). Trainer (2010: 139) has a high regard for indigenous and peasant lifestyles, which are "typically highly collectivist, and their economies are governed by customs and traditions, not profit and gain." He maintains that appropriate development recognizes the "value of many aspects of traditional cultures" (Trainer 2010: 139). Trainer cites three illustrations of appropriate development, two of which ironically have been guided by socialist rather than anarchist principles. The first is the state of Kerala in southern India and the second is Cuba in the wake of the collapse of the Soviet Union. The third example is the Ladakh region of India, near Tibet, where people engage in farming and pastoralism at an altitude of about fourteen thousand feet. Trainer reports:

> The Ladakhis are kind and generous.... Their production is labour-intensive, yet the pace of work and life in general is relaxed, with much time for ceremonies and religious observance. No one is isolated or lonely, they recycle everything and waste nothing. They have no interest in power, domination or competition. They are very conscious of their dependence of their dependence on nature, they are multi-skilled and practical, and they live simply. There is no crime and no poverty and no drug problem and no social breakdown. Above all they are notoriously happy people (Trainer 2010: 146).

While it is possible that Trainer is romanticizing the people of Ladakh, his characterization of them is consistent with the ethnographic observations that anthropologists have made about many indigenous societies and peasant communities around the world.

Anthropologist John Bodley (2003: 17) makes the following poignant observations about the highly egalitarian and environmentally sustainable lifestyle of tribal or indigenous societies, particularly foraging ones, simple horticultural ones, and simple pastoral ones:

> [The] combination of a universal human rights entitlement, supported by an emphasis on economic cooperation and equity together with specific social features that reduce conflict, all worked together to reduce the natural incentive for rivalry between groups of individuals. Ideally, there was no cultural incentive for households to produce more than their immediate material needs. Instead, people were free to focus on minimizing energy expenditures and risk, and maximizing household security and well-being. There was also no incentive for either economic growth, or population growth. Even more remarkable, in this kind of world there was no economic scarcity because everyone's material needs could be readily satisfied (Bodley 2003: 17).

While obviously it would not be possible to return to the "forces of production" practiced by these people on a global scale, the question becomes how we can maintain some level of technological sophistication, but one that is environmentally sustainable, while adopting their "social relations of production." I explore Trainer's ideas on how to transition to a sustainable and just world in chapter 6 as part of an effort of achieving a democratic eco-socialism world system.

Trainer (2014: 8) maintains that while socialists and anarchists essentially agree on the need to ultimately achieve communism, in other words, a system in "which there are no classes and no domination of some by others, no relations of power or privilege, and in which things are done cooperatively, all are cared for, and in which there is 'no alienation'," they disagree on how such a social system is to be achieved, with the former advocating capturing the state and the latter building it through grassroots efforts that prefigure the future by local communities running themselves.

Maria Mies and Veronika Bennholdt-Thomsen have developed a "subsistence perspective" that blends together elements of feminism, eco-anarchism, and perhaps even eco-socialism, and bears some resemblance to Trainer's notion of a Simpler Way, although it places more emphasis on the need for equitable gender relations. Some of the main features of their perspective include the following:

- Subsistence production has priority over commodity production.
- Socially and materially useful work has priority over wage work.
- Men should do as much unwaged work as women.
- The economy should recognize the limits of nature and must be regional and decentralized.

- Local markets should function to satisfy the subsistence needs of all.
- The notion of the commons should be revitalized in order to resist the commercialization and privatization of the nature.
- Money should be a means of circulating goods rather than accumulating them (Mies and Bennholdt-Thomsen 1999: 62–63).

Mies (2010: 194) points to *Nayakrishi Anolon*, a peasant movement in Bangladesh that assisted many villages to create "poison free zones," which denied access to multinational corporations in their efforts to sell the villagers artificial fertilizers, high-yielding seeds, and pesticides, as a living example of the application of the subsistence paradigm in a developing country. Terisa E. Turner and Leigh S. Brownhill present a succinct overview of the basic parameters of a *subsistence political economy*:

> Subsistence is life-supporting activity in which values predominate. That is, people produce primarily for use and while they may trade items or services, the production is not primarily for exchange or for the appropriation of money demand. A subsistence way of life may be pre-capitalist or it may co-exist in the interstices of the capitalist political economy. Depending on the power relations which exist, the subsistence political economy may be more or less subsumed by the commodified political economy or autonomous from it. Those producing sustenance have an interest in working to make the society sustainable through conversation, restitution, relative regional autonomy and planning (Turner and Brownhill 2001: 107).

When Mies and Bennholdt-Thomsen introduced their subsistence perspective to the attendees at the congress on "Women and Ecology" in Cologne, Germany, in 1986, which in their view made "evident that women in the industrialised countries and the middle-class women in the South are not only victims but also beneficiaries of the international exploitative system," they encountered resistance from some of the women present on a wide variety of grounds, including that it allegedly reinforces traditional women's roles and is a throw-back to a more rudimentary way of life (Mies and Bennholdt-Thomsen 1999: 13). However, Mies and Bennholdt-Thomsen maintain that the subsistence perspective has been adopted by many women in the developing world who have rejected the modernization development project that inevitably will marginalize them as well as many children and men.

Eco-anarchists tend to subscribe to certain prefigurative visions that are very much compatible with eco-socialism, including "decommodification; alternative development schemes; movements for ethical consumption; new forms of labour internationalism; and organization of marginal workers and other excluded groups" (el-Ojeili 2014: 456). In keeping with the eco-anarchist and sufficiency perspectives of Trainer and Mies and Bernnholdt-Thomsen, Robinson and Tormey (2012: 144) assert that anarchists and particularly eco-anarchists use anthropology as a "as means to imagine alternatives to oppressive social orders such as neoliberal capitalism." In a similar vein, Nugent (2012: 209) maintains that anthropology and anarchism exhibit an affinity for each other in that ethnographic research, such as in lowland South American foraging and horticultural village societies, has "revealed possibility for a utopian society based in part on what has been possible in past."

Can Eco-Socialism and Eco-Anarchism Dispense with Money?

In an anthology edited by two Australians, Anrita Nelson and Frans Timmerman (2011: 3), various eco-socialists and eco-anarchists revisit the notion that money could be abolished in a "moneyless, marketless, wageless, classless, and stateless planetary society." Such an approach is obviously quite at odds with various propositions for market socialism. Indeed, party elites during the early Soviet era seriously discussed the possibility of creating a moneyless economy (Nelson 2011: 11). This debate re-emerged with encouragement from Che Guevara in Cuba during the mid 1960s. In reality, all postrevolutionary societies in the past or present have operated or still do operate with money, which generally is not officially recognized by capitalist countries. However, the need to obtain "hard currency" has long forced postrevolutionary societies to operate within the parameters of the capitalist world system.

Ariel Salleh delineates fourteen features of a "non-monetised society-nature metabolism" that are characteristic of "meta-industrial communities," some of which I list below:

- "The consumption footprint is small because local resources are used and monitored daily with care."
- "Meta-industrial labour is intrinsically precautionary because it is situated in an intergenerational time frame."
- "Regenerative work patiently reconciles the time scales of humans and other species, and readily adapts to disturbances in nature."
- "It is an empowering work process, without a division between the worker's mental and manual skills."
- "The labour product is immediately enjoyed or shared, whereas the industrial worker has not control over his or her ability" (Salleh 2011: 100).

From an eco-anarchist perspective, Terry Leahy envisions a "gift economy" in which: "[P]roducers in all sectors would decide what to produce and how to distribute it based on the needs of other groups. The status of the givers would depend on genuine needs being met by the gift. Producers in each sector would be aware of their dependence on the services of other sectors" (Leahy 2011: 114).

He argues that "hybrid strategies" or practices employed by producers or community members could serve as transitions from the existing system to a gift economy. Actually existing examples of hybrids include support for environmental taxes, willingness to pay a higher price for more environmentally sustainable products, and volunteering to work in community gardens (Leahy 2011: 122). Leahy (2011: 125) argues that a "'revolution by stealth' might happen as the pervasiveness and influence of hybrid organisations came to dominate the economy, nationally, regionally or internationally." Conversely, any gift economy would probably "have to achieve either a near universal spread or the military capacity to defend attacks from capitalist forces" (Leahy 2011: 125).

The Concept of Democratic Eco-Socialism

The concept of democratic eco-socialism obviously constitutes a merger of the earlier existing concepts of democratic socialism and eco-socialism and draws on my work with Merrill Singer and Ida Susser in critical medical anthropology or critical health anthropology (Baer, Singer, and Susser 1997, 2003, 2013; Baer, Singer, Long, and Erickson 2016) as well as my work with Merrill Singer on developing a critical anthropology of climate change that posits the following premises:

- Social systems do not last forever, whether at the local, regional, or global level.
- The capitalist world system or global capitalism has been around for about 500 years but has come to embody so many inherent contradictions that it must be transcended to ensure the survival of humanity and animal and plant life on a sustained basis.
- There is a need for an alternative global system, one that is committed to meeting people's basic needs, social parity and justice, democracy, and environmental sustainability.
- Anthropologists and other progressive social scientists are too small a group to act as a vanguard in the struggle against global warming (or climate change) and capitalism. They must form links not only with anti-systemic movements, including the labor, anti-corporate globalization or social justice, peace, indigenous and ethnic rights, and environmental movements, but also the climate justice movement (Baer and Singer 2009, 2014).

Democratic eco-socialism entails the following principles:

- An economy oriented to meeting basic social needs—namely, adequate food, clothing, shelter, education, health, and dignified work
- A high degree of social equality
- Public ownership of the means of production
- Representative and participatory democracy
- Environmental sustainability (Baer, Singer, and Susser 2013: 398–403).

Democratic eco-socialism rejects a statist, growth-oriented, productivist ethic and recognizes that humans live on an ecologically fragile planet with limited resources that must be sustained and renewed as much as possible for future generations.

The vision of democratic eco-socialism closely resembles what world systems theorists Terry Boswell and Christopher Chase-Dunn (2000) term *global democracy*, a concept that entails the following components: (1) an increasing movement toward public ownership of productive forces at local, regional, national, and international levels; (2) the development of an economy oriented toward meeting social needs, such as basic food, clothing, shelter, and health care, and environmental sustainability rather than profit making; (3) the eradication of health and social disparities and the redistribution of human resources between developed and developing societies and within societies in general; (4) the curtailment of population growth that in large part would follow from the previously mentioned conditions; (5) the conservation of finite resources and the development of renewable energy resources; (6) the redesign of settlement and transport systems to reduce energy demands and greenhouse gas emissions; and (7) the reduction of wastes through recycling and transcending the reigning culture of consumption.

Fred Magdoff (2014) delineates twenty-two principles in order to create a socially just and ecologically sustainable economy and social system that are wholly compatible with democratic eco-socialism. Too numerous to list here, they touch on issues such as economic democracy, social equality, community and regional self-reliance to the extent possible, reliance on renewable energy sources, conservation and sparing use of nonrenewable resources, sustainable agriculture, situating people near workplaces to which they can commute on efficient public transportation, adequate leisure time to pursue personal interests, and educational practices that encourage compassion, cooperation, reciprocity and sharing, and an awe of nature. Magdoff (2014: 34) fears that "barbarism may be the fate that awaits our grandchildren and their grandchildren unless we change the way of thinking and start to envision, and begin to work towards, an economy and society under truly democratic social control with the very purpose being to satisfy basic human needs, which ... includes a healthy and living environment."

Socialism for the Twenty-First Century

The term "twenty-first-century socialism," more often referred to as "socialism for the twenty-first century," is attributed to Heniz Deiterich, a German scholar of Marxism (Kennemore and Weeks 2011: 267). It seeks to transcend the mistakes of both late capitalism or neoliberalism and postrevolutionary societies, or "twentieth-century socialism," by adhering to democratic principles. At any rate, various progressive scholars have tried to predict from which countries or regions efforts seeking to transcend global capitalism are likely to occur.

Chase-Dunn (2010: 52) argues that it is unlikely that in the foreseeable future such efforts will emanate from the core countries because a "sizable segment of the core working classes lack motivation because they have benefitted from a less confrontational relationship with core capital." Despite the fact that workers and peasants in peripheral countries are among the most exploited and oppressed members of the world proletariat, he argues that they generally "have either been overthrown by powerful external forces or forced to abandon most of their socialist programs" (Chase-Dunn 2010: 52). These realities prompted Chase-Dunn to turn to the semi-peripheral regions as the weak link in the capitalist world system as the locus of a shift toward "democratic socialism" because they, "especially those in which the territorial state is large[,] have sufficient resources to be able to stave off core attempts at overthrow and to provide some protection to socialist institutions if the political conditions for their emergence should rise."

Venezuela starting out with the rule of Hugo Chavez has been a semi-peripheral country that has proven to be a challenge to global capitalism and has provided a protective umbrella to various peripheral countries, particularly Bolivia and Ecuador, by raising the hope of creating "socialism for the 21st century" (Katz 2007). Indeed, overall Latin America is the region where the anti-imperialist struggle has been the most pronounced in the early twenty-first century. However, Chase-Dunn (2010: 53) mentions South Africa, Brazil, India, Mexico, and perhaps even South Korea as potential candidates to make "socialist bids for state power in the semiperiphery," although at the present time such bids appear to be rather unlikely.

At the present time or for the foreseeable future, the notion that democratic eco-socialism may be eventually be implemented in any society, developed or developing, in a number of societies, or at the global level may seem absurd. Boswell and Chase-Dunn (2000: 227) assert that the "emergence of democratic collective rationality (socialism) at the world-system level is likely to be a slow process." In keeping with Trotsky's notion of a "permanent revolution," they also assert that just as capitalism is a world system, "socialism must be organized at the global level as well" (Boswell and Chase-Dunn 2000: 86). Indeed, Chase-Dunn (2010: 55) views a world state as a more direct and stable way to prevent nuclear holocaust but while initially it may be dominated by the transnational capitalist class, the "very existence of such a state will provide single focus for struggles to socially regulate investment decisions and to create a more balanced, egalitarian and ecologically form of production and distribution."

More recently, drawing on a world systems theory perspective, Chase-Dunn along with Lerro have delineated three possible scenarios for the next several decades (Chase-Dunn and Lerro 2014: 363–372):

- Scenario 1: A second round of U.S. hegemonic domination
- Scenario 2: Collapse entailing interstate rivalry, eco-catastrophe, and deglobalization
- Scenario 3: The creation of a global democratic and sustainable commonwealth.

In scenario 1, the United States would rejuvenate itself by capitalizing on various new technologies, such as biotechnology, nanotechnology, and green energy, and its scientific expertise situated in its vast network of research universities. A more enlightened political leadership would distance itself from military adventurism such as in Iraq and Afghanistan and make an even greater commitment to achieving greater global social parity. In scenario 2, the United States would continue its present hegemonic decline and the capitalist world system would experience an ongoing financial and ecological crisis, including climate change. Germany, Japan, China, and India would function as serious rivals of U.S. hegemony.

I focus on scenario 3, which is the most optimistic of Chase's three scenarios and the one that most closely corresponds to the notion of global democratic eco-socialism that I espouse as a vision for a more egalitarian, democratic, and environmentally sustainable future. Chase-Dunn's and Lerro's (2014: 367) vision of a "global democratic and sustainable commonwealth" could potentially emerge out of the efforts of the "New Global Left"—a "subgroup of global civil society that is critical of neoliberal and capitalist globalization, corporate capitalism, and the exploitative and undemocratic structures of global governance," which he juxtaposes with the "larger global civil society ... which includes defenders of global capitalism and of the existing institutions of global governance as well as other challengers of the current global order." The New Global Left includes social movements, various global political parties, and even progressive national regimes. A new global democratic government would be supported by a majority of the world's population, would attempt to try to redistribute the global wealth and resources, would seek to achieve environmental sustainability, would operate a global multilateral military apparatus, and would exert democratic control over global financial institutions.

History tells us that social changes can occur very quickly once economic, political, and social structural changes have reached a tipping point, a term that has become popular in climate science. As indicated earlier, Wallerstein (1979) some time ago asserted that the transition from the capitalist world system to a socialist world government—if it were to occur, which he did not see as inevitable—would take 100–150 years, but in light of the global economic and ecological crises, he has more recently argued that it would have to be achieved in thirty to forty years in order to assure the survival of much of humanity (Wallerstein 2008). According to Foster (2009: 265), the "transition from capitalism to socialism is a struggle for sustainable human development in which societies on the periphery [and the semi-periphery] have been leading the way."

The Case of Venezuela

Perhaps the leading example of such a transition is Venezuela, where Hugo Chavez declared his commitment to socialism of the twenty-first century at the World Social Forum in Brazil by calling for the creation of the United Socialist Party of Venezuela. He had first assumed power in 1998 and managed to thwart an attempted military coup in February 2002. Chavez's vision of an alternative Venezuela is referred to as the Bolivarian Revolution, named after Simon Bolivar, who fought for independence from Spain in the early 1800s. It also stresses not only a commitment to socialism but also endogenous development or "development from within" the country, which emphasizes relying primarily on Venezuela's internal resources, including agrarian production with the aim of achieving food sovereignty and participatory democracy. By 2009 Venezuela had over thirty-five thousand community councils that monitored and administered food needs (Schiavoni and Camacaro 2009). Due to its heavy reliance on oil production and export, agricultural production declined from a third of the GDP in the 1920s to less than 10 percent of the GDP by the 1950s, and to about 6 percent in 2005 (Wilpert 2005: 9).

The Venezuelan government extended its commitment to a "social economy" by promoting the redistribution of wealth vis-à-vis land reform and other social policies, and promoting cooperatives, industrial eco-management, social production enterprises, and renationalization of the national telephone company (Webber 2011: 191). Chavez issued a decree in 2007 that required all foreign oil companies to relinquish the majority control of their operations in Venezuela to PdVSA (Petroleos de Venezuela S.A.), the state oil company, a stance that prompted many of these companies to leave the country. Nevertheless, Chevron and its partners along with various other oil companies, including several Chinese ones, want to develop the rich oil deposits in the Orinoco Belt. Venezuela now exports more oil to China than the United States, making it in some observers' eyes the Saudi Arabia of Latin America (Schwartzman and Saul 2015: 17). Oil accounts for 90 percent of Venezuela's exports and over half of the government revenues in Venezuela (Webber 2011: 199). Although for the present Venezuela does constitute a progressive player in national efforts to resolve the ecological crisis, including the climate crisis, it finds itself in a complex and embattled situation that forces it to act in contradictory ways. Conversely, as Foster observes, "Venezuela has been using its surplus from oil to transform its society in the direction of sustainable human development, thereby laying the foundation of a greening of its production. Although there are contradictions to what has been called Venezuelan 'petro-socialism,' the fact that an oil-generated surplus is being dedicated to genuine social transformation rather than feeding into the proverbial 'curse of oil' makes Venezuela unique" (Foster 2009: 274).

While still alive, Chavez developed Plan Paria, an eco-socialist program, which included training thousands of cadres in agroecology, bioremediation, sustainable land and water management, and urban agriculture. Despite these efforts, Venezuela has the highest per capita murder rate in Latin America (Schwartzman and Saul 2015: 19).

In addition to reining in the nationalized oil industry, which according to Chavez, operated as a "state within a state," the Chavez government nationalized the "Banco de Venezuela, the Exito supermarket chain, Venezuela's four electricity companies, the telephone company Entel, the cement industry, and the steel industry" (Escobar 2010: 13; Harnecker 2015: 116). In fact, it nationalized many companies of varying size, with some them being turned over to cooperatives that often "lacked the skills and experience to run them properly (Buxton 2016: 11).

In 2003–2004, the government initiated a bottom-up strategy that sought input from both communities and workplaces. This resulted in the creation of Local Councils of Public Planning and Communal Councils, part of a program of People's Power. The Communal Councils draw on input from urban residents and peasant communities and are "financed directly by the State and its institutions, thus avoiding major interference from the municipalities" (Azzellini 2010: 25). The Oil Sowing plan in Venezuela invites local communities to design their own development projects that are funded by state oil company revenues and have helped to reduce the incidence of extreme poverty, reduce infant mortality, and essentially eradicate illiteracy. However, there exist numerous contradictions in the People's Power initiative due to the symmetry of power between the state and the Communal Councils, "meaning that initiatives and movements from below risk being co-opted, in the sense of reproducing the logic and patterns of constituted power, such as hierarchical structures, representative mechanisms, divisions between leaders and led, and bureaucratization" (Azzellini 2010: 26).

Poverty and extreme poverty rates in Venezuela fell from 48.6 and 22.2 percent, respectively, in 2002 to 28.5 and 8.5 percent, respectively, in 2007 (Webber 2011: 196). In contrast, poverty rates in Chile stood at 13.7 percent in 2006, 18.6 percent in Costa Rica in 2007, and 18.1 percent in Uruguay in 2007. Venezuela surpassed Chile and Costa Rica in 2008 to become the Latin American country with the second-lowest level of inequality, only surpassed in this regard by Cuba (Leech 2012: 126). Nevertheless, as Webber observes (2011: 324), "[h]uge concentrations of personal wealth and privilege remain untouched by the Bolivarian process," a situation that requires a radically progressive taxation system that would redistribute the wealth in Venezuela. Capitalism continues to be well-entrenched in Venezuela as is illustrated by the fact that the share of private sector GDP increased over 6 percent between Chavez's election in 1998 and the third quarter of 2008, in spite of the nationalization of various industries (Giri 2012: 24). A small group of very wealthy families "have insulated themselves from the Bolivarian Revolution (Buxton 2016: 11).

Venezuela has implemented three state-owned firm arrangements: (1) a worker-managed firm that seeks to maximize the income per worker and is modeled on self-managed enterprises in the former Yugoslavia, (2) profit-oriented state capitalist firms, and (3) "statist" firms in which productive units are "directed by the state to achieve specific targets, e.g., of output of revenue" (Lebowitz 2014: 2). State capitalist companies, however, can easily exploit workers and engage in environmentally damaging activities in their drive for profits. In reality, Venezuela has a mixed economy in that banking, commerce, and foreign trade largely remain in private hands (Petras 2011: 15). Unfortunately, since the death of Chavez in 2013, Venezuela has experienced extremely difficult times under the Maduro government, including the threat of the opposition party Mesa de la Unidad Democratica (MUD), shortages of many consumer items, the world's highest inflation rate at 70 percent, and a crisis in the health-care system, including 16,000 beds out of 45,000 hospital bed being empty due to shortages (Saunois 2015: 8). Furthermore, a 97 percent drop in earnings from oil due to a large drop in oil prices worldwide has contributed to economic problems. Oil revenues account for about 95 percent of export earnings, 60 percent of budget revenues, and 12 percent of Venezuela's GDP (Buxton 2016: 7).

In December 2015 the government lost the parliamentary election, only the second electoral defeat by the Chavistas since the election of Hugo Chavez in 1998. Whereas the opposition Mesa de lad Unidad Democratica (MUD) Party garnered 56 percent of the popular vote, the Partido Socialista Unido de Venezuela (PSUV) won only 41 percent of the popular vote (Buxton 2016: 6). Maduro has felt compelled to install a new management at the state oil company, in which minority shareholders have been granted disproportionate influence; it has had very low support in opinion polls. In early 2016 the Maduro government implemented economic reforms that include changes to Venezuela's "multi-tiered exchange rate, an increase in domestic petrol prices, a new tax system and expansion of community control over food distribution" (Boothroyd 2016: 18). Despite a 6,000 percent increase in oil prices for consumers, the cost is extremely inexpensive, about one cent per liter, one of the lowest prices in the world. Unfortunately, Venezuela has been experiencing extremely high inflation, possibly over 700 percent in 2016 (Buxton 2016: 14) and has incurred huge debts, including to China. At the present time, the future of the Bolivarian Revolution, hangs in the balance as the country finds itself under assault from the more affluent sectors of its population as well as U.S. imperialism.

The Case of Bolivia

In Bolivia Evo Morales and his governing Movement Towards Socialism (MAS) party have adopted a policy of "communitarian socialism" and "proposed a 'Law of Mother Earth', which requires that human society live in harmony with the earth and that the earth has the right to maintain the integrity of its ecosystems through the preservation of clean air and water, and the maintenance of diversity and equilibrium" (Leech 2012: 144). Morales called for a Socialism of the South or Indigenous Socialism that includes ten "commandments." These call for the eradication of the capitalism in order to "save the planet"; ending wars that have benefitted empires, multinational corporations, and selected families; developing "relations of coexistence" among the countries of the world; treating access to water as a human right; developing "clean energies"; respecting Mother Earth; treating "basic services, such as water, electricity, education, health care, communications, and collective transportations" as human rights; consuming only what is necessary and seeking to consume local products; promoting "cultural and economic diversity"; and developing a "communitarian socialism that is in harmony with Mother Earth" (Derber 2011: 135). The Bolivian legislature established eleven new rights for nature, including the right to life and to existence, the right to clean water and air, the right to be free from pollution, and the right to "not be affected by mega-infrastructure and development projects that affect the balance of ecosystems" (Bell 2016: 81).

The MAS emerged as the political arm of the indigenous-peasant movement based in the department of Cochabamba in the mid 1990s and models itself on an "assembly-style, rank-and-file democracy of peasant unions in region," which consisted in large part of coca growers who opposed the U.S.-supported "drug war" in the region Webber (2011: 3). Over the course of its ascent, the party has become more moderate as an urban, mestizo middle-class intelligentsia has assumed influence. The MAS has also sought to appease the capitalist class in the eastern section of Bolivia in terms of the assembly's governance structure, in part to stave off a secession movement. Garcia Linera, whom Morales chose as his vice-president in 2005, developed the notion of "Andean-Amazonian state capitalism." He asserts: "The State is the only actor that can unite society. It is the State that takes on the synthesis of the general will, plans the strategic framework and steers the front carriage of the economic locomotive. The second carriage is Bolivian private investment. The third is foreign investment. The fourth is small business. The fifth is the peasant economy and sixth, the indigenous economy" (quoted in Touissant 2010).

In 2006 Morales signed a law that nationalized Bolivia's natural gas reserves and required all foreign energy companies to renegotiate their contracts so that they would benefit the Bolivian people (Derber 2011: 138). His administration embarked on a program of "nationalization without expropriation," which nationalized oil and natural gas reserves within Bolivia but did not want multinational corporations in the hydrocarbon sector to cease operations, but rather wanted them to enter into joint ventures with the government (Kaup 2015). Under this arrangement, foreign direct investment in Bolivia increased from $281 million in 2006 to

$1.75 billion in 2013. In addition to providing financial support for social programs, the government's profits from the hydrocarbon sector have substantially increased international monetary reserves in Bolivia.

Bolivia has embarked on a program of food sovereignty by promoting local and sustainable food production, which seeks to allow peasants to obtain ownership of seeds, thereby freeing them from dependence on agrobusinesses that dominate seed production. Morales withdrew Bolivia from the World Bank and the International Monetary Fund and undertook nationalizing the oil, natural gas, and lithium resources, a move that required altering the constitution.

In January 2009 the Bolivian people in a national vote of 61.4 percent approved a new constitution that declared that Bolivia constitutes a "pluri-national state, communitarian state" that is "based on respect and equality for all, with principles of sovereignty, dignity, complementarity, solidarity, harmony and equality in the distribution and redistribution of social goods, where the quest for the common good predominates; with respect for economic, social, juridical, political and cultural plurality of the inhabitants of this earth; in collective coexistence with access to water, work, education, health, and housing for all" (quoted in Kennemore and Weeks 2011: 270).

Despite the best of intentions, the movement toward implementing Morales's vision of socialism has been slow and patchy: "Morales's first four years in office ... witnessed the hollowing out of the left-indigenous demand for a revolutionary constituent assembly—the assembly introduced by the MAS government in 2006 purged the process of all revolutionary and participatory potential by seeking to appease the eastern-lowland bourgeoisie concerning its rules of conduct and content. The combined liberation-struggle for indigenous liberation, with socialist transformation is only a remote possibility, 50 to 100 years in the future" (Webber 2012: 329).

Bolivia under Morales has preserved private property rights, has not undergone serious land reform measures, and has experienced fiscal austerity. In 2011 Morales slashed fuel subsidies in order to reassure foreign investors. Also, the Morales government provides hundreds of millions of dollars in loans, export subsidies, and tax incentives to wealthy agricultural exporters and has expelled landless indigenous squatters from large agricultural estates (Petras 2011: 15). In essence, Webber (2011: 232) maintains that despite MAS's stated commitment to socialism over the long run, Bolivia at least between 2005 and 2010 followed a "reconstituted neoliberalism, one that abandoned features of neoliberal orthodoxy, but retained its core faith in the capitalist market as the principal engine or growth and industrialization"—one based largely on exports of natural gas, oil, and mineral resources. Furthermore, while Morales initially embraced direct democracy, many coca growers on whose support he relied in his ascent to the presidency feel that he has "slipped into old fashion patronage politics, rewarding friends and pushing opponents" (Grisaffi 2013: 49). Despite Morales's capitulation to neoliberal pressures and retreat from participatory democracy, there still exists a left-indigenous collection of critical thinkers who are committed to the transformation of Bolivia into a highly equitable and environmentally sustainable society.

On the positive side, poverty in Bolivia has been reduced from 60.6 percent of the population in 2005 to 43.5 percent in 2012 (Fidler 2014: 18). The Morales government has reduced infant mortality and illiteracy and expanded rural electrification and access to employment, education, and health care. The Bolivian government has created small state-owned enterprises that allow local producers input into their operation, and has allocated more than 35 million hectares of land as communal land or indigenous territories. It has reduced greenhouse gas emissions since 2010 and made improvements to local transportation systems, including the construction of a new cable car public transit system in La Paz that has enabled particularly poor people to more easily access the central business district (Bell 2016: 82).

Perhaps the most profound contradiction of the "socialism for the twenty-first century" experiment in Bolivia is, on the one hand, ongoing reliance on fossil fuels for generating revenue, much of it for needed social programs, and, on the other hand, the government's defense of Mother Earth. In December 2009, Morales denounced the UN Copenhagen Climate Conference for failing to recognize that climate change is ultimately rooted in global capitalism. In April 2010 he convened the World's Peoples Conference on Climate Change and the Rights of Mother Earth in Cochambamba, where he told a gathering of over thirty-five thousand people that "either capitalism dies or Mother Earth dies." While Ecuador has continued oil extraction, it has invested heavily in renewable energy, particularly hydroelectric schemes, in joint ventures with China and Russia (Anderson 2015: 115).

Unfortunately Bolivia finds itself in the awkward position of remaining "dependent on a volatile global economy and the finicky interests of foreign investors," thus seriously damaging the "credibility of the government as it seeks to project an environmental image in the international arena" (Kennemore and Weeks 2011: 273–274). Furthermore, the Morales government in its attempts to function within the parameters of constitutional democracy finds itself forced to make "ugly compromises with racist ranchers, concessions to foreign capitalists, and distorted by an economic dependence on exploiting gas" (Parenti 2014: 50). The case of

Bolivia is a profound illustration that socialism cannot be achieved in one country, especially one on the periphery of the capitalist world system, and that the construction of socialism for the twenty-first century will have to be a process occurring in numerous countries. As Webber (2011: 41) so aptly observes, "over the long term, constructing popular power from below in a socialist direction throughout the region is a necessity for the survival of any transformative change within Bolivia."

The Case of Ecuador

Under the leadership of Rafael Correa, a U.S.-trained economist who became president in 2006, Ecuador has embarked on its own version of socialism for the twenty-first century. Shortly after being elected, he promised to convene a constituent assembly to implement a "citizens' revolution." Becker reports:

> The new document that voters overwhelmingly approved in an October 2008 plebiscite rejected neoliberalism and embraced increased resource allocation to education, social services, and healthcare. It expanded democratic participation, including extending the vote to 16-year-olds, to foreigners residing in the country from more than five years, and to emigres.... The constitution also defended the rights of nature, the *sumak kawsay* (an Andean concept of living well rather than living better, thus favouring sustainability over material accumulations and the commodification of resources), Indigenous languages, and, in a highly symbolic gesture, plurinationalism as a way to incorporate Indigenous cosmologies. (Becker 2012: 73)

The Correa government doubled poverty assistance payments and credits available for housing loans, subsidized electricity rates for low-income consumers, and rechanneled funds into social programs (Kennemore and Weeks 2011: 275). Correa increased the royalty tax on foreign oil companies from 50 to 90 percent.

Correa has spoken of utilizing "extractivism" as a means of overcoming poverty or a short-term strategy to obtain funds needed for social programs, but some Ecuadorians have not accepted this argument (Harnecker 2015: 148). Indigenous people and environmentalists have criticized Correa for his alleged lack of commitment to protect the five million hectares of biodiversity in the Ecuadorian rainforest by allowing plans to drill for oil in the area to move forward. According to Kennemore and Weeks (2011: 279), although his "Citizen's Revolution promised a corrective to the economic, political and social crises brought by neoliberal structural adjustments, policies such as mining and water legislation continue to alienate those groups that have been most affected by its doctrine." Indeed, the Ecuadorian state has suppressed indigenous and environmental protests against developing new copper and gold mines and withdrew legal status from Accion Ecologica, the largest environmental organization in the country (Bonds and Downey 2012: 178). Hollender (2015: 81) asserts that cases such as Bolivia and Ecuador "make it clear that escaping dependence on natural resources will involve more than individual country efforts" to do so.

Developments in Cuba

While not part of the Bolivarian Revolution as such, Cuba has worked in solidarity with it in a number of ways, including sending health workers to Venezuela and importing oil from that country. Despite predictions that the revolutionary experiment in Cuba would end soon after the collapse of the Soviet Union, this beleaguered island state of some eleven million continues to push on over two decades later. The United States has persisted in continuing its economic embargo on Cuba. During the Special Period (1989–1993), Cuba's GDP decreased 35 percent due to fall of markets for its products and its exports fell 75 percent (Murphy and Morgan 2013: 333). Severe food shortages resulted from a decline in food imports. Despite a decrease in daily energy intake from 2,899 calories to 1,863 calories per person, the shortage in food forced people to walk or cycle, translating into an increase in physically active adults from 30 percent to 67 percent with the average adult losing 9–11 pounds, or 5–6 percent of body weight (Murphy and Morgan 2013: 333). Oil shortages forced a transportation revolution with trucks being converted into buses, the manufacture of local buses, and the extensive reliance on horse-drawn carriages and carts. The government implemented an extensive taxicab service.

In addition to a food-rationing program, which favored children, elderly people, and pregnant and lactating mothers, Cuba shifted away from industrial agriculture with its reliance on oil and chemical fertilizers to a more labor-intensive agricultural system that relied on organic fertilizers and organic pesticides, oxen traction, mixing cropping. The government responded to the agricultural crisis by shifting collectivized land to private cultivation. Whereas Cuban agriculture had been conducted on 80 percent state-operated farms, this shifted to 80 percent on employee-owned cooperatives (Heinberg 2004: 107).

Murphy and Morgan report: "In 1993 Cuba's legislature passed the National Energy Sources Development Program. Its goals include increased energy efficiency (the first priority), reduced energy imports, and maximized domestic energy sources. Cuban began a drive to save energy and use more renewable sources. Off-grid schools, health clinics, and social centers were electrified with solar panels" (Murphy and Morgan 2013: 235).

The Cuban Energy Revolution (CER) initiative of 2005 emphasized economic and energy sustainability and resulted in the distribution of 9.4 million compact fluorescent light bulbs and the replacement of energy inefficient appliances, fans, and air conditioners. CER has resulted in extensive changes in the production, transmission, and utilization of electricity, which has included a gradual shift to reliance on wind farms, hydropower, solar photovoltaic panels, solar water heaters, and biomass fuel from reforestation and sugarcane. Cuba has also been exporting its CER model to other countries, including Bolivia, Honduras, Lesotho, Mali, South Africa, and Venezuela (Murphy and Morgan 2013: 237). In 2006 the World Wildlife Foundation rated Cuba in its *Living Planet Report* the only country in the world exhibiting "sustainable development." As Heinberg (2004: 108), "Cuba managed to power down dramatically and quickly, relocalizing its economy with little increase in internal violence, and with relatively little sacrifice in terms of many basic measures of social welfare."

Despite its numerous achievements and resilience, Cuba is not a socialist paradise situated on a relatively large Caribbean island. It is a place filled with contradiction and pathos in that it constitutes a peripheral country functioning on the margins of the capitalist world system, and one that at least the United States continues to treat by and large as a pariah, even a "rogue state." Numerous reforms were implemented in Cuba particularly the period between 1989 and 1991. Paz maintains: "These reforms can be characterized as liberalization in the economic sphere and (to a lesser degree) democratization in the political arena, in a combination that, despite its gradual and careful implementation into a society that is more differentiated, more acquisitive, more participatory, and more efficient. The challenge will be to navigate those changes while maintaining its qualities of equity, frugality, and solidarity" (Paz 2007: 105).

Without romanticizing Cuba, it is a country that is trying to its utter best under extremely difficult conditions. Instead of often sitting idly around as is the case for many U.S. military personnel, Cuban soldiers often carry out a wide array of economic activities and many military officers hold government positions, with some of them serving as delegates to the National Assembly (Levins 2010: 10). The Special Period stimulated the expansion of educational programs at levels ranging from elementary school to university. Levins reports: "Art education was expanded, schools for art teachers established, and special programs were organized for handicapped students. Higher education was expanded by the establishment of university centers in all municipalities. Study as salaried work became an option for sugar workers displaced by the closing of some of the sugar mills" (Levins 2010: 11).

In terms of UN Human Development Index, in 2016 Cuba ranked 68th, below Costa Rica, Panama, Argentina, Barbados, and Chiles, but ahead of Brazil ranked 79th, Peru ranked 87th, and well ahead of China, ranked 90th, and India, ranked 131th (United Nations Development Program 2016: 222–225).

After the revolution, the Cuban health system was in shambles, in large part because doctors left the country, many of them to come the United States. Despite this, the physician/population ratio in Cuba increased from 9.2 per 10,000 in 1958 to 58.2 per 10,000 in 1999. The Family Doctor Program provides a physician-nurse team who live in a designated community and serve about 140 families or 700 individuals (Kath 2010: 24). The family physician as the gatekeeper to the rest of the healthcare system is generally is situated within walking distance from patients' homes. In her ethnography of community health care in Cuba, sociologist Christina Perez (2008: 270) discusses "[a] moderately poor developing country like Cuba. One that is cut off from the largest founders of health care in the world (the United States, World Bank, and IMF), can create health bodies because the government decided that nothing is more important than the lives of the population." The infant mortality rate of 6.33 per 1,000 in 2006 was the lowest in the Americas, along with Canada. In 2006, the life expectancy at birth was 77.23 years, comparable to that of the United States. However, it is important to note that numerous problems exist within the Cuban health-care system, including low pay for doctors, many hospitals and clinics in disrepair, poor provision of biotechnology, frequent absence or shortage of essential medicines, and reliance on medical tourism, which generates some $40 million annually in "hard currency."

Whereas many developing countries export health workers to developed countries, another example of unequal economic exchange between the core and both the semi-periphery and periphery of the capitalist world system is that Cuba has been exporting health workers, at least for periods of time. There are some 52,000 Cuban health workers who care for more than 70 million people in 92 countries, over six times the population of Cuba itself. The Bolivarian Revolution in Latin America has drawn on health-care assistance and inspiration from Cuba. More than 20,000 Cuban health workers are assisting the effort to restructure the Venezuelan health-care system. Cuba has established eye care centers in Bolivia with a capacity to treat more than 100,000 patients free of charge per annum (Webber 2011: 41). Cuban health workers provided relief in the aftermaths of the earthquakes in Pakistan and Haiti and were prepared to go to New Orleans to treat the victims of Hurricane Katrina in 2005 but were denied entry by the George W. Bush administration. Shortly after the Ebola epidemic broke out in West Africa in 2014, none other than the *Wall Street Journal* recognized that Cuba was "at the forefront" of fighting Ebola when its first contingent of 165 Cuban doctors arrived in Sierra Leone in early October of that year.

Aside from its remarkable achievements in health care, Cuban society continues to be highly bureaucratic, with too much of daily life being "constrained by rules and procedures often applied in rigid and inhumane way" (Levins 2010: 22–23). Cuba as a society purporting to be committed to a socialist democracy needs to be more democratic, including allowing multiple parties to function within its borders. However, creating more democratic structures in a besieged society, particularly one that has been besieged for over half a century by the most powerful country is the world, has been a very daunting and even perilous endeavor. Fortunately on 17 December 2014, U.S. President Barack Obama and Cuban President Raul Castro announced to the world that their two countries would resume normal diplomatic relations, which had ended in January 1961. Pope Francis and Canadian Prime Minister Stephen Harper had assisted in facilitating this monumental agreement, which still entails details to be hammered out. Wallerstein (2015: 2) views Obama's action as the "single most positive foreign policy decision he had made in his term of office, amidst a record that has been otherwise rather dismal."

Despite the U.S. economic embargo, Cuba has managed to survive, in part by becoming an international tourist destination, including for medical tourism, and in part by "Venezuelan money in return for Cuban educational and medical aid" (Therborn 2008: 11). Cuba is committed to maintaining the state sector as the dominant part of its economy has expressed openness to increased workers' participation in workplace decision making and increased nonstate employment, up to 40–50 percent, including in the cooperative sector (Campbell 2016: 14). It is also open to limited private capital investment, including the creation of small private restaurants. The Cuban currency has been stable at 22–25 CUP for a "convertible" peso or CUC (Anderson 2015: 112). Luxury goods often require CUC currency but many goods and services, including basic foods, arts, public transportation, home and personal services, books, and selected goods and services can be purchased with CUP currency. Anderson argues: "Major challenges for Cuba involve reconciling the dysfunction of some measure adopted during previous crisis (*e.g.* the dual currency system) and deepening another, the drive for food self-sufficiency. There is also a new challenge: that of maintaining the sense of mission: the morale of the socialist project, in context of greater commercialisation and the proliferation of small business" (Anderson 2015: 113).

Delinking from Global Capitalism

Samir Amin (1985: 55) developed the concept of *delinking* to refer to the process that follows socialist-oriented revolutions, which constitutes "one of the indispensable aspects of the emergence of a new social mode—whether socialist or not." He views delinking as part of the "'transition'—outside of capitalism and over a long time—towards socialism" (Amin 1985: 55). Delinking allows a postrevolutionary society to achieve national development outside of the imperatives of the capitalist world system. Under Stalin, the Soviet Union embarked not on the path of socialism but the "path of statism, a new class mode of production (and non-restoration of capitalism" (Amin 1985: 77). Through a rather awkward process, China was able to undergo rapid economic development in contrast to India because it pursued a policy of delinkage during the Maoist era. Amin (2013: 143) has recently reintroduced the notion of *delinking*, which he views as synonymous with *de*globalization, as a "strategic reversal in the face of both internal and external forces in response to the unavoidable requirements of self-determined development" and as a mechanism for "reconstruction of a globalization based on negotiation, rather than submission to the exclusive interests of imperialist monopolies" and contributing to the "reduction of international inequalities." He asserts that delinking can only be "initiated in the context of states/nations with advanced radical social and political struggles, committed to a process of socialization of the management of their economy" (Amin 2013: 143).

In contrast to Amin, Paul Baran asserted that:

> total self-sufficiency or economic autarky was impractical, even for large states like the U.S.S.R. and China. And in small states, in particular, a unilateral attempt to achieve the degree of diversification or production necessary to industrialize in an autarkic fashion would impose prohibitively high costs, while the ultimate object would remain forever out of reach. Moreover, the shortage of resources in small countries demanded considerable involvement in the world economy.... Baran's answer was that the main hope lay in mutually beneficial trade and economic collaboration among socialist states. (Foster 2014a: 198)

Aside from the matter of whether societies such as the Soviet Union and China were ever socialist or whether they were something else,

COMECON as a semi-autarkic structure no longer exists and China has more or less become fully integrated into the capitalist world system, to the degree that some would argue that it is a state capitalist society.

In light of these realities, perhaps the notion of delinking as a transitional strategy should be revisited as a strategy to break the stranglehold of the capitalist world system that relies on countries such as Venezuela, Bolivia, and Ecuador, which seek to create socialism for the twenty-first century, for oil, natural gas, and various minerals. Perhaps in time these countries associated with the Bolivarian Revolution, along with Cuba, and other Latin American countries, such as Argentina and Uruguay, could form a delinked bloc. Only time will tell whether blocs of delinked countries in the developed world could form, including one in southern Europe consisting of Greece, Italy, and Spain, which have increasingly found themselves to be marginalized partners in the European Union.

While it may appear as farfetched at this point in time, another delinked bloc among the developed countries might include Australia and New Zealand, both of which have an abundance of mineral resources and agricultural land. Although Australia relies heavily on coal for domestic energy, it could easily shift to reliance on renewable energy sources, particularly solar, wind, and geothermal, particularly if a new left party were to come to power that had the political will to nationalize the energy sector and orchestrate such a shift. While the manufacturing sector has gone into decline in Australia, it has a highly educated and skilled labor force that could manufacture a wide variety of products for domestic consumption and rely less on imported products, ranging from cars to clothing and electronic appliances. At any rate, blocs of delinked countries seeking to shift toward socialism could form a super-bloc that would stand in opposition to the capitalist world system, which is integrated by structures such as the World Bank, IMF, WTO, the World Economic Forum, G7, and G20.

Ultimately, delinking would constitute only a short-term strategy until there would be a groundswell of socialist-oriented regions and countries in the world to achieve a socialist world system that would have edged out the existing capitalist world system. Under a global eco-socialist system, as Sarkar observes, "[T]he minimal long-distance trade between its various regions that would be necessary and possible within the limits of sustainability would be just, fair exchange. Perhaps it would then be conceivable to form a world economic council that would plan such trade. One of the objectives of such planning would be to distribute the world's resources equitably, transferring resources from richly endowed regions ... to poor endowed regions" (Sarkar 1999: 221–222).

Conclusion

Socialism does not ensure utopia in a complete sense but rather constitutes but provides a vision for achieving a real utopia in which humans learn to live in harmony with each other and nature. Sanderson (2010) believes that the "human nature" problem seems to serve as an impediment to developing a viable socialism. Indeed, there is a common view in modern societies that human nature is acquisitive, greedy, and self-serving. Indeed, according to Peter J. Taylor (1996: 214), most working- and middle-class people are "enemies of capitalism because they see what the system does for people better off than themselves" and that the system that exploits them today may allow them to be affluent tomorrow, a reason "why socialists have so much trouble selling their message as anti-capitalist." Capitalist hegemonic ideology has diffused itself through supposedly commonsense thought that humans have an innate need to acquire more and more things, whether they need them or not. Ultimately, critical pedagogy demands that radical educators demonstrate that human nature is highly malleable. Numerous ethnographic and historical accounts of indigenous or tribal peoples, particularly foraging or hunting and gathering societies, demonstrate that most people in those societies emphasize sharing and reciprocity (Gowdy 1999: 391–392). While in the nineteenth century anthropologists viewed cultural evolution as progression from savagery to barbarism to civilization, with the latter being not only technological superior but morally superior to the former stages, in the twentieth century many anthropologists became more ambivalent about civilization. In making a distinction between "primitive society" and "civilization," White's assessment for "primitive" or indigenous societies was always positive: he asserted that because they place human relationships above property relations, they are humane, ethical, and personal systems that focus on meeting human needs. His assessment of civilization, which implicitly included capitalist civilization, was quite critical:

> The economic systems of civil society ... subordinating as they do human rights and human welfare to property rights, are impersonal, nonhuman, and nonethical in character. These traits translated into terms of human experience are impersonal, inhuman, inhumane, and unethical. All the suffering, indignities, and degradation that come from slavery, serfdom, prostitution, usury, dependence upon wages, unemployment, wars of conquest and expropriation, colonial rule, and exploitation are inherent in the economic systems of civil society. In commercial systems, anything may be bought or sold: a woman's

chastity, a judge's honor, a citizen's allegiance. There is no crime however heinous that men will not commit in obedience to their economic systems. And the unethical nature of civil society's economic systems corrupts and blights the higher classes as well as the lower. (White 1959: 330).

White later in his life modified his optimism—one shared by most conservatives, liberals, social democrats, and socialists—about the capacity of technology to provide a more secure life for humanity: "Cultural systems are moving rapidly to make the earth uninhabitable. And over all hovers the specter of annihilation by two lethal cultural devices: nuclear bombs and national sovereignty" (White 1975: 11).

Because the zenith of his career coincided closely with the Cold War era, including McCarthyism, White opted not to make explicit his intellectual debt to Marx and Engels, both of whom, like White, were impressed by Morgan's work (Carneiro 1981). Indeed, using a pen name, White wrote articles for a socialist newspaper and even visited the Soviet Union.

During the late 1960s and 1970s, many anthropologists and other social scientists, particularly sociologists, turned to Marxian perspective and contributed to a neo-Marxian analysis. For a period of time, there was even a Council for Marxist Anthropology within the umbrella of the American Anthropological Association. The disillusionment with the demise of the Soviet bloc did much to shift anthropology and the social sciences away from historical/dialectical materialism toward poststructuralism and postmodernism, but the former perspective still exists, although in somewhat muted form, in anthropology and the social sciences. Fortunately, postmodernism may be on the wane as perhaps a growing number of scholars recognize that it leaves us at a political impasse in comprehending and challenging the global socioeconomic and ecological crises, which are interrelated.

While I stress in this book the need to develop a critical anthropology of the future, I maintain along with various anthropologists that we can gain insights about possible pathways to sustainability based on sustainable practices in indigenous and tribal societies in the past and even to some degree today. Without romanticizing indigenous societies, anthropologist John H. Bodley (2013) proposes "the small nation solution" whereby indigenous small-scale societies can serve as living models for complex societies and world society can address global problems of conflict, poverty, and environmental degradation. Small-scale indigenous or tribal societies often exhibit economic systems, particularly if not highly distorted by the outside world, oriented to meeting basic human needs, such as food, clothing, and shelter, produce until such needs are met, reward generosity with prestige rather than material objects, discourage outward signs of differential material wealth, make decisions by consensus, seek to live in harmony with their local ecosystems, and have very low ecological footprints, particularly compared to developed capitalist societies. Bodley observes:

> Indigenous peoples are unique in the contemporary world because they share a way of life that is focused on family and household and is organizationally small scale and more sustainable than life in urban-based societies organized by political centralization, market exchanges, and industrial mass production.... Small-scale societies have enormous human advantages, especially because people living in large, dense populations, where they usually are divided sharply by differential access to vital resources, wealth, and power. In small-scale societies, where all households have assured access to food and shelter and to the rewarding experiences offered by their culture, there is less cultural incentive to accumulate and concentrate wealth. Likewise, there is little incentive for population and resource consumption to expand (Bodley 2015: 3).

Many indigenous societies practice communal ownership of resources, a practice anathema to global capitalism, and view themselves as stewards of their land (Fenelon 2012). The Ifugao in the highlands of the Philippines have developed a highly sustainable agricultural practice entailing a sophisticated system of terraced irrigation streams that have provided them with a diversity of crops without the use of industrial fertilizers and chemical pesticides (Jacques and Jacques 2012). Indigenous or tribal societies exhibit much smaller ecological footprints than most modern nation-states. Whereas the average ecological footprint was 2.2 hectares per person worldwide in 2001, Bodley (2008: 74) found that tribal footprints ranged from 0.08 to 1.1 per hectare based on his "estimates for Australian foragers, African herders, and tropical forest gardener-foragers."

In seeking inspiration for developing a renewal of socialism in the twenty-first century, radicals need to take care that they do not simply revive the Third World-ism characteristic of the New Left in the 1970s in seeking inspiration from calls for a renewed socialism in the guise of the Bolivarian Revolution. As Bronner (2014: 172) observes, "[o]rganizational examples from less economically developed nations with fewer consumer expectations like Bolivia and Venezuela cannot be arbitrarily lifted out of their context to instruct more economically developed nations in a transhistorical and reified manner." However, the socialism of the twenty-first-century experiment in Latin America for the most part has "not had a galvanizing effect on politics in North America or Northern Europe, beyond specific circles of solidarity activists and individuals and groups with ties to those regions experiencing mobilizations" (Sears 2014: 98). Progressive governments in Latin America, as has been the case with the Maduro government in Venezuela, "must prepare to confront and defeat elite maneuvers meant to the block way toward socialism" (Harnecker 2016: 28). Aside from the efforts to create socialism for the twenty-first century in Latin America, there are some modest signs that antipathy to socialism and even calls for a renewal of socialism have been occurring in various parts of Europe. However, creating a democratic eco-socialist world system at this point in time remains a most elusive endeavor. As Li observes, "The collapse of capitalism and the establishment of a post-capitalist society will not automatically guarantee the solution of the climate change crisis and a successful transition to ecological sustainability. However, without the compulsive competitive demands imposed by the global capitalist market, humanity will be freed from the constant and intense pressure of ceaseless accumulation" (Li 2009: 1058).

CHAPTER 5

The Role of Anti-systemic Movements in Creating a Socio-ecological Revolution

The term "social movement" reportedly was coined by Lorenz von Stein in referring to the "working class movement for socialism in Europe that culminated in the mass socialist parties of the nineteenth and early twentieth centuries" (Hammond 2012: 215). Prior to this time, at least within the context of state societies, spontaneous rebellions have come and gone, most of them "only partially efficacious at best"; in some cases they "forced the oppressors to reduce the pressure or the exploitation," but in other cases they failed at doing so (Arrighi, Hopkins, and Wallerstein 1989: 29). Around 1848 or so, social movements came into being in which groups of people involved in activity opposing some aspect of the capitalist world system "began to create a new institution: the continuing organization with members, officers, and specific political objectives (both long-run and short-term)" (Arrighi, Hopkins, and Wallerstein 1989: 29). Wallerstein states: "There were two main varieties—social movements and nationalist movements—as well as less strong varieties such as women's movements and ethno/racial/religious movements. These movements were all antisystemic in one simple sense: They were struggling against the established power structures in an effort to bring into existence a more democratic, more egalitarian historical system than the existing one" (Wallerstein 2014: 160).

Obviously, efforts to create socialism in Russia, China, and other developing countries came out of anti-systemic movements that resulted in new political regimes that in part created more egalitarian societies, but unfortunately not more democratic ones. Various world systems theorists view 1968 as a time of "world revolution" in many parts of the world, not only Western Europe and the United States, but also Eastern Europe. This revolution rejected U.S. hegemony and the Old Left in both its communist and social democratic guises. While the revolutionaries were not able to come to power anywhere, the "world revolution of 1968 transformed the geo-culture" in the sense that the "radical left and the conservative right re-emerged as fully autonomous actors on the world scene" (Wallerstein 2014: 164). According to Silber (1994: 6), "movements to replace capitalism with a more equitable (and, yes—despite the Soviet debacle—a more efficient) social arrangement are sure to be a permanent feature of the world's political landscape so long as capitalism rules." While the Global Left briefly went into a period of dormancy during the late 1970s and early 1980s, anti-systemic movements representing a wide diversity of voices ranging from indigenous peoples (e.g., the Zapatistas in Chiapas under the leadership of Commandte Marcos), landless people, peasants, anti-corporate globalization activists (e.g., the anti-WTO demonstrations in Seattle in 1999 and the World Social Forum) have been challenging the parameters of a neoliberalism around the world.

Furthermore, various voices have been challenging the environmental unsustainability of capitalism and the anthropogenic climate change that it engenders. In that corporations and most governments have not been acting in a responsible manner in terms of serious climate change mitigation, despite much rhetoric to the contrary, much of the collective effort will have to be spurred by anti-systemic movements, including a burgeoning international climate movement, which is quite variable in terms of addressing social justice or equity issues (Baer and Singer 2009; Baer 2012; Connor 2012; Dietz and Garrelts 2014). Powerful anti-systemic movements will be needed to counteract the resistance of the powers-that-be in the capitalist world system). Where social movements, such as the environmental, peace, women's rights, community, and identity movements, including ethnic rights movements in developed countries, tend to be middle-class in their composition, social movements in developing countries tend to be populist and working class. According to Hardt and Negri (2009: 94–95), "only movements from below" possess the "capacity to construct a consciousness of renewal and transformation"—one that "emerges from the working classes and multitudes that autonomously and creatively propose anti-modern and anticapitalist hopes and dreams." They assert that while Lenin and Trotsky's perspective about a vanguard paving the way to revolution may have been a "realistic and pragmatic means to address the realities of late 19th and early 20th century Russia," a revolutionary strategy in the contemporary period necessitates an "organizational process that establishes revolutionary decision making and the overthrow of ruling power from within, not above, the movements of the multitude" (Hardt and Negri 2009: 351). Hardt and Negri (2009: 365) draw inspiration from Antonio Gramsci's notion of "war of position" rather than "war of movement" that revolutionary activity be conducted through an "extended series of battles in cultural and political spheres in an effort to wrest hegemony away from the bourgeoisie." They note that the anti-corporate globalization or alter-globalization movement has devised new strategies of opposition, democratic decision making, which includes concerns for social justice, peace, and environmental sustainability (Hardt and Negri 2009: 368).

It is important to point out that social movements can be progressive or reactionary. While I focus on the former, both right-wing or reactionary movements have had in the past and have today profound impacts on human societies, not only within specific nation-states, but also in geographic regions and even globally. Reactionary social movements sometimes evolve into political parties that may come to power and play a profound role in altering human history, such as in the case of the National Socialist Party or the Nazi Party in Germany. Whereas McCarthy and Zald (1977: 1217–1218) define a social movement as a "set of opinions and beliefs in a population which represents preferences for changing some elements of the social structure and/or reward distribution of a society," a *countermovement* constitutes a "set of opinions and beliefs in a population opposed to the social movement." Nazism evolved in part as a reaction of the reparations imposed by the Allied powers on Germany after World War II but also against socialist movements and political parties and labor unions. In fact this fascist movement even co-opted May Day, a traditional left-wing working-class celebration of international solidarity, by declaring it a workers' holiday on which numerous Germans marched in honor of the Fatherland.

Other examples of countermovements in the past and today include the Ku Klux Klan; the anti-abortion movement; the Religious Right in the United States; the climate denialist movement; the Tea Party, which has gained a powerful foothold in the U.S. Republican Party; right-wing anti-immigration groups in Europe, the United States, and Australia; and Islamic State in the Middle East, to name only a few.

Progressive social movements exhibit degrees of radicalism in critiquing capitalism, conceptions of democracy and gender relations, and environmental perspective (Amin 2008a: 79). Indeed, despite an "exhilarating new flourishing of movement activity, a slowly resurgent opposition to the onslaughts of neoliberalism in crisis by a globally expanded and recomposed working class, and the expression of widely popular ideological challenges to the fundamental principles of capitalist society," according to Barker, Cox, Krinsky, and Nilsen (2013: 12), the current time period may constitute the "first time since 1948 when specifically

Marxist ideas are not the natural *lingua franca* of a rising movement" or an array of movements. In a similar vein, John Holloway (2005: 214) observes: "The shift from politics of organisation to a politics of events is ... taking place: May 1968, the collapse of the regimes of Eastern Europe too; more recently, the development of the Zapatista rebellion, for all its organizational formality, has been a movement through events, and the wave of demonstrations against global neoliberalism (Seattle, Davos, Washington, Prague, and so on) is obviously event-centred." The same characterization can be applied to more recent events, such as the Occupy movement, the Arab Spring, and numerous climate rallies around the world. Some of these movements that are centered on events peter out and others take on organizational forms or coalitions of organizations.

Chris Dixon identifies a strong "anti-authoritarian" orientation in current social movements, which operates according to four core principles:

1. "Struggling against all forms of domination, exploitation, and oppression";
2. "Developing new social relations and forms of social organization in process of struggle";
3. "Linking struggles for improvements in lives of ordinary people to long-term transformative visions";
4. "Organizing that is grassroots and bottom-up" (Dixon 2014: 6–7).

She asserts that anti-authoritarianism draws from a diversity of social movements, including anarchism, "anti-racist feminism, prison abolitionism, autonomous movements in the global South, and Indigenous struggles for self-determination" (Dixon 2014: 65).

While traditional forms of left organizing still exist in abundance, ranging from political parties and militant political sects to labor unions to environmental and peasant movements, as David Harvey (2012: 119) observes, "they all seem to swim within an ocean of more diffuse oppositional movements that lack overall political coherence." He also asserts that:

> The left as a whole is bedevilled by an all-consuming "fetishism of organizational forms." The traditional left (communist and socialist in orientation) typically espoused and defended some version of democratic centralism (in political parties, trade unions, and the like). Now, however, principles are frequently advanced—such as "horizontality" and "non-hierarchy"—or visions of radical democracy and the governance of the commons, that can work for small groups but are impossible to operationalize at the scale of a metropolitan region, let alone for the 7 billion people who now inhabit planet earth. Programmatic

priorities are dogmatically articulated, such as the abolition of the state, as if no alternative form of territorial governance would ever be necessary or valuable. (Harvey 2012: 125)

More recently, Harvey (2014: 145) has argued that "much of the anti-capitalist as opposed to social democratic left prefers in these times to wage its war at the micro scale, where *autonomista* and anarchist formulations and solutions are most effective, leaving the macro level almost bare of oppositional powers." A New Global Left has been theorized by autonomists such as Hardt and Negri as a "multitude" of workers who have been casualized by emergence of flexible specialization (Chase-Dunn and Lerro 2014: 369).

Wallerstein (2013: 33) delineates two broad possible organizational strategies for change that started sometime in the 1970s and will probably play themselves out until around 2040 or 2050, the first of which he terms "the spirit of Davos" and the second "the spirit of Porto Alegre," both of which are internally divided. In the case of the former, "One group favors immediate and long-term harsh repression, and has invested its resources in organizing a network of armed enforcers to crush opposition. There is however another group who feel that repression can never work over the long term. ... They talk about meritocracy, green capitalism, more equity, more diversity, and an open hand to the rebellious—all in the spirit of heading off a system premised on relative democracy and relative equality" (Wallerstein 2013: 34).

"The spirit of Porto Alegre" consists of a camp that favors "horizontalism," which entails a maximization of debate and a "search for relative consensus among persons of divergent backgrounds and immediate interests" and a rejection of economic growth, and another camp favors a "vertical organization of some kind" that seeks political power but also argues for "short-run economic growth in the 'less developed' areas of the present-day world in order to have the wherewithal to redistribute benefits" (Wallerstein 2014: 34). Wallerstein (2014: 35) is reluctant to definitively predict the outcome of these four strategies, the first two of which are essentially pro-systemic and the latter two of which are anti-systemic, but conjectures that "we have at best a 50-50 chance of getting the kind of world-system we prefer." At any rate, anarchism of one sort or another has become more fashionable than socialism among anti-capitalist systemic movements or progressive social movements in general. Anarchists eschew forming linkages with political parties and labor unions. While "anarchist and autonomist hostility to 'organisation' is comprehensible as a response to the real dangers of co-optation and bureaucratisation," it avoids the need to capture the state and move it in a more progressive direction on a number of levels and "makes the formation of a different world almost unthinkable, for how—to take an obvious example—is the world to be pulled back from a climate-induced catastrophe without a major exercise of *popular power*" (Barker, Cox, Krinsky, and Nilsen 2013: 21).

Amory Starr (2000) has developed an elaborate typology of international anti-corporate social movements that consists of three broad types and fifteen ideologically distinct types that are depicted in Figure 5.1 below:

Figure 5.1. Types and Subtypes of Anti-corporate Social Movements

1) Contestation and reform movements
    - Fighting structural adjustment
    - Peace and human rights
    - Land reform
        - Explicit anti-corporate movements (e.g., which oppose genetically engineered agricultural inputs and the construction of roads)
        - Anti-roads opposition
        - Cyberpunk, which critiques technology while using it
2) Globalization from below
    - Environmentalism (e.g., Greenpeace, the environmental justice movement in the United States)
    - Labor
    - Socialism (e.g., numerous parties and institutions)
    - Anti-FTA
    - Zapatismo
3) Delinking/relocalization
    - Anarchism
    - Sustainable development (e.g., voluntary simplicity, urban sustainability movements)
    - Small business (e.g., informal enterprises, institutionalized public markets, small business associations, campaigns against malls, community currency, punk do-it-yourself)
    - Sovereignty movements (e.g., indigenous groups)
    - Religious nationalisms

A profound challenge has proven to be and will continue to be how to create cooperation among the variegated progressive social movements in challenging global capitalism, how for example to resolve tensions between the environmental and labor movements and how to create a democratic global governance regime when some groups emphasize local autonomy (Chase-Dunn 2005: 184). Chase-Dunn (2013: 179) argues that while the New Global Left is becoming increasingly strong, it still is very much in formation, with a major divide between the *horizontialists* (who have been rising in influence) and the more traditional activists (who are committed to electoral politics and capturing state power). The World Social Forum as a central actor in the New Global Left contains numerous contradictions, including the fact that in 2005 more than half of its participants came from the host country, that participants from peripheral and nonlocal semi-peripheral countries are underrepresented, and that participants tend to be relatively privileged in terms of class, race, ethnicity, and education (Chase-Dunn and Reese 2007: 78). Furthermore, the alter-globalization movement is divided between those who wish to reform capitalism and those who wish to replace capitalism with a more just economic system.

In the remainder of this chapter, I focus on various progressive movements, namely, the labor movement, the women's movement, the anti-corporate or global justice movement, the ethnic and indigenous rights and peasants' movements, the environmental movement, and the climate movement, which have tendencies or segments that are explicitly anti-systemic. To some degree, these various movements overlap with each other in that they may collaborate on mass actions or have members who belong to two or more movements. Some progressive movements, however, are not clearly anti-systemic. An example of this would be the human rights movement, which has no clearly delineated critique of corporations, believing that they can either be reformed or shamed into adopting human rights principles.

The Labor Movement

Despite its anti-systemic origins, the labor movement to a large degree became domesticated during the twentieth century and probably is more so than it ever has been. Alain Touraine (1974), a prominent social movement theorist, asserted some time ago that whereas the labor movement once constituted *the* foremost social movement, it has evolved into an institutionalized political actor. In a similar vein, Hammond argues:

> In the United States today, the labor "movement" is called a movement only by historical memory. Organized unionism, weakened by the changing industrial structure, the globalization of the economy, and the political attack it has suffered under increasingly conservative federal administrations, has also been weakened by its own organizational sclerosis. Large unions are ossified bureaucracies: most do not work for social change. Their steadily shrinking membership is enrolled through voluntary (whether or not unwanted) due checkoffs from their paychecks and has no active association or activity within the "movement." (Hammond 2012: 216)

The same characterization could be applied to trade unions in most, if not all, developed societies and probably many in developing countries. Trade union officials have a tendency to confine the class struggle to "reformist reforms" in their efforts to improve wages and working conditions, all too often making concessions to the corporate class during management-labor negotiations. These officials often earn substantial salaries and receive other lucrative perks that they justify as compensation for working long hours under stressful conditions. Aside from the existence of a "labor aristocracy" in developed societies, one that has become much decimated under neoliberalism, the working class today, most of which remains unorganized around the world, includes "immaterial workers, agricultural workers, migrants, the unemployed, the underemployed, and the poor" (Bencievenni 2006: 30).

Nevertheless, in their effort to seek a wide base of support, various scholars assert that social movements must find a natural ally in the labor movement (Glaser 2007: 206). Wallerstein (2007: 385) asserts that the environmental movements will face an impasse in the next twenty-five to fifty years if they fail to form a "symbiotic relationship with other kinds of antisystemic movements." In contrast to the corporations, which have increasingly globalized themselves, labor organizations for the most part remain national and local. Organized labor's response to neoliberalism and corporate control has been muted, possibly due to allegiances to supposedly left-center parties, such as the Democrats in the United States, the Social Democrats in Germany, the Labor Party in the Britain, and the Australian Labor Party, which by and large have made their peace with corporate agendas. While left-center or social democratic parties were able to win gains for the industrial working class in various developed countries, the deindustrialization in these countries subsequently resulted in a rift between the old working class from growing segments of the working class, namely, the "salaried professionals, the 'feminized' service-sector employees, and the 'ethnicized' unskilled or semi-skilled labor force" (Arrighi, Hopkins, and Wallerstein 1989: 88). Conversely, Sernau maintains: "International labor organization will be a difficult task, but forces of globalization may make it easier. Certainly, communication among workers across national borders is becoming easier than before. Further, falling industrial wages in wealthier countries are bringing these workers closer to their fellow workers in middle-income countries such as those in Latin America. Industrial plants themselves are becoming more similar internationally...." (Sernau 2011: 302).

Creating a labor-green alliance has proven difficult in that most unions "make no critique of industrialization, centralization, standardization, consumption, ecological limits or growth" (Starr 2000: 92). Nevertheless, it is imperative that adherents of other anti-systemic movements make alliances with the labor movement because of the sheer numbers of people who are exploited by capitalism that it represents and its potential to mobilize against global capitalism.

Indigenous, Peasant, and Ethnic Rights Movements

According to various estimates, indigenous peoples number 250 to 350 million people worldwide and represent some five thousand distinct cultures (Hall and Patrino 2012). Particularly in developing countries, various indigenous rights and peasant movements have appeared in recent decades, which, while seeking to preserve local cultures and traditions, have done so by challenging the activities of multinational corporations and their political allies. Indigenous protesters are often connected to a wide array of NGOs, even UN organizations, and social media but often retain their own autonomous groups. The adherence to communal ownership of resources often constitutes their most basic challenge to global capitalism. While indigenous movements may not constitute serious threats to global capitalism and may not even seek to transcend it per se, they engage in a "war of position" in Gramscian terms in their relationship to it.

The Zapatistas constitute the best example of an indigenous rights/peasant movement in the world today. On 1 January 1994, the very same day that the North American Free Trade Agreement (NAFTA) went into effect, some three thousand armed Mayan Indians under the leadership of Subcommandte Marcos seized San Cristobol de las Casas, the capital of the Mexican state of Chiapas, and various surrounding municipalities. The Zapatistas demanded adequate land, employment, adequate food and shelter, good health care, and a more democratic social order in which to participate as enfranchised citizens. Lynd and Grubacic (2008: 10) regard the Zapatistas as a synthesis of Marxism, for example in its analysis of capitalism, and anarchism in that it "rejects the goal of taking state power and sets forth the objective of building a horizontal network of centers of self-activity." The Zapatistas resemble other indigenous rights movements in the world in that they also incorporate elements of spirituality, in their case primarily Native American. Marcos had been a member of Marxist-Leninist group in Mexico City before moving on a permanent basis to the Lacandon Jungle in the 1980s.

An example of a clearly anti-systemic labor movement is the Brazilian Landless Workers Movement, which has occupied land in order to provide a subsistence base for peasants. La Via Campesina, an organization consisting of some two hundred thousand small farmers around the world, has become part and parcel of the climate justice movement, and often asserts that "agroecology is the solution to solve the climate crisis" and that "small farmers cool the planet" (quoted in Klein 2014: 134). Indigenous movements that are at least partially anti-systemic are found in many countries and regions and include international organizations such as the International Indian Treaty Council, the Indigenous World Association, the International Land Coalition, Earth Peoples, Survival International, and Cultural International.

The Women's Movement

The women's movement or the feminist movement has a long history in Western societies but has also exploded over the past several decades in developing societies. It has taken on mainstream versions and more radical ones, mystical feminism, lesbian-feminism, eco-feminism, and socialist eco-feminism. While all versions of feminism in one way or another oppose patriarchy, only some are anti-systemic per se in that they mount critiques of capitalism, imperialism, and class relations. Radical feminists in focusing on patriarchy as the main source of oppression have been criticized for downplaying or even ignoring class and racial/ethnic forms of oppression. Conversely, socialist-feminism and anarchist-feminism are explicitly anti-capitalist and aware of class as an important force contributing to oppression, along with patriarchy.

Radical women of color during the 1970s organized groups such as the Third World Women's Alliance, Women of All Red Nations, and the National Black Feminist Organization (Dixon 2014: 35). The Women's International Coalition for Economic Justice in 1999 issued a Declaration for Economic Justice and Women's Empowerment, which demanded "macro-policies designed to defend the rights of women and poor people and protect the environment, rather than expand growth, trade, and corporate profits exclusively" and to redefine "economic efficiency to include measuring and valuing women's unpaid as well as paid work" (quoted in McMichael 2012: 193). Other radical women's groups include DAWN (Development Alternatives with Women for a New Era), which calls for "gender and economic justice, IFI [International Financial Institution] policy, reproductive rights and health—especially for poor Southern women, redistribution of global wealth"; WIDE (Women in Development Europe), which advocates "feminist alternatives to economic theory and development policies of the North"; and WEDO (Women's Environment and Development Organization), which calls for sustainable solutions to global problems" and "democratization of the WTO" (McMichael 2012: 196).

In terms of eco-feminist principles, Chipko in India is particularly well known. While including men within its ranks, its leaders and activists are primarily women. It seeks to protect Indian forests, which for eons have provided rural peoples with subsistence. Chipko, which is based on Gandhian principles, promotes resistance to the state's promotion of forestry for its own revenues as well as exploitation by private companies of traditional forest common areas (Nhangenge 2011: 369–379). One branch of Chipko, called Uttarkhand Sanharsh Vahini, was started by a man, namely, Chanti Prasad Bhatt, and functions essentially as an eco-Marxian group that opposes capitalism, not only in terms of its exploitation of forests but also in terms of its promotion of mining and liquor sales (Nhanenge 2011: 371–372).

Mies and Bennholdt-Thomsen, whose subsistence perspective I described in the previous chapter, discuss the transition that the women's movement had made by and large from its earlier incarnation of the 1960s, 1970s, and early 1980s to the form that it had taken by late 1980s and 1990s. They argue that during its earlier incarnation, "most feminists still knew that issues like the exploitation of the Third World, ecological issues, militarism and peace, and the economy were feminist issues" (Mies and Bennholdt-Thomsen 1999: 182). Mies and Bennholdt-Thomsen (1999: 189) maintain that in its later incarnation much of the women's movement shifted to seeking "integration into mainstream economics and politics" where "women would participate in existing power structures." In the case of their native country of Germany, they witnessed this transition play itself out within the corridors of the Green Party. They assert: "It is not sufficient to struggle for 50 percent participation of women in a political and economic system that, as a whole, is based on oppression and exploitation. Unless women produce a vision of another economy and society this sharing of power is of no avail" (Mies and Bennholdt-Thomsen 1999: 194).

The Global Justice Movement

The global justice or anti-corporate globalization movement drew inspiration from the Zapatistas and at the 1996 International Encounter for Humanity and against Neoliberalism in Chiapas, developed a unique organizing tool known as affinity groups, which are small collections of like-minded individuals that "enabled protestors from all over world to situate themselves both within the wider movement and within an unfamiliar urban landscape" (Ruiz 2014: 103). The global justice movement captured international attention at the "Battle of Seattle" in 1999 when some fifty thousand labor and environmental activists managed to shut down the World Trade Organization meeting for a day, leading to its eventual collapse. Numerous protests opposing corporate power represented by the WTO, the World Bank, and the IMF have occurred since then in various parts of the world. David Graeber, an anthropologist who has been part of the movement for some time, maintains:

> The real origins of the movement ... lie in an international network called People's Global Action (PGA). PGA emerged from a 1998 Zapatista *encuentro* in Barcelona, and its founding members include not only anarchist groups in Spain, Britain and Germany, but a Gandhian socialist peasant league in India, the Argentinian teachers' union, indigenous groups such as the Maori of New Zealand and [indigenous federations] of Ecuador, the Brazilian landless peasants' movement and a network made up of communities founded by escaped slaves in South and Central America.... It was PGA that put out the first calls for days of action such as J18 and N30—the latter, the original call for direct action against the WTO meetings in Seattle. (Graeber 2005: 169)

The World Social Forum as a later incarnation of the global justice movement started out with a conference in Puerto Alerge in 2001 and according to some "constitutes an effort to organize global political parties" (Chase-Dunn 2010: 55). Chase-Dunn and Reese (2007) conducted a survey of attendees at the 2005 WSF meeting in Porto Alegre, Brazil. Out of 625 respondents, 69 percent came from South America, 10.5 percent from North America (excluding Mexico), 7.5 percent from Asia, 1.1 percent from Central America and the Caribbean, and 0.3 percent from Australia and New Zealand. Three-quarters of the respondents were situated in middle-class occupations or endeavors, including professionals, technicians, artists, students, activist/organizers, and NGO workers. Respondents came from a wide range of social movements, including human rights, environmental, global justice, peace, socialist, communist, anarchist, and feminist movements. In terms of how to deal with capitalism, whereas 53.5 percent of the respondents said they would like to "abolish it and replace it with a better system," 38.7 percent said they would like to "reform it" (Chase-Dunn 2007: 81). Humphrys (2013: 371) asserts that the WSF in the post-Iraq era "seemed to drift, its constitutionally enshrined refusal to take political positions leading to significant doubts about its effectiveness, perhaps best exemplified by infighting over 2005 attempts by prominent activists to forge a Porto Alegre Manifesto."

In a sense the Occupy Wall Street (OWS) movement that emerged in the wake of Global Financial Crisis of 2008–2009 and the Arab Spring can be viewed later incarnations of the global justice movement. OWS blended social media with face-to-face interactions, encampments, and rallies but also struggled with the matter of the digital divide among their ranks. Ultimately, word-of-mouth communication may have played the stronger role in the organizing (Buechler 2014: 258). Sociologist Richard Flacks, a longtime observer of social movements, views OWS as "one expression of a rising, and very long delayed, national and global class struggle" that formed alliances with the labor movement and a wide array of working-class grievances. Conversely, Chomsky (2012: 58) argues that in contrast to the strong labor input into the Arab Spring rebellions in Tunisia and Egypt, the influence of the labor movement in the Occupy movement in the United States will matter because it "has been decimated" in the latter. OWS's "'we are the 99%' formulation obviates race, occupation, religious difference, and poor/middle-class categories," thus providing the basis for broad solidarity, but perhaps it important to note that there is a 19 percent, namely, the managers and professionals, who are much more closely aligned with the 1 percent than they are with the remaining 80 percent (Tabb 2014: 511).

Colours of Resistance (CORE) is a North American network that seeks to "develop feminist, multiracial, anti-racist, anti-authoritarian, anti-capitalist politics in the global justice movement," in which women of color take the lead in organizing and the proportion of whites and men participating is capped (Dixon 2014: 10).

The global justice movement in its various manifestations has a strong anarchist tendency as opposed to a socialist tendency within it (el-Ojeili 2014). Anthropologist David Graeber (2005: 171) rejects the frequent assertion on the left that the global justice movement lacks a coherent ideology or action plan, noting that in reality it is a "movement about creating and enacting horizontal networks instead of top-down (especially, state-like, corporate or party) structures, networks based on principles of decentralized, non-hierarchical consensus democracy." More recently, Graeber (2013: 277) has argued that the global justice movement "peaked between 1998 and 2003, but essentially experienced a lull beginning with the rally against the IMF in 2002 as a result of the security apparatus that the George W. Bush administration had erected in the aftermath of 9/11."

The Peace Movement

By and large, the peace movement is characteristic of various developed countries and has historically waxed and waned, but has experienced a lull since exuberant periods in the 1980s during the last days of the Cold

War and then again in 2001–2003 with the invasions of Afghanistan and Iraq on the part of a U.S.-led coalition of military powers. According to Mary Kaldor (1999: 198), with some five million people demonstrating in many cities and towns across Europe in October 1981 and October 1983, it "was probably the largest transnational movement in history" up until that time. While the peace movement is anti-systemic in that it "opposes 1st World imperialism and excessive militarization of 1st world societies," as Starr (2000: 54) observes, "few versions of the peace movement put forth critiques of capitalism." Theoretically, the peace movement could have been more explicitly anti-systemic but perhaps had not been because of the strong presence of middle-class people, and often religiously oriented ones at that, in it. While I was involved in the Arkansas Peace Center in the 1980s, I found some of its faith-based members were more oriented to praying for peace than participating in peace rallies. Conversely, the U.K. religious group the Prince of Peace Ploughshares viewed "global corporate domination" as among the "madnesses" of the world (quoted in Starr 2000: 54).

The peace or anti-war movement had gone into a lull by 2005. According to Dixon (2014: 46–47), the "protests against the 2004 Republican National Convention in New York City were arguably the last gasp of hope for carrying out the momentum of the global justice movement into a large-scale anti-war movement." Small peace initiatives persist, such as Iraq Veterans Against the War and Courage to Resist in the United States and the War Resisters Support Campaign in Canada (Dixon 2014: 47).

The Environmental and Climate Movements

The environment movement that emerged in the late 1960s and early 1970s globally has evolved in many directions, ranging from mainstream environmental organizations to radical environmentalists of different sorts, including deep ecologists, eco-socialists, and eco-anarchists. In the United States, the "Group of Ten" environmental organizations consists of the Environmental Defense Fund, the Environmental Policy Institute, Friends of the Earth, Izaak Walton League of America, National Audubon Society, National Parks and Conservation Association, the National Wildlife Federation, Natural Resources Defense Council, the Sierra Club, and the Wilderness Society (Merchant 2005: 166–167). Mainstream environmental groups in Australia include the Australian Conservation Foundation, the Wilderness Society, Environment Victoria, and various state nature conservation societies. Various environmental groups have been more willing to engage in civil disobedience and even direct action against industrial and forestry operations. Earth First! is well-known for engaging in strategic ecotage or "monkeywrenching"; it proudly proclaims in its journal that it makes "no compromise in defense of Mother Earth." Greenpeace, which has national chapters in many countries, is also known for engaging in acts of civil disobedience and direct action, including promoting nuclear-free seas, saving whales and seals, and protesting the dumping of toxic chemicals and coal mining. In the United States, the environmental justice movement has achieved much currency among people of color, particularly African-Americans, Latinos, Americans, and Native Americans. In terms of parliamentary politics, Green political parties have become highly visible in many countries in the world, particularly Germany but also the U.K., Australia, Canada, and New Zealand. The Green Party in the United States became particularly visible during the 1996 and 2000 presidential elections when consumer advocate Ralph Nader, who actually did not belong to the party, ran as its candidate. Many Democrats accused him of being a spoiler in 2000 election when George W. Bush defeated Al Gore. Conversely, many progressive Americans opted to vote for Nader in order to protest the shift of the Democratic Party to the right.

Over the past decade or so, many environmental groups and Green parties have in one way or another become involved in the climate movement as have many other political actors, including grassroots climate action groups, socialist political parties and groups, and even labor unions. Charles Derber (2010: 205) maintains that in order to address the climate crisis, social movements must acknowledge the existential reality of climate change, struggle for regime change within nation-states and internationally, allow labor to take the lead, integrate personal and systemic changes, and both cooperate with the heads of state and push them to take more radical stances on systemic problems. He argues that the "labour movement has begun to advocate strongly for green economic solutions to the current crisis, as well as to remake the economy and ensure full employment for the long term" (Derber 2010: 209). Many environmentalists, however, may not want put labor in a position of leadership because many labor union leaders are committed to providing their members with a high material standard of living with little thought of the environmental consequences associated with it. Conversely, environmental groups all too often have been willing to cooperate with corporate interests, even the more progressive ones. As James Anderson observes, "National and international NGOs, such as Friends of the Earth, Greenpeace, Oxfam, and other international aid and development organizations, play a crucial role in mobilizing scientific expertise, and some are important players in anti-globalization. But despite this progressive role, they generally occupy an ambiguous and often highly constrained position in the power game and hence are vulnerable to incorporation and ideological subversion by established powers" (Anderson 2006: 252).

In that corporations and most governments along with the United Nations have not been acting in a responsible and effective manner in terms of serious climate change mitigation, despite considerable rhetoric to the contrary, much of the collective effort will have to be spurred by anti-systemic movements, including a burgeoning international climate movement that is quite variable in terms of addressing social justice or equity issues. The climate movement, both internationally and nationally, is a broad and disparate phenomenon that draws in part from the earlier movements, particularly the environmental movement but also the anti-corporate globalization movement and even the labor movement. Countries with active climate movements include the United States, Germany, the U.K., Canada, Australia, New Zealand, South Africa, Brazil, and India (Dietz and Garrelts 2014).

Many climate action groups in North America, Europe, and Australia tend to focus on ecological modernization as their primary climate change mitigation strategy, thus either ignoring or downplaying social justice issues. The climate movement both at the international level and within specific countries is quite fragmented, with some of its segments, such as the Climate Action Network, Greenpeace, and Al Gore's Climate Reality Project, trying to work within the parameters of global capitalism, and others, such as Friends of the Earth, Rising Tide, Climate Justice Action, and various socialist and anarchist groups, challenging it and even calling for transcending it, thus prompting the byline, "not climate change, system change." Despite internal differences, climate activists of varying political orientations tend to converge on specific campaigns, such blockades against the Northern Gateway Pipeline in North America, the construction of the Keystone Pipeline (which was slated to run from the oil tar sands in northern Alberta to the Gulf of Mexico until President Barack Obama blocked its completion), and what was reportedly the world's largest climate change rally attended by an estimated half million people from some 166 countries on 21 September 2014 in New York, two days before the UN Climate Summit there. Giacomini and Turner (2015: 30) observe that two broad tendencies were represented at the New York rally, one between the supporters of green capitalism that view climate markets as being essential to mitigating climate change and the defenders of life-affirming solar communing who "seek an end to capitalist relations (meaning the demise of corporate value chains)" and an elaboration of "commoners' value chains (better described as a web or matrix)."

Indigenous groups, such as Indigenous Environment Network and Native Alaskans, have become part and parcel components of the climate movement. Tokar maintains that climate change is linked with problems addressed by other social movements, particularly the anti-war and global justice movements: "This [climate justice] movement is sharply focused on the social justice implications of the global climate crisis, highlighting the voices of those already massively affected by the heating of the earth. It is linked to antiwar efforts, demonstrating how continuing US military adventures, including the wars in Iraq and Afghanistan, are without question the most grotesquely energy-wasting activities on the planet" (Tokar 2010: 10).

Andreas Malm (2014: 29–38) delineates the following five theses on the role that climate change may play in fostering revolutionary change, of which obviously the climate justice movement could be part and parcel.

- Thesis 1: Climate change increases the likelihood of revolution.
- Thesis 2: Climate change will cause victorious revolutions to degenerate.
- Thesis 3: Revolution improves the prospects for adaptation to climate change.
- Thesis 4: No revolution can survive business-as-usual, because no one can.
- Thesis 5: The only revolution that can save humanity is the climate revolution.

In terms of thesis 1, he acknowledges that climate change cannot serve as a "sufficient cause for revolution but it can be one ingredient in a powder keg, and it can, at least potentially light the fuse" (Malm 2014: 29). For example, between 2010 and 2012, a series of extreme weather events very likely related to climate change adversely impacted the global food system in places such as Russia, Ukraine, Canada, Australia, Argentina, the United States, and China. In terms of thesis 2, there is the danger that "victorious revolutions in the era of climate change will give birth to bulky bureaucracies or even blood-soaked dictatorships" because climate change is likely to create shortages of a wide variety of basic goods, starting with food (Malm 2014: 33). In terms of thesis 3, a progressive revolutionary regime could potentially ensure that everyone would be granted enough resources to survive the ravages of climate change. In terms of thesis 4, "any social formation, whatever the character of its relations, will succumb to the forces of global warming, because it cannot do without a biophysical resource base" (Malm 2014: 37). Finally, in terms of thesis 5, Malm (2014: 41) highlights that any serious effort to mitigate climate change before humanity embarks beyond the 2-degree threshold will require nothing less than a "war on the accumulation of fossil capital," but unfortunately at the moment no one has yet taken the lead in instigating a climate revolution that will move humanity beyond business-as-usual.

While the overwhelming claims of climate science that the planet has in recent times been experiencing anthropogenic climate change and the acceptance of this reality on the part of the United Nations, the European Union, governments, NGOs, and even many corporate leaders, although not all, a counter–climate change movement in the form of climate skepticism or climate denialism continues to play a prominent role in many spheres of public life. As Jacques et al. summarize:

> Climate skepticism comes as an anti-reflexive counter-movement to beat back the ontological threats to Western modernity, organized through conservative think tanks, most in the US, with some in the UK [as well as Australia].... The true ideological and material objectives of the counter-movement are camouflaged by several tactics that confuse fair-minded citizens but empower those ready to deny climate change.... While most climate deniers do not have substantial climate expertise ..., well-credentialed contrarians serve as spokesmen ... to media forums outside peer outside peer-reviewed journals. Thus, it appears to policy elites, journalists, and of course the general public that there are two equally legitimate "sides" and that each should receive equal attention. (Jacques et al. 2008)

The key features of climate change denial are that (1) it is led by a coordinated movement; (2) it seeks to defend the notion that capitalism in not only a productive economic system but is characterized by a moral and beneficent agenda; (3) the funding and organizational origins of the movement are largely hidden; and (4) it seeks to sow confusion in the media, among policy makers, and in the general public (Antilla 2005).

Social Movements as Prefigurative Social Experiments

Anarchists are more apt to use the concept of prefigurative actions or communities, which entails "'building the new world in the shell of the old" or "creating local institutions to meet communal needs" (Starr 2005: 120). In a somewhat different vein, John L. Hammond (2012: 224) argues that many social movements "attempt to anticipate in their own social relations the future society they aim to create" in three principal ways, namely, "community, democracy, and decommodification." He views Occupy Wall Street as having been in essence a prefigurative movement, which I discussed briefly in the section on the global justice movement. Numerous prefigurative social experiments have emerged out of social movements, which, while not clearly anti-capitalist per se, are seeking to operate on the margins of the capitalist world system or skirt around it in various ways. Examples include the Bruderhof, cooperatives, the voluntary simplicity movement, the Transition Town movement, eco-villages (both rural and urban), community centers, libraries, technology centers, craft production, free medical clinics, and alternative media production.

Eco-socialist Joel Kovel cites the six Anabaptist Bruderhof communities, six of which are situated in the United States and two in England, as examples of "prefiguration." Historically, they are indirectly related to the Hutterites in the Plains states of the United States and the Prairie colonies, on one of which I conducted my initial ethnographic research intermittently in 1970–1973 (Baer 1976). The Bruderhof communities, which limit their size to 300–400 individuals, "thrive in the capitalist market" by manufacturing "fine and useful objects, using sophisticated machinery, computers, and a functioning distribution and sales network," while simultaneously adhering to the principle of "all things in common," eschewing cars, DVD players, and designer jeans and eating communal meals and providing health care and education for their members (Kovel 2007: 208–209). In essence, the Bruderhof observe what eco-anarchist Ted Trainer (2010) terms a "Simpler Way," as I found to be the case for the four Hutterite colonies that I visited in South Dakota.

With respect to cooperatives, Mary Mellor maintains: "[they] are well placed to offer a model for a democratic provisioning economy. They are based on member participation on the principle of one person, one vote. The most important aspect of economic democracy in co-operatives is that they do not have to meet the demands of shareholders or chase share value. Finance was constrained in the early co-operatives by a limit on interest paid and in many later co-operative constitutions limits on the nature and role of any capital invested" (Mellor 2012: 109).

Alternative economic institutions or cooperatives tend to limit the size of each enterprise and start new enterprises instead of expanding existing ones but under pressure to grow in order to stave off competition from capitalist enterprises (Schmitt 2012: 204). They also are caught in the dilemma of purchasing their raw materials from capitalist companies and selling their products to them. Perhaps the best-known cooperative venture is the Mondragon cooperatives, which began in 1956 in the Basque city of Mondragon in Spain with the company Ulgor and consisted of twenty-four workers who manufactured paraffin heaters and gas stoves. The Mondragon Cooperative Corporation functions as a network of cooperatives that are directly owned by worker-members. The managers of Mondragon have stepped up the capitalist nature of their operation due to perceived competition from corporate globalization. As a result, as of 2007, "of the roughly 100,000 workers in the various cooperative firms of MCC, somewhat less than 40 percent were owner-members of the cooperatives," with the remaining workers being ordinary employees, some of them temporary (Wright 2010: 244). Unfortunately, MCC is hostile to its ordinary employees unionizing. In the case of the United States, Ratner (2015: 18) maintains that the co-op movement is "tied to U.S. federal agencies whose agenda is promoting neoliberalism, both domestically and abroad." In contrast, village-scale cooperatives that have been emerging in China appear to operate along genuinely socialist principles. Conversely, socialism ultimately cannot be created within a single communal village and ultimately "would require a general movement toward socialism and communism in Chinese society," not to speak of the world system (Ratner 2015: 30).

The voluntary simplicity movement sponsors annual "buy nothing boycotts, websites which assist people in the sharing and bartering of unneeded goods, and Common Security Clubs in the US which to connect people within the informal economy" (Assadourian 2012). The simplicity movement also encourages people to "downshift" from high-income, stressful professional jobs to more basic jobs that people find more meaningful. Sociologist Mary Grigsby (2004: 12) reports that her research on the movement indicates that participants "don't generally talk about policy initiatives, instead focusing on the individual as the primary mechanism for change." Based on a multinational online survey of 2,268 movement participants, Alexander and Ussher (2011: 15) contest this assertion, noting that their survey found that "67% of participants report that they are involved in a community organization and, more specifically, 41% report that they are engaged in a community or political organization related to simple living." Alexander, Trainer, and Ussher (2012: 17) list numerous suggestions for living simply on a wide range of issues, including mindfulness, money, work and time, food consumption, transportation, housing, energy utilization, clothing, "stuff," water consumption, waste reduction, appropriate technology, socializing and entertainment, community relations, social activism, and the politics of sustainability. They argue that participants in the simplicity movement need to adopt lifestyle choices that challenge the "consumer society" and must free themselves of its social structures, an endeavor that includes actions such as relocalizing food production; promoting bicycle paths and public transportation; reducing working hours; encouraging governments to price CO_2 emissions, invest in renewable energy, and cease subsiding the fossil fuel industry; promoting "post-growth economics"; reducing poverty and socioeconomic inequality; voting progressively, and assuming responsibility for one's actions.

Eco-villages, which in some ways are a continuation of the countercultural or hippie movement of the late 1960s and early 1970s, are found in many parts of the world, particularly developed countries, both in rural and urban areas. In one way or other, eco-villagers are committed to living in an environmentally sustainable manner. Christina Ergas conducted ethnographic research in an urban eco-village in a medium-sized city in the Pacific Northwest, which she did not view as necessarily typical of urban eco-villages. She reports:

> The primary intention or goal for this community is achieving ecological sustainability. However, members do not have a monolithic vision of how to achieve this goal. Some only care to live their everyday lives in ways they view as sustainable, whereas others push for action outside the confines of the village. Whatever their individual avenues toward sustainability, villagers communicate, think of collective actions, and define and redefine sustainability in the collective identity process. Additionally, how actors act on their goals is dependent on how they negotiate the political opportunity structure, sanctioning institutions, infrastructure, dominant ideologies, bureaucrats, and neighbors in the greater community. (Ergas 2010: 39–40)

Some members of the eco-village in question were "hoppers" in that they networked with our intentional communities, sharing ideas with others in terms of living sustainably. Some members attempted to make as many articles as possible on their own rather than purchasing them from the larger market. Many members opted out of owning or driving cars or limited their use of cars. Some members purchased food with government food stamps or obtained food from city food boxes or by dumpster-diving.

More recently, political scientist Karen T. Litfin (2014) published a book based on her visits to and interviews with members of fourteen eco-villages in Europe, Africa, Asia, and Australia, each of which she visited for at least two weeks. Table 5.1 below summarizes some of the details of the eco-villages in her study.

Drawing generalizations from the broad range of eco-villages and other intentional communities in Liftin's study, most of them situated in developed societies but some in developing societies and most of them situated in rural areas but some in urban areas, is rather difficult to do. She observes, however, "their average footprints are 10–50 percent less than their home country averages" (Liftin 2014: 35). While she does not see eco-villages as panaceas for achieving environmental sustainability, Liftin (2014: 187) views them as "seedlings" or, one might say, prefig-

Table 5.1. Eco-villages in Karen Liftin's Study

Eco-village Location	Year Established	Number of members or social composition
Earthhaven rural North Carolina (USA)	1994	55
EcoVillage Ithaca (New York)	1991	160
Findhorn rural Scotland	1962	600
Svanholm rural Denmark	1979	125
UfFabrik West Berlin	1979	35
ZEGG rural western Germany	1991	80
Sieben Linden rural eastern Germany	1997	140
Damanhur rural Italy	1995	1,000
Coluffifa Senegal	1964	network of 350 villages
Crystal Waters rural northeastern Australia	1984	200+
Auroville rural Tamil area in Sri Lanka	1968	spiritual community
Lagoswatte Sri Lanka	2004	part of network of 15,000 villages working
Konohana Family rural Japan	1994	80
Los Angeles USA Eco-Village	1992	ethnically diverse community

Source: Adapted from Liftin (2014).

urative communities, that point humanity in the direction of a "viable future."

The Transition Town movement, which bears some strong resemblances to the eco-village concept, was started by Rob Hopkins (2008), a permaculture teacher, in 2005 and started out with Transition Town Totnes (est. 2006) in Totnes, Devon, England. The movement seeks to reduce community energy use and relocalize communities and food systems in order to build up resilience to counter anticipated peak oil and climate change. The official Transition Town Initiatives consist of nearly 400 communities in thirty-four countries (Assadourian 2012: 35). In some ways, the Transition Town movement resembles the countercultural back-to-the-land movement of late 1960s and early 1970s in that its groups "see their role as creating sustainable livelihoods outside the formal economy, through self-provisioning and the creation of alternative currencies" (Scott-Cato and Hillier 2010: 882). For the most part, the Transition Town movement has been a phenomenon of the developed world with its initiatives presently based primarily in Anglophile countries, particularly the U.K., the United States, Canada, Australia, and New Zealand. While sympathetic toward the movement, Trainer (2009) is concerned that its initiatives will operate very much within the parameters of global capitalism in that they obtain many of their resources from it. Hopkins (2008), in his extensive discussion of the political ecological premises of the Transition Town paradigm, tends to downplay how global capitalism contributes to the ecological crisis, climate change, and social disparities around the globe.

While various community-based social experiments, such as cooperatives, eco-villages, intentional communities, the simplicity movement, and the Transition Town movement, seek to sever themselves from the worst aspects of the capitalist world system, at best they can only be "semi-autonomous" endeavors and generally do not advocate transcending global capitalism per se. At the same time, they do illustrate that even within the constraints imposed by global capitalism, people are seeking out alternatives or prefigurative social experiments that attempt to create more meaningful and less alienated lives (Hahnel 2007: 70).

Finding Cracks in the System and Dilemmas Facing Anti-systemic Movements

The prospect of getting a critical mass of humanity to engage in anti-systemic movements and efforts is a most daunting one. Anti-systemic social movements are a crucial component of moving humanity to an alternative social system but the process is a tedious and convoluted one with no guarantees. With respect to climate change, as Victor Wallis so forcefully states: "Until large numbers of people are well organised and thoroughly aware of their long-term interests, the idea of reducing carbon emissions by 80% will appear totally unreal. Only with the social transformation well underway will everyone be able to see through the false dilemma—'either' protect the economy, 'or' preserve the environment—propounded by those who resist even the most minimal international accords on global warming" (Wallis 2010: 52).

Thus, John Holloway (2010: 11) maintains that the "only way to think of changing the world radically is as a multiplicity of interstitial movements" situated within mainstream society. Ordinary people constitute the carriers of social movements and they often come in and out of them and move from one movement to another. Ultimately, "Social change is

... the outcome of the barely visible transformations of the daily activities of millions of people. We must look beyond activism, then, to the millions and millions of cracks that constitute the material base of possible radical change" (Holloway 2010: 12). In trying to survive in the cracks of the capitalist world system, people have developed alternative social relations and even communities, as was the case in El Alto, an indigenous settlement on the outskirts of La Paz that played a central role in the Bolivian response to neoliberalism under the leadership of Evo Morales. A crack provides an opening for deeper systemic changes and the transcendence of global capitalism will require the "recognition, creation, expansion and multiplication here and now of cracks in the structure of domination" (Holloway 2010: 35).

To a greater or lesser degree, social movements have historically focused on relatively limited objectives, whether they were better wages and working conditions in the case of the labor movement, voting rights and better economic and educational opportunities in the case of the women's movement and the civil rights or ethnic rights movements, environmental protection in the case of the environmental movement, and so on. It is important to note that social movements, including anti-systemic movements, often undergo a process of institutionalization, in which they evolve into "social movement organizations" (SMOs) in order to "act with greater efficiency in the areas of recruitment, fundraising, political lobbying, and, of course, mobilization" (Johnston 2011: 72). As they become professionalized, bureaucratized, and more efficient, SMOs often become less democratic, less radical, more domesticated, and even co-opted to the point that they make concessions to the capitalist world system, something that is manifested in many NGOs that to a large degree have their roots in various types of social movements. Indeed, many of the largest environmental NGOs, which by and large came out of the environmental movement, receive core funding from some of the world's largest and most polluting multinational corporations and obtain contributions from very wealthy individuals.

Nevertheless, a strong anti-systemic orientation appears to have resurfaced, albeit in disparate ways, even as old social movements become institutionalized. Social movements struggle internally with how much hierarchy and centralization to adopt in an effort to be efficient, coordinated, and capable of rapid action or how decentralized and democratic to be in order to maintain a sense of commitment among their followers. Although anti-systemic movements have not yet managed to displace capitalism with some sort of global democracy, aside from whether it is based on socialist or anarchist principles or some other notions of democracy, they have not been inconsequential in getting the attention of the powers-that-be. As Buechler observes, "The best indicator is that national governments and corporate interests can no longer ignore them and have been forced to respond. The responses have included violence against protesters and repression of dissent was well as attempts to co-opt and pre-empt movement initiatives without substantive change" (Buecher 2014: 278).

Conversely, one of the problems that social movements have is that many of their members, including organizers, lack a concrete theory of social activism and "turn out to be sprinters rather than long-distance runners" (Lynd and Grubacic 2008: 42) who move from one protest to another protest, sometimes around the world (and in the process burn up many carbon miles.) From the very start, certain social movements have manifested an anti-systemic quality, although in today's world, social movements, including the labor movement, are not necessarily anti-systemic per se or may be only partly anti-systemic or have anti-systemic wings. As a scholar-activist who has been over the years been involved in various social movements, including the labor, peace, anti-apartheid, environmental, climate, and socialist movements, I know firsthand that social movements, even those with some semblance of anti-capitalist tendencies, can be quite disparate, contradictory, and factionalized. However, in today's world they are the site where the greatest potential to transcending global capitalism lies. While anti-systemic movements often espouse democratic decision making, all too often a small clique of professional activists "can quickly dominate an organisation and exclude alternative means of communication" (Ruiz 2014: 36).

In recent times, social movements have come to rely heavily on the internet and social media for organizing and mobilizing their activities. This has been the case for the Arab Spring in Egypt and the Middle East, the Los Indignados in Spain, and the Occupy Wall Street movement around the world. Manuel Castells (2012: 2) views digital social networks as "spaces of autonomy largely beyond the control of governments and corporations that have monopolized channels of communication." He believes that they "offer the possibility for largely unfettered deliberation and coordination of action" (Castells 2012: 10). Whereas as in the past social movements relied on word of mouth, pamphlets, leaflets, newspapers, and conferences for organizing protest rallies, they have increasingly come to rely on multimodal, digital networks for this purpose. Castells (2012: 233) maintains that contemporary social movements are "largely made by individuals living at ease with digital technologies in the hybrid world of real virtuality." Of course one of the dangers that social movements face in relying too heavily on digital networks is that the powers-that-be can shut them down, such as was the case for the

2007 Saffron Revolution in Myanmar. China has created sophisticated techniques to censor material on the Internet that potentially could provoke social dissent. Thus it is imperative that social movements continue to rely on more traditional forms of communication. Furthermore, digital networks can destroy or interfere with convivial relationships among individuals who isolate themselves in their homes with their laptops, smartphones, and so forth, rather than interacting with each other face-to-face, whether in each other's homes, cafés, classrooms, or a wide assortment of public places. Ultimately, anti-systemic movements need structures, albeit democratic ones, to coordinate their efforts, to follow up on protest rallies, which all too often fail to go anywhere concretely in terms of creating another world.

Conclusion

Perhaps the biggest challenge that anti-systemic movements face is how they can they unite in their struggle against a common foe, namely, global capitalism. Ethnic, national, religious, and cultural differences all too often serve to divide subalterns around the world and prevent them from forming a united revolutionary movement. Some time ago, Wallerstein (1990: 52) maintained that social movements need to devote "considerably more energy than has been historically the case to intermovement diplomacy." That recommendation continues to apply today. Chase-Dunn (2005: 184) maintains that humanity "needs both more democratic global governance and more local autonomy, and that the globalization-from-above movements should work together with the local-autonomy movements, or at least with those who are progressive and willing." In contrast, the corporate class and its political allies are able to use a number of transnational organizations, such as the World Bank, the IMF, the WTO, the G8 and G20, the World Economic Forum, and even the UN, to further their various agendas.

Harvey (2012: 127) maintains that the formation of a viable anti-capitalist movement will have to re-evaluate many anti-capitalist or anti-systemic movements of the past as well as the present in terms of "what can and must be done, and who is going to do it where" and needs to address three critical questions. He asserts that the "first is that of crushing material impoverishment for much of the world's population," which requires an "anti-wealth politics and to the construction of alternative social relations to those that dominate within capitalism" (Harvey 2012: 127). We often hear parties ranging from the World Bank to superstars make an appeal for "eradicating global poverty" or "making poverty history." However, following Harvey, we can make a case for "making wealth history" and the eradication of poverty will follow. For him, the "second question derives from the clear and imminent dangers of out-of-control environmental degradations and ecological transformations," which will require "major shifts in consumerism, productivism, and institutional arrangements" (Harvey 2012: 127). And the third question for Harvey (2012: 127–128) "derives from a historical and theoretical understanding of the inevitable trajectory of capitalist growth," which "requires the abolition of the dominant class relation that underpins and mandates the perpetual expansion of surplus value production and realization."

At the moment, while there is no clear organizational model for anti-capitalist movements to follow, Holloway maintains that a key organizational form that has had some success historically is: "the council or assembly or commune: a feature of the rebellions from the Paris Commune to the Soviets of Russia to the village councils of the Zapatistas or the neighbourhood councils of Argentina. The ideas of council organization are also present in many of the current attempts to respond to the crisis of the party as a form of organisation. Necessarily, such attempts are always contradictory and experimental, always in movement" (Holloway 2005: 223).

Nick Srnicek and Alex Williams (2015: 5) lament that progressive social movements tend to rise quickly, mobilizing large numbers of people, and then fade away, resulting in apathy, melancholy, and a sense of defeat, a fate that has befallen by and large the anti-corporate globalization movement, the anti-war and environmental coalitions of the early 2000s, and the more recent Occupy movement. In their view, the contemporary New Left tends to be hindered by an overemphasis on horizontalism, localism, and withdrawal from representative politics, possible due to a strong anarchist orientation. When it comes to localism, for example, "in attempting to reduce large-scale systemic problems to the more manageable sphere of the local community, it effectively denies the systematically interconnected nature of today's world" (Srnicek and Williams 2015: 40). While this characterization has not applied so much to the anti-corporate globalization, climate justice, and even Occupy movements, it certainly has applied to the Transition Town and permaculture movements. Srnicek and Williams (2015: 108) maintain that one way to unify the agendas of many progressive social movements is to build a "post-work society on the basis of fully automating the economy, reducing the working week, implementing a universal basic income, and achieving a cultural shift in the understanding of work." I will return to some of these ideas in the following chapter on system-changing transitional reforms. While global capitalism has certainly been moving in the direction of automation, a shorter work week, a universal basic income, and a new understanding of work are not components of its modus operandi.

It is imperative that anti-systemic movements collaborate with each other. Obviously specific individuals may over the course of their lifetime or at any given movement be participants in two or more anti-systemic movements. For example, Yotam Maron, an Occupy Wall Street organizer based in New York: wrote in July 2013: "The fight for climate isn't a separate movement; it's both a challenge and an opportunity for *all* of our movements. We don't need to become climate activists, we *are* climate activists. We don't need a separate movement; we need to seize the climate *moment*" (quoted in Klein 2014: 152).

My hope is that somehow democratic eco-socialism will serve as a vehicle for addressing Harvey's three questions and serve as an integrative focus for anti-systemic movements within nation-states and cross-nationally, although I recognize that achieving such a stance would be a daunting one, perhaps even utopian one. Sarkar (1999: 227) notes that an eco-socialist movement "would tell workers—undoubtedly in the first world, and perhaps also in the organised sector in the third world—that their standard of living is too high for the health of the environment, for the survival of other species, and for the fights of future generations." He maintains that an eco-socialist party would have difficulty getting its candidates elected to a parliament or legislature in the foreseeable future. Aside from the matter of eventually getting its candidates elected, Sarkar delineates two more immediate objectives for an eco-socialist movement: (1) a struggle for job and social security, job sharing, reduced working hours, in labor-intensive industries and (2) pushing for state control of the market, which would provide protectionism for workers from capitalist firms relying on cheaper labor in other countries. He maintains that the eco-socialist movement would have great difficulty gaining a foothold in a small country, although in "large countries such as India, Brazil, the USA or Australia, it might be less difficult" (Sarkar 1999: 230). Ultimately, Sarkar (1999: 230) argues that the "best strategy would be to build an eco-socialist movement first in large regions, such as western Europe, and not to try to come to power before the pressure for change in the direction of eco-socialism has built up in several countries."

CHAPTER 6

Transitional System-Challenging Reforms

Historically Marxists or socialists have engaged in intense debates as to whether the transition from capitalism to socialism will occur through revolutionary change or more gradual change in the form of reforms in various parts of the world. After all, Marx himself had stated in one of his early writings that "without revolution, socialism cannot be made possible" (Marx 1975: 420). Revolutions involve more sudden and radical social transformations and are often associated with much violence, as was the case with the American, French, Chinese, and Cuban revolutions, but ironically was not the case for the Bolshevik Revolution in Russia in October 1917.

Reforms, despite the best of intentions, are often problematic in that they may serve to stabilize capitalism, as has been the case around the world repeatedly. In light of this reality, Andre Gorz (1973) differentiates between *reformist reforms* and *nonreformist reforms*. He uses the term reformist reform to designate the conscious implementation of minor material improvements that avoid any alteration of the basic structure in the existing social system. Between the poles of reformist reform and complete structural transformation, Gorz identifies a category of applied work that he labels nonreformist reform. Here he refers to efforts aimed at making permanent changes in the social alignment of power. In reality, these two types of reforms are sometimes hard to distinguish, but one way might be whether they are initiated by the powers-that-be or whether they are initiated by the working class or various other subaltern groups or anti-systemic social movements. Gorz (1994: 40) argues that the welfare state particularly characteristic of developed societies constitutes "more or less humanized capitalism" rather than "democratic socialism" as the Scandinavian countries are often described.

The transition toward a democratic eco-socialist world system is not guaranteed and will require a tedious, even convoluted, path, one in which anti-systemic movements will have to play a central role. Amin (2013: 2013) argues that achieving socialism as a "higher level of civilization" will entail a "very long" (albeit hopefully not too long, given the seriousness of climate change) road for which humanity needs to specify "intermediate strategic objectives." Perhaps bearing Marx's eschewal of a blueprint pointing the way to socialism in mind, sociologist Raewyn Connell (2011: 166) argues: "We do not want blueprints any more: we expect to feel our way into the future." Indeed, Marx viewed blueprints as a distraction from the political tasks that needed to be undertaken in the present moment and it is important to note that these indeed are paramount. But history tells us that there always will be immediate struggles that must be addressed. Wallerstein (2008: 51) also expresses skepticism about efforts to create "utopian models," which he argues "will not have much impact on what actually emerges," adding that the "most we can probably do is to push in certain directions that we think might be helpful." Perhaps in the latter spirit I seek to delineate various proposals for system-challenging transitional reforms. I often find that when people ask me what it would take to make a transition to a democratic eco-socialist world system, they are seeking some basic guidelines on how to move forward beyond merely bumbling along haphazardly a step at a time. As Foster observes, "The transition from capitalism to socialism is the most difficult problem of socialist theory and practice. To add to this the question of ecology might, therefore, be seen as unnecessarily complicating an already intractable issue" (Foster 2009: 265).

Proposals for System-Challenging Reforms

More and more, scholars and activists have been exploring how to move beyond the existing capitalist world system and toward an alternative world system based on social justice, democratic processes, and environmental sustainability. David Schweickart maintains that there is a need to devise a "plan" for an alternative economic system at this juncture in history: "I think we do need, not simply a vision of a better world, but a sense of how that world might be structured institutionally. But before defending this claim, let me be clear: the numerous, often disparate efforts now underway to build a better world are *not* counterproductive, diverting energy from the revolutionary struggle to replace capitalism with a humane order. To the contrary, they are essential" (Schweickart 2012: 60).

In this spirit, sociologist Erik Olin Wright (2010: 8) makes a case for envisioning *real utopias* as part and parcel of a "general framework for systematically exploring alternatives" to capitalism. Some time ago, Howard J. Sherman (1995: 327–334) proposed immediate programs and medium-term programs for the left designed to increase political and economic democracy in the United States. In terms of a long-term program for the left, he, in a similar vein to Wright, asserts: "Humans need the dream of a future utopia or at least a vision of the road to travel" (Sherman 1995: 334). Below is a summary of Sherman's long-term vision for a future world society:

- A world government with the continuation of national, state, and local governments and provision for a "wide range of public and cooperative ownership"
- The "voluntary withering away of private enterprise"
- Workers' control of the economy
- The provision of free goods and services as the market withers away (Sherman 1995: 336–337).

In his "anti-capitalist manifesto," Alex Callinicos (2003: 132) envisions a transitional program that includes immediate cancellation of the Third World's debt, introduction of a Tobin tax on international currency exchanges, restoration of regulations on capitalist operations, the implementation of a universal basic income, reduction of the work week, support of public services, renationalization of privatized industries, progressive taxes in order to finance public social services and redistribute wealth and income, and abolishing the military-industrial complex that has been an integral component in profit making, maintaining the military hegemony of powerful nations, and facilitating the grab for resources around the world. Derek Wall delineates ten transitional policies that he hopes will promote discussion:

- Supporting indigenous control of rainforests and other vital ecosystems
- Empowering workers to assume control of bankrupt businesses
- Utilizing government bailouts to support mutualistic resources
- Converting arms and SUV production for public programs
- Promoting open-source patenting
- Land reform
- Large-scale for libraries and other social services
- A tax and welfare system to support common ownership

- "Competition reform to transform ownership"
- Public ownership of pharmaceuticals and health care (Wall 2010a: 68).

Radical economist Juliet Schor (2010: 2) proposes a "plenitude" paradigm that is committed to transitioning to a "sustainable economy" that "places ecological and social functioning at its core," and includes green technologies, working less, the art of slow spending, reciprocity, and share ownership.

In their discussion of how to extend democracy in U.S. society, Wright and Rogers (2011: 468–469) delineate the following reforms: a redefinition of the boundary between government regulation and private rights in the market, a massive public investment in public transport, public control of energy development, the creation of a comprehensive public health-care system, an expansion of government regulation of the labor market and workplace, an expansion of public sector employment, and the implementation of taxation and redistribution policies that diminish social inequality. David Pepper (2010: 37–39) juxtaposes *utopian eco-socialism* to the practical side of eco-socialist theorizing and strategizing by listing various nonreformist reforms, including "local, soft renewable energy for businesses"; community farming, including in urban areas; producer and consumer cooperatives; local currencies that can be used in local employment and trade systems; and community councils.

As these various proposals loosely suggest, the transition from a capitalist world system to a democratic eco-socialist world system will entail numerous transitional or nonreformist reforms at the local, national, regional, and global levels. Whereas the achievement of democratic eco-socialism constitutes the long-term vision, the struggle over the short term will entail more modest reforms, but ones that pave the way for deeper systemic changes. As Victor Wallis so astutely observes,

> The long-term dimension will encompass the total reorganization of society, including a dramatic change in the definition of both individual and common goals. The short-term dimension will reflect the need to immediately arrest the headlong drive to environmental collapse. It will inescapably require steps that can engage capitalist participation: the challenge will lie in not letting corporate interests define the bigger picture. This will entail developing a high level of mass consciousness, so that when ecological advances are attained, popular constituencies will be ready to claim the credit for them, and not allow corporate adaptations or concessions—such as renewable energy and "green jobs"—to be repackaged as a triumphant achievement of capital. (Wallis 2009: 96–97)

Thus the trick becomes the extent to which radical activists work within the system, be it, for instance, with private renewable energy companies or government agencies, such as the U.S. Environmental Protection Agency, that espouse progressive reforms without being co-opted by them.

For purposes of discussion, and debate, this chapter focuses on the following transitional reforms: (1) the creation of new progressive, anti-capitalist parties designed to capture the state; (2) the implementation of greenhouse gas emissions taxes at the sites of production that include measures to protect low-income people; (3) social planning and increasing public ownership, socialization, or nationalization in various ways of the means of production; (4) increasing social equality within nation-states and between nation-states and achieving a sustainable global population; (5) the implementation of workers' democracy; (6) the creation of meaningful work and shortening the work week; (7) the creation of a new zero-growth economy; (8) the adoption of energy efficiency, renewable energy sources, and green jobs; (9) the expansion of public transportation and massive diminution of a reliance on private motor vehicles and air travel; (10) the implementation of sustainable food production and forestry; (11) resistance to the culture of consumption and adoption of sustainable and meaningful consumption; (12) the implementation of sustainable trade; and (13) the implementation of sustainable settlement patterns and local communities. These transitional steps constitute loose guidelines for shifting human societies or countries toward democratic eco-socialism and a safe climate, but it is important to note that both of these phenomena will entail a global effort, including the creation of a progressive global climate governance regime. While I suggest a litany of possible transitional reforms, I am not calling for a definitive blueprint for achieving a socio-ecological revolution. In calling for an alternative to the capitalist world system that is "relatively democratic and relatively egalitarian," Wallerstein (2013: 33) cautions that the latter has "never yet existed" and is "only a possibility." He goes on to argue: "Of course, none of us can design either alternative in institutional detail. Such a design will evolve as the new system begins its life" (Wallerstein 2013: 33).

My litany of proposed transitional reforms is a modest effort to contribute to an ongoing dialogue and debate as to how to move forward from the present impasse in which the world finds itself today. The application of my suggested transitional reforms will have to be adapted for many countries, both developed and developing, around the world. Furthermore, my suggested transitional reforms do not exhaust possible transitional reforms necessary for creating an alternative world system.

New Left Parties Designed to Capture the State

Multinational or transnational corporations to a greater or lesser extent have come to dominate governments and politicians in countries around world. National governments have found much of their power taken over by transnational organizations in which corporate elites and politicians often confer, such as the United Nations, the World Bank, the International Monetary Fund, the World Trade Organization, the G7, the G20, the World Economic Forum, and the European Union. These bodies constitute components of what Robinson (2014: 77–95) terms the "transnational state" (TNS), a structure that facilitates the accumulation efforts of the "transnational capitalist class" (TCC). As Scott Sernau observes, "The real powerhouses of the twenty-first century, however, are not transnational governmental organizations but multinational corporations. In part, this simply reflects their enormous research and financial clout. The assets of the largest corporations often exceed the gross national products of many countries, including large countries. In dollar terms, many of the largest entities in the world are not nations but corporations" (Sernau 2011: 202).

Althusser (2014: 109) maintains that revolution is "unthinkable without the capture of state power." In a similar vein, Dean (2016: 206) argues that "gaining control of the state remains an important goal because it presents a barrier to political change." Obviously the shift to a democratic eco-socialist world will require a revolution of some sort that will have to be played out in various ways depending on the national context. Obviously the capitalist class and its political allies around the world will be resistant to such as revolution. The larger question is whether a democratic eco-socialist-oriented revolution can be achieved largely through peaceful measures or whether it will entail violence or perhaps a mixture of both, depending on the country. As Arrighi, Hopkins, and Wallerstein (1989: 32) observe, regardless of whether state power is achieved through a "legal path of political persuasion" or through the "illegal path of insurrectionary force," in reality the terms reform and revolution "have become so overlaid with polemic and confusion that today they obscure more than they aid analysis." After all, one hears mention in various circles of a wide variety of revolutions, for instance, the Agricultural Revolution, the Industrial Revolution, the Internet Revolution, and the Ecological Revolution. Althusser (2014: 107) asserts that achieving socialism through parliamentary means has been a nonexistent path, although various attempts, such as in Chile in the early 1970s when Salvador Allende was elected President, have been pursued. This effort, however, failed as the result of a U.S.-supported military coup that brought General Augusto Pinchot to power. Indeed, as support for the Popular Unity movement increased between 1970 and the 1973, "so did the *reaction* to these advances" (Hoffman 1984: 161).

Nevertheless, while Marx and Engels indeed envisaged an armed overthrow of capitalism in some situations, they also gave attention to achieving reforms within the bowels of capitalist societies and viewed such efforts as a "school for evoking the political consciousness of the proletariat and building up its political organization," a strategy for "weakening the class domination of the bourgeoisie in all of its aspects," and a means for "potential peaceful transformation of capitalist society into a socialist one" (Wlodzimierz 1967: 87). Marx believed that socialist-oriented revolutions might occur in countries such as England, the Netherlands, and even the United States through peaceful and presumably parliamentary means. As Eagleton observes, "He did not dismiss parliament or social reform. He was an enthusiastic champion of reformist organs such as working-class political parties, trade unions, cultural associations and political newspapers. He also spoke out for specific reformist measures such as the extension of the franchise and the shortening of the working day. In fact at one point he considered rather optimistically that universal suffrage would itself undermine capitalist rule. His collaborator Friedrich Engels also attached a good deal of importance to peaceful social change, and looked forward to a nonviolent revolution" (Eagleton 2011: 192).

Ultimately achieving most of the transitional reforms in my listing may require that new left parties or socialist-oriented parties come to power and in a sense "capture the state" and ensure that there is a political will that will guarantee their implementation. For example, nationalization of the means of production would be difficult to achieve without a leftist political party in power. Martin Ryle (1988) maintains that an environmentally friendly progressive state will be necessary to challenge corporate power. Dean (2016: 259) argues that the Communist Party needs to be "revised, renewed, rethought, reimagined" so that it functions as a vehicle with the "will and capacity to bring an egalitarian world into being."

Until the election of Syriza in Greece in early 2015, the possibility of new left parties coming to power in developed capitalist countries had appeared to be remote. But given the gravity of both the global economic and ecological crisis, including climate change, one should not rule out the possibility of political tipping points just as climate scientists speak of tipping points that had set off a number of irreversible climatic events. At the same time, it is important to note that various new left parties have emerged on the word scene. In addition to Syriza, perhaps the most prominent examples are the German Left Party (Die Linke), the Left Front in France, Left Unity in the U.K., and Podemos in Spain.

DIE LINKE

DIE LINKE is a merger of the Party of Democratic Socialism (PDS), the successor party of the former ruling Socialist Unity Party of Germany in the GDR, and the Jobs and Social Justice—the Election Alternative (WASG), which was created in West Germany by Social Democratic dissidents, members of the German Communist Party based in the West, disaffected left-wing Greens, and left trade-unionists. During the early 1990s, PDS "had adopted a role as 'societal opposition,' using its parliamentary positions to campaign for East German grievances and for pacifist, ecological, feminist, and social justice themes" (Oswald 2002: xiv). In the late 1990s, in contrast to the more than two million members who had belonged to the SED, the PDS had 280,882 members and still was losing members (Oppelland 2012: 433). Nevertheless, as David F. Patton observes, "[T]here are signs that the PDS was becoming an eastern catchall party (*Volkspartei*) that appealed across class, regional, and generational lines. The party did well among young voters, many of whom had but childhood memories of the SED state" (Patton 2011: 8).

Indeed, in the 1998 elections the PDS obtained 5 percent of the nationwide vote, thus granting it membership for the first time in the Bundestag. Also in 1998, for the first time it entered a coalition government in Mecklenburg-West Pommerania.

The PDS renamed itself Left Party.PDS in 2005 and had been involved in governing coalitions with the Social Democratic Party (SPD) at the provincial, namely, in Mecklenburg-West Pomerania in 1998–2002, Berlin in 2002–2011, and in Brandenburg since 2009. In contrast, the WASG arose in 2004–2005 and drew on "trade union and SPD malcontents with Schroeder social reform policies on the one hand and extreme leftists on the other" (Oppelland 2012: 435). Former SPD chairperson Oskar Lafontaine joined WASG in 2005 and played a pivotal role in its merger with Left Party.PDS. However, the merger process, which was completed in 2007, was not an entirely smooth one with the WASG proving to be the more radical partner. As Patton (2013: 222) observes, "Even as the Left Party.PDS and WASG prepared to merge, Oskar Lafontaine and numerous WASG politicians criticized the Left Party.PDS's record in regional governments and crossed swords with the internal grouping Forum Democratic Socialists (FDS), which was closely identified with eastern politicians in office or seeking office." The WASG was the smaller partner in the merger with a membership of 8,944 members at the end of 2006 as compared to 60,338 members in Left Party.PDS.

Ingar Solty (2008: 2) maintains that DIE LINKE is unique in that it constitutes the "first (party-) political leftist articulation of the contradictions of neoliberalism in the core capitalist countries." Further, he maintains that DIE LINKE has prevented a significant portion of the working class to be attracted to right-wing extremism in Germany as has been the case in various other European countries, such France, Italy, Denmark, the Netherlands, and Switzerland (Solty 2008: 3). Solty argues: "While the Greens were made up of the so-called 'post-material' enlightened, culturally left, upwardly mobile citizens and accordingly today have the highest average income of all parties (vying with the FDP in this respect), the originally strongly white-collar LINKE has in recent years steadily acquired more of an unequivocal class basis. In this respect, Die LINKE is increasingly winning support among workers and the unemployed" (Solty 2008: 32).

DIE LINKE has become a force not only in elections in the new eastern states but also in the federal Bundestag or parliament. In the 2009 Bundestag election, DIE LINKE won 8.3 percent of the Western vote (including West Berlin), 28.5 percent in the East and 11.9 percent overall. Table 6.1 below indicates the Left.Party.PDS's share of votes in Bundestag elections in 2005 and 2009, clearly indicating an overall strengthening of its position at the federal level.

In the 2010 election the Left Party.PDS obtained 10 percent of the vote for the Bundestag. DIE LINKE approached Frigga Haug "for advice to improve its competence in women's issues, in order to address the low percentage of women among new recruits to the party (18%)" (Haug and the Dialektikfrauen 2012: 143). In 2010 DIE LINKE implemented a policy whereby at least one of the two chairpersons have to be women. In its Erfurt party program, DIE LINKE specified the conditions under which it would be willing to enter a governing coalition. "We will not participate in a government that goes to war, supports military engagement of the Bundeswehr abroad, promotes rearmament and militarisation, pursues the privatisation of public services and cuts in social welfare and whose policies undermine the ability of the public sector to fulfil its tasks" (DIE LINKE 2011).

Despite its relative success in the Bundestag as well as in various provincial parliaments, DIE LINKE continues to be a fragile alliance of

Table 6.1. Die Linke.PDS/DIE LINKE Vote Share in Bundestag Elections, 2005–2009

	National votes, %	West Germany, %	East Germany, %
2005	8.7	4.9	25.3
2009	11.9	8.3	28.5

Source: Adapted from Patton (2011: 141).

eastern and western forces, in part illustrating the East-West divide that lingers in Germany a quarter of a century after the unification. As Table 6.2 below indicates, this East-West divide was illustrated in the national capital in the 2011 election, which revealed distinctly different voting patterns in the two Berlins.

In 2013 the parties in the federal Bundestag order were the Christian Democratic Union with 255 seats, the Social Democrats with 193 seats, DIE LINKE with 64 seats, the Greens/Alliance 90 with 63 seats, and the Christian Social Union (a sister party of the CDU in Bavaria) with 56 seats. Whereas the far-right-wing Republican Party has dropped below the 5 percent threshold required to for representation in the federal Bundestag, the Alternative in Germany Party is positioning itself to be the next far-right-wing party in that body.

In early November 2014, DIE LINKE found itself in the first-time situation of heading the government in Thuringia in coalition with the SPD (Fertl 2014: 18). At the federal level, the SPD has repeatedly refused to form coalitions with DIE LINKE. In 2013, the SPD, the Greens, and DIE LINKE could have formed a "red-red-green" ruling coalition but the SPD opted to enter into a Grand Coalition with the CDU. Conversely, Lafontaine has opposed the willingness of DIE LINKE to cooperate with the SPD at the local and provincial levels because "it interfered with his strategy to demonize the SPD and to force it out of power" (Spehr 2012: 164).

The larger question is to whether some time down the road, DIE LINKE can form a coalition government with the Greens, just as at one time the Greens formed a coalition with the Social Democrats to form government under the leadership of Gerhard Schroeder, or even enter into a red-red-green coalition at the federal level. Even if DIE LINKE were to adopt a more explicitly eco-socialist program, a dilemma of forming a coalition is the latter's tendency in Germany as well as elsewhere, including Australia, not to "take into account the intrinsic contradictions between the cap-

Table 6.2. Voting Patterns in the 2011 Berlin Land Election—West and East

Party	West	East
Social Democrats	27.9	28.8
Christian Democratic Union	29.5	14.2
Greens	20.3	13.5
DIE LINKE	4.3	22.7
NPD	1.6	2.9
Free Democrats	2.3	1.2

Source: Adapted from McKay (2012: 235).

italist dynamics of the unlimited expansion of capital and accumulation of profits, and the preservation of the environment" (Loewy 2006: 295). Even though DIE LINKE has expressed a willingness to work with the Social Democrats within parliamentary politics, William Tabb (2014: 515) maintains it is "about pushing for non-reformist reforms." Furthermore, Wolf (2015: 88) reports: "As to the participation of DIE LINKE in a federal government coalition (electorally possible with the Social Democrats and *Gruene* (Greens), on the basis of the last federal elections, the SPD and *Gruene* have formulated conditions—which would exclude any participation of DIE LINKE. Whether this exclusion can be lifted will be a central issue for electoral politics in Germany on the federal level for at least the next decade."

In terms of the larger European community, DIE LINKE, along with "Blockupy," has initiated mobilizations in Germany opposing the EU's austerity measures imposed on Greece (Wolf 2015: 91). For the most part, however, DIE LINKE and the radical left in Germany have been limited in challenging the dominant neoliberal agenda being promoted by the Christian Democratic government headed up by Angela Merkel. In part this has occurred in the form of transformative projects (*Einstiegsprojeckte*), such as publicly funded employment and participatory budget planning in cities "designed to demonstrate a socialist-type approach to organizing inside a capitalist society" (Spehr 2012: 167).

At the present time, DIE LINKE, while clearly identifying itself as a "democratic socialist" party, does not appear to have a strong explicit commitment to eco-socialism per se except in a nominal sense. On its website, it states: "We have united in a new political force that stands for freedom and equality, fights resolutely for peace, is democratic and social, ecological and feminist, open and plural, militant and tolerant" and also notes its commitment to the "conservation of nature" (http://en.dielinke.de/die-linke/welcome/). In its statement of its policies, DIE LINKE states that it is committed to a "social energy policy":

> An entirely different energy policy has to be geared to the aim of a renewable and at the same time democratized energy supply. Energy and climate policy has to be designed in a socially fair way. Every person has the right to an affordable energy provision. DIE LINKE demands an effective supervision of current and gas prices plus consumer advisory boards safeguarding the energy buyers' insight in and co-decision on pricing. The prompt introduction of social tariff could quickly relieve low-income households. In the medium and long-term prospect, the shift from the fossil-nuclear energy industry to energy supply from renewable sources coupled with the sparing and efficient use of energy is the only way to guaranteeing affordable energy prices. (http//:en.die-linke/die-linke/ policies/)

For the most part, other than the reference to the need for a "democratized energy supply," this energy policy closely resembles the commitment to ecological modernization expressed by Green parties around the world as well as by many mainstream environmentalists and climate activists.

Syriza

Syriza formed in large part due to the severe austerity crisis experienced by Greece since it was admitted by EU leaders into the Eurozone economy in 2001, in part spurred by the cost spirals resulting from the 2004 Olympic games in Athens (Sayer 2016: 369). Even before Greece's admission into the Eurozone, German and other bankers had advanced large loans to Greece during the 1980s and 1990s, despite the fact that, as anthropologist Michael Herzfeld (2016: 12) observes they "must have realized that Greece would never be able to meet such heavy accumulation." The Eurozone economy has been regulated largely by the German state and its various allies, including France as a junior partner (Jessop 2016: 203). The International Monetary Fund has reportedly earned 2.5 billion Euros of profits from its loans to Greece since 2010 (Sayer 2016: 370). Furthermore, German and French arms companies had supported loans to Greece so that previous governments could purchase their products. Greek debts skyrocketed from 133 percent of the country's GDP in 2010 to 174 percent in 2014 (Sayer 2016: 371). While various German power brokers often argue that Greece is a country filled with corruption (Herzfeld 2016: 12), the same characterization applies to Germany, as has been evidenced by the fact that "Daimler, Siemens, and the German national rail system (Deutsche Bahn) have been embroiled in major bribery scandals in Greece."

Syriza (Coalition of the Radical Left) (established 2004) is an alliance or coalition of a broad array of political groups, some of them socialist and green, and independent politicians. The political party SYNASPISMOS (Coalition of the Left, of Movements and Ecology) was the largest party in the newly formed coalition. While a large portrait of Rosa Luxemburg hangs in the Thessaioniki office of Nikos Samanidis, a founding member of Syriza, this relatively new left party certainly is not as radical as Luxemburg and her Spartacist League were nearly a century ago. Syriza has become a political force to be reckoned with in the little over a decade since its establishment, both in Greece and now in the European Union. Syriza ran in the Greek general elections of 2004, 2007, 2012, 2014, and 2015. Syriza began to come into its own when the Greek government announced in October 2009 that the country's debt exceeded 12 percent of the national income and appealed to the Eurozone for financial assistance (Varoufakis 2013: 206–207). As the leading actor of the Eurozone, German chancellor Angela Merkel issued her three "neins" or "no's" to the possibility of a bailout, interest relief, or to default from the debt for Greece. Instead, the Eurozone, the European Common Bank, and the International Monetary Fund agreed to extend Greece a loan of 110 billion Euros, at high interest rates, in order to resolve the debt crisis, which instead drove the country deeper and deeper into insolvency.

Prior to 2012, Syriza never exceeded 5 percent of the vote in Greek elections. However, in June 2012, it obtained nearly 27 percent of the vote and outflanked the social democrats by becoming the second largest party in Greece. The neofascist Golden Dawn increased its vote share from 0.28 percent in 2009 to 6.92 percent in 2012, reflecting a different sort of dissatisfaction with the government's austerity measures. Syriza had gained its strength in large part due to the protest against the austerity measures that the consortium, often referred to as the Troika, consisting of the European Bank, the European Commission, and IMF, has attempted to impose on Greece. Since the Global Financial Crisis of 2008, the Troika has imposed severe austerity measures on Greece, contributing to an unemployment rate of 31 percent (about 50 percent for youth) and a 20 percent poverty level among employed workers (Munckton 2015: 5).

Syriza delineates Four Pillars in its National Reconstruction Plan:

- Confronting the humanitarian crisis
- Restarting the economy and promoting tax justice
- Regaining employment
- Transforming the political system to deepen democracy.

In the 2014 elections for the European Parliament, Syriza obtained 26.58 percent of the votes, more than any other Greek parties, including New Democracy and PASOK (Panhellenic Socialist Movement).

Syriza won the 25 January 2015 election, in which it obtained 36.3 percent of the vote. Syriza's inability to obtain two more seats in the Greek parliament, leaving it short of an absolute majority of the 151 seats, forced it to form a government with Independent Greeks (ANEL), a right-wing nationalist party that formed as a split from the former ruling New Democracy Party and that holds thirteen seats in the 300-seat parliament. This "unholy" alliance prompted concern on the part of Exarchaia, an anarchist/radical Left group near the Polytechnic in Athens (Rundle 2015: 5). Prime Minister Alexis Tsipras at age forty became the youngest political leader in Greek history. He rose to leadership of Syriza in 2008 and was elected to the Greek Parliament in 2009. Syriza and the Independent Greeks both oppose the EU's austerity measures but differ significantly on policies relating to immigrants, refugees, and gay rights. Only time will tell how this alliance will work out but it certainly attests to the saying that "politics makes for strange bedfellows." On a positive note, Ecologist Green formed an alliance with Syriza during the election.

The new Syriza-led government announced a wide array of anti-austerity measures, including debt relief, blocking privatizations, raising the minimum wage, and reinstating unemployed workers (Munckton 2015: 5). Syriza sought relief for a half of Greece's 321 billion Euro debt and to tie repayments to economic growth. The new government also abolished fees for prescription drugs and hospital visits and restored the right of unions to engage in collective bargaining. Members of the thousands of cooperatives—medical cooperatives, co-op cafes, farmers markets, etc., that emerged in response to the austerity crisis "are making plans about how they will access Syriza's plans for a robust 'third sector', a real post-capitalist socio-economic space, not an add-on" (Rundle 2015: 5).

German finance minister Wolfgang Schauble rejected the idea of restructuring the Greek debt on 5 February 2015 and the previous evening the European Central Bank "announced it would not accept Greek government debt as collateral for loans to Greek private banks after February 11" (Nichols 2015: 13). When Syriza found the EU unwilling to accept its demands, Tspiras called for a referendum on whether or not to reject the EU's terms. On 18 September 2015, the no vote won by an overwhelming majority, but the EU did not make any concessions and offered worse terms to Greece, which Tspiras felt compelled in large measure to accept, prompting the left caucus in the Syriza to form a new party. Despite this split, Syriza managed to win the election of 20 September 2015, thus retaining its power despite much dissatisfaction among the voters, who felt they had no other option under the circumstances in which their country constituted a small player in the German-dominated EU. In the meantime the Troika has been sending its officials to Greece to monitor and micromanage repayment of its loans, thus altering the regulations governing the ownership of pharmacies, allowable shelf life of milk, and the weight of bread loaves sold by bakeries (Sayer 2016: 370). How to resolve the dilemma that Syriza faces is extremely complex, especially given Greece's peripheral status with a German-dominated EU. Rasmus (2016: 17) asserts that "Syriza's strategy grossly underestimated the ability of the Troika, which included the ECB, to economically pummel Syriza into submission over the course of 2015 negotiations by virtually shutting down Greece's banking system to force Syriza to submit to Troika demands and concessions." In essence Syriza hoped to return Greek workers, small businesses, and farmers to social democratic practices that were opposed both by the neoliberal agenda of the Troika as well as by Greek bankers, investors, and wealthy people who were financially prospering under an arrangement that allowed them to avoid taxes, speculate in the Greek government's bonds, and send their incomes to banks and shadow banks offshore (Rasmus 2016: 117).

Some leftists have advocated that Greece withdraw from the Eurozone, refuse to pay its debt, re-establish its old currency, and nationalize its banking system (Flassbeck and Lapavitsas 2015). Hahnel (2016: 18) maintains that rather than having submitted to the EU's draconian austerity measures, the Syriza-led government should have launched "its own economic recovery plan[;] its economic program could have built worker and neighborhood councils and the beginnings of participatory planning." Other leftists argue that the Syriza-led government needs to stay the course and continue to leave the exit possibility open while negotiating for better fiscal terms within the EU.

Other New Left Parties

In the U.K., Left Unity emerged from a call by filmmaker Ken Loach in 2013 for a new party to the left of Labour. Left Unity has attracted disaffected Labour Party members, including councillors, along with feminists, environmentalists, and socialists of various stripes, including Trotskyists. Kate Hudon, Left Unity's secretary, told *Green Left Weekly* in an interview published on 29 October 2014: "We include people with revolutionary politics, but we also include those who consider themselves left-of-Labour. Everyone is welcome, but we don't want people trying to make it something that it is not" (Rogatyuk 2014: 20).

Left Unity has put forward two platforms, the first titled "The Left Party" platform and the second the "Socialist Platform." Left Unity has over fifty branches around the U.K., cooperated with the Green Party at the time of the European Parliament elections, and supported Peter Cranie, the Green Candidate for the North-West. Green Left, an eco-socialist network in the Green Party of England, has been seeking to move the party further to the left since its creation in 2006. Derek Wall (2010b: 109) maintains that the "Green Party of England and the U.K. has a reputation as the most left-leaning Green Party in Europe, and the U.K. has seen ecosocialists organize with increasing effectiveness both inside and outside the Green Party." Furthermore, Socialist Resistance, the British section of the Fourth International, has adopted eco-socialist principles and "cooperates where possible with Green Left to promote ecosocialism" (Wall 2010b: 114).

In Spain Podemos started out as a social movement responding to anti-austerity measures, became the third major political party in January 2014, and draws its inspiration from both Syriza and the Bolivarian revolution in Latin America. At the end of 2013, austerity measures suggested by the International Monetary Fund and the European Committee resulted in some six million people unemployed and massive cuts in social and health services in Spain (Iglesias 2015: 119–120). Podemos has called for fiscal reform, an audit of Spain's debt, increased employment, and regulation of financial institutions and seeks to restore social democratic policies that were commonplace in much of Europe, including somewhat belatedly Spain, some thirty to forty years ago.

The Notion of a Global Left Party

Aside from the issue of new left parties in specific countries, in their discussion of the World Social Forum, Chase-Dunn and Reese observe:

> World party formation is once again on the agenda of the transnational social movements confronting capitalism and global injustice. Before real global parties can emerge, contemporary transnational social movements need to come to an agreement about organizational practices and structures, as well as democratic and effective forms of leadership, education and recruitment. The forms taken by earlier efforts, especially the Comintern, need to be carefully studied, critiqued and improved upon. New communications technologies make horizontal forms of communication and decision-making far easier than in the past. Party formation (or party-network formation) can shape new structures and processes using digital technologies that did not exist when the revolutionaries of earlier eras tried to change the world. (Chase-Dunn and Reese 2007: 84)

Unfortunately, thus far, the concept of a global far-left-wing political party exists primarily in science fiction, as we saw in chapter 4 and, namely, in W. Warren Wagar's (1992) *A Short History of the Future*, in which in the World Party comes to power and forms a bureaucratic, technocratic, socialist-oriented world government in 2062, following a tumultuous struggle between the Leninists and Ghandians within the party. While the Commonwealth is governed by a democratically elected world parliament and built on environmentally sustainable principles, it is replaced by a federation of small political communities, some of which live in outer space, and manage to reach farther into space by 2300.

Stephen Gill develops five theses on global political parties that provide food for thought:

1. "The study of global parties requires a long-term view, e.g. grounded in what Ferdinand Braudel called the longue duree of history."
2. "All political parties are doomed to obsolescence unless they remain fully relevant to and engaged in both immediate and longer-term political questions."
3. "We need to avoid the fallacy of assuming that all forces of opposition of progressive forces can be or should be unified into a single form of agency which we call a political party."
4. "We should reimagine political agency in terms of movement rather than organization."
5. "Successful political movements need to connect their forms of organization to feasible utopias or mobilizing myths" (Gill 2007: 115–118)

In terms of principle 3, any observer of far-left-wing politics in various countries around the world will notice the diversity of socialist parties that may exist in a single country, let alone in a single city. For example, in Melbourne, a city of some four million people where I have resided since 2006, socialist parties and groups vary in size and influence and include the Socialist Alliance, the Socialist Alternative, the Socialist Party of Australia, the Communist Party of Australia, Solidarity, the Socialist Equality Party, and the Freedom Socialist Party. There were discussions a few years ago to merge or "fuse" the Socialist Alliance and the Socialist Alternative, the two largest socialist parties in Australia, but they collapsed, in part because the latter did not regard the former as being sufficiently "revolutionary" in its politics. Perhaps what is needed is what Samir Amin (2007: 139) in his call for a Fifth International maintains "should be a socialist/communist international open to all who want to act together to create convergence in diversity" and would "not exclude the formulation of theoretical concepts for the society to come."

Emissions Taxes

Along with various types of eco-taxes in general, emissions taxes and various other eco-taxes potentially can serve as a progressive climate change mitigation strategies given the seriousness of the climate crisis (Loewy 2015: 11). Meinshausen et al. (2009) have argued that between 2010 and 2050 only 750 billion tons (0.75×10^{12} tons of 750 Gt) of CO_2 can be emitted in order to guarantee that the global average temperature will not increase by over 2°C. At the present rate of burning of fossil fuels, the "carbon budget" will be depleted before 2050. Thus, it is imperative that humanity figure out ways to reduce greenhouse gas emissions very quickly to keep the planet in a relatively safe climatic state. However, various other climate scientists, including James Hansen, are arguing that a more realistic safe level of CO_2 is in the order of 300–350 ppm. Olli Tammilehto states:

> If we choose 1.5 degrees instead of 2 as the maximum temperature rise, then the carbon budget gets smaller. One study calculated the difference as human emissions of 420 Gt instead of 750 Gt CO_2 between 2010 and 2050.... This would require emissions to decline by about 10 percent annually. Fairness demands that most of the allowable increase should be allocated to the global South or the Majority World, because per capita emissions have been and still are small compared to the North. In order to stay within the 420 Gt emissions target, the old industrial countries would have to cut their emissions much more than 10 percent per year, a prescription that collides directly with continuing economic growth (Tammilehto 2012: 80).

Much ink has been spilled on how to reduce greenhouse gas emissions, including weighing the pros and cons of emissions taxes and emissions trading schemes. The latter, thanks in part to the advice of Al Gore when he was vice-president in the Clinton administration, is embodied in the Kyoto Protocol—which includes a "cap and trade system" that imposes national caps of CO_2 emissions, which the vast majority of countries that have ratified the agreement consistently fail to meet, in part because there is no legal mechanism for sanctioning failure to meet the targets. Under the dictates of the Kyoto Protocol, developed countries can circumvent emissions reductions by either investing in forestation as a form of sequestration or soil conservation or investing in Clean Mechanism projects in developing countries that supposedly result in the transfer of more environmentally sustainable technologies to developing countries. Larry Lohmann delineates five reasons why emission trading constitutes a flawed approach to mitigation:

> *First*, in order to work, greenhouse gas trading has to create a special system of property rights in the earth's carbon-cycling capacity. This system sets up deep political conflicts and makes effective climate action exceedingly difficult. *Second*, pollution trading is a poor mechanism for stimulating the social and technical changes needed to address global warming. *Third*, the attempt to build new carbon-cycling capacity is interfering with genuine climate action. *Fourth*, global trading systems for greenhouse gases can't work without much better global enforcement regimes than are likely in the near future. And *fifth*, building a trading system reduces the political space available for education, movement-building and planning around the needed fair transition away from fossil fuels (Lohmann 2006: 72).

Various trading schemes, including those in the United States, the Kyoto Protocol, and the EU Emissions Trading Scheme, initiated in 2005, essentially grant corporations and developed countries property rights to emit greenhouse gas emissions. The carbon permit prices under the EU ETS have fluctuated wildly, from a high of 30 Euros in April 2006 to 0.03 Euros at the end of 2007, to 35 Euros during 2008, then down to 6.04 Euros in April 2012, and up to 9.80 Euros in August 2012.

Conversely, a carefully crafted emissions tax has the potential to serve as a transitional reform. However, emissions taxes or carbon taxes are fraught with danger given the tendency on the part of corporations to find loopholes and pass the burden of the tax onto working-class people, thus compounding the primary injustice of climate change with secondary injustices. Progressive British journalist George Monbiot advocates the global implementation of a "carbon tax and carbon swap program" (Chase-Dunn and Lerro 2014: 373). A carbon tax has several advantages over an emissions trading scheme, including that "taxes are generally less complicated, more transparent and less expensive to administer than emissions trading" and it "does not allow speculators to create financial instruments that would provide money-making opportunities and hence increase the costs of the carbon price" (Diesendorf 2014: 216–217). Sweden and Finland were the first countries to implement a carbon tax in the early 1990s. Only a few other countries have done likewise, but several U.S. states and Canadian provinces have followed suit. Boulder, Colorado, became the first U.S. city to implement a carbon tax in 2007 and extended it in 2012. Norway has a carbon tax of about $50 per ton of CO_2, Sweden one of about $40 per ton, and Denmark one of about $14 per ton of CO_2 (Metz 2010: 229). France imposes a modest tax on chemical industries for emitting nitrous oxide. Efforts to impose a carbon tax in the EU as a whole met with failure, because certain member states, such as the U.K., caved to various industries' claims that they would not be able to compete with non-EU industries.

James Hansen, a retired NASA climate scientist, has called for a steep carbon tax as a strategy for quickly reducing greenhouse gas emissions. He has proposed a "fee and dividend" strategy for the United States that includes the following recommendations (see Foster 2013). Fossil fuel companies would be charged an easily implemented carbon fee imposed at the well head, mine shift, or point of energy.

- 100 percent of the revenue collected would be distributed monthly to the population on a per capita basis as dividends, with up to two half shares for children per family.

- Dividends would be sent directly via electronic transfers to bank accounts or debit cards.
- The carbon fee would be a single, uniform amount in the form of dollars per ton of CO_2 emitted from the fuel.
- The carbon fee would then gradually and predictably be ramped up so as to achieve the necessary reductions.
- Current subsidies to the fossil fuel industry would be eliminated.

It is important to note that a carbon tax can be ineffective if it is poorly designed, as was the case in Australia under former Julia Gillard's Australian Labor Party (ALP) government. In reality, this carbon tax was intended to serve as an interim measure that would transition into an emissions trading scheme three years after its implementation. The scheme, which was called the Carbon Price Mechanism (CPM), was crafted by the Multi-Party Climate Committee, consisting of ALP and Greens representatives, which actually took effect on 1 July 2012, after passing both the House and the Senate, despite much opposition from the Coalition (Liberal and National parties), whose leader Tony Abbott accused the government of trying to impose "a great big new tax" on the Australian people. The CPM was designed to result in a 5 percent cut in Australia's CO_2 emissions, based on 2000 levels, by 2020, and an 80 percent cut on those levels by 2050. Some five hundred big polluters were to pay a tax to be set at AU$23 for every ton of CO_2 emissions, rising to $24.15 in 2013–2014 and $25.40 in 2014–2015. Agricultural emissions were exempted. The CPM during its brief existence was an extremely complex scheme that included numerous subsidies for industries and low- and middle-income households, and free permits for some industries. To encourage renewable energy, the CPM included a Clean Energy Finance Corporation, which was to invest in renewable energy projects and finance research and development of renewable energy. In a complicated political coup Kevin Rudd, who had preceded Julia Gillard as the ALP Prime Minister, managed to orchestrate her ouster as prime minister and be reinstated as the prime minister. Rudd's government whittled the now unpopular carbon tax from $23 per ton of CO_2 emissions to the then-current EU carbon price of $9 per ton of CO_2 emissions. However, Tony Abbott managed to win the 2013 federal election in large part because he promised that his government would abolish the carbon tax. Indeed, with support from Palmer United Party, the Abbott government managed to orchestrate the abolition of the "carbon tax" in the Parliament on 17 July 2014. The Coalition Party's insider removed Abbott as prime minister and replaced him with Malcolm Turnbull, a successful banker and businessman, who despite having favored an emissions trading scheme while serving as the leader of the Opposition when Kevin Rudd was prime minister, for the moment does not plan to overturn Abbott's climate change policies. At the UN COP21 conference in early December 2015, which Turnbull and Foreign Affairs Minister Julie Bishop attended, Australia was considered to be a leading "climate laggard" along with Saudi Arabia and Kazakhistan.

Ultimately, the record on the extent to which the few existing emissions tax schemes found in various countries have been effective is modest or mixed in terms of curtailing emissions or promoting a shift to renewable energy sources. Sociologist Anthony Giddens, a proponent of emissions taxes, reports: "In Finland, without the CO_2 tax, emissions would probably have been 2–3 per cent higher by the year 2000 than they turned out to be; in Sweden, Norway and Iceland the figure was 3–4 percent. The absolute level of emissions, however, increased across the 1990s in all of these countries. Only in Denmark did the absolute volume of CO_2 emissions fall. The reason is that the Danes directed the tax revenue to environmental ends—it was used to subsidize energy-saving practices" (Giddens 2009: 151–152).

Emissions taxes at best are only a short-term solution and a market mechanism at that and would perhaps not be necessary if ownership of energy production were publicly owned rather than privately, which is generally the case today around much of the world. Public ownership of utilities as well as mining could serve in rapidly transitioning from reliance on fossil fuels to renewable energy sources. What are needed are governments that exist not to prop up corporate endeavors but seek to achieve social parity and environmental sustainability.

Public Ownership of the Means of Production

In an era of increasing privatization of social and health services, and even military activities and prisons, raising the specter of public ownership, nationalization, or socialization of the means of ownership is taboo in conventional economic and political circles. Privatization is often justified in terms of economic efficiency. While state or government enterprises or services can be terribly inefficient for complex reasons, this does not necessarily have to be the case. There are numerous examples of publicly owned enterprises that operate relatively efficiently. As Tanuro (2010: 124) observes, ultimately eco-socialism requires "expropriation of monopolies in the energy sectors and confiscation of their assets" as well as a radical extension of the public sector in general, particularly in the "domains of transport and housing" and social services, which would be free. In calling for socialization of the ownership of monopolies,

Amin (2013: 136) maintains that "[m]onopolies are institutional bodies that must be managed according to the principles of democracy, in direct conflict with those who sanctify private property." Public ownership could consist of a number of social arrangements, including state ownership, collective ownership, worker-owned enterprises, and cooperatives. State ownership in term can and already does occur at various levels of government, namely, the federal or national, state or provincial, and the municipal or local level.

It is important to note that public ownership or nationalization of the means of production does not in and of itself constitute socialism, despite the fact that people have often assumed that it does. For example, after World War II the British state nationalized heavy industry that had been in decline for over fifty years, but retained previous owners in managerial positions. According to Rogers (2014: 106), "it appears that the government socialized the costs of post-war reconstruction and British relative economic decline, before returning industries to private hands, which in turn allowed those who were already at the top end of the income pyramid to reap the financial benefits of government's investment in declining industries while sharing the cost." Australia historically exhibited extensive public ownership of various productive forces, not only utilities but also banks, manufacturing operations, communication networks, Qantas Airlines, and transportation systems. Robert Menzies, a conservative prime minister, reportedly stated in 1943, "Few people would have any quarrel with government control of railways, or tramways, or water supply, or such other great public utilities" (quoted in Wettenhall 1965: 428). Even in the United States, some utility companies are publicly owned.

Nevertheless, nationalization or socialization of private wealth would constitute an essential step toward the creation of a democratic ecosocialist society and would reduce the power of the corporate class and wealthy individuals to influence elections around the world through the support of selected candidates via campaign contributions, favorable media coverage, and even bribery. Wall (2010a: 57) maintains that "ecosocialism is founded on the principle of common property rights, which allow individuals [and groups] free access to a resource as long as they don't damage it." What is also needed is preservation and extension of the common property system that still exists throughout much of the world. According to Burkett (2006: 314), "The range of resources that are still being managed as common property is impressive: Southeast Asian wetlands, forests in parts of India and Nepal, fisheries in Japan and other Pacific Island nations, dams in Sri Lanka, canals in India, rangelands in various African countries, grazing lands in Britain and continental Europe, and groundwater in numerous countries."

What needs to be guarded against is the increasing privatization of water resources, supposedly in the form of public-private partnerships under which a small number of multinational corporations, such as Suez, Veolia, Thames, America Water, Bechtel, and Dow Chemical, assert that they are not buying or selling water per se, simply managing its delivery. These corporations along with "U.N. representatives, international development specialists, and heads of state from around the world" converge at the World Water Forum, which occurs every three years (Piper 2014: 2). The corporations are being assisted not only by national governments, but, not surprisingly, the World Bank and the International Monetary Fund. Fortunately, they are being opposed by the French-based Alternative World Water Forum (Forum Alternatif Mondial de l'Eau or FAME), which argues that no one owns water, but that it constitutes a common good.

Increasing Social Equality and Achieving a Sustainable Global Population

The United Nations Development Programme (UNDP) has consistently recognized widespread social inequality in the world. While organizations such as UNDP and the World Bank claim to be committed to the eradication of global poverty, they rarely, if ever, call for a redistribution of wealth or increasing social equality. Instead such organizations, along with mainstream economists and most of the governments in the world, call for more economic growth, never acknowledging that wealth and poverty, development and underdevelopment, are interwoven. While some redistribution of wealth has been achieved under capitalism at various historical junctures and particularly in developed societies with strong labor unions and left-of-center governments, social inequality is an inevitable dimension of the capitalist world system.

Ultimately, a shift toward greater social equality or parity will require transcending global capitalism and moving toward a democratic eco-socialist world system. Socialists have over the years engaged in intense debates about what sort of wage differentials should exist under socialism. Stilwell (2000: 130) argues that a 3:1 ratio of the highest to lowest incomes would be a tolerable standard for a socialist society. In reality, there are other compensations for work than material rewards, such as the intrinsic rewards of intellectual and even physical stimulation and the sense that one has contributed to the greater good. Ozlem Onaran (2010: 19) argues that in order to be socially acceptable, maintaining "long-term economic growth at zero or low levels" will require a "guarantee of high employment and an equitable distribution of income," all of which are incompatible with capitalism. Needless to say, as long as rich people and corporations exist, progressive taxation that does not allow for tax loopholes constitutes an important mechanism for redistributing wealth.

As I note in chapter 1, population growth is largely a symptom of a deeper problem, poverty, which would be eradicated in a postcapitalist or democratic eco-socialist world. As Fred Magdoff (2013: 22) argues, while population can be an environmental problem, it is "usually not the main one, given that economic growth generally outweighs population growth and environmental degradation arises mainly from the rich rather than the poor." Many middle-class environmentalists who posit population growth as the principal environmental problem appear to want to maintain more or less their present material standard of living, albeit on a planet with far fewer people. However, in reality bringing down population growth will entail an improvement in quality of life for the poor, or essentially the elimination of poverty, which from an eco-socialist perspective should go hand-in-hand with creating a high degree of social equality. Authoritarian measures to reduce population growth have historically either met with "open rebellion, as in the case of Indira Gandhi's India, or evasion and subversion, as in the case of Deng's China under the one-child family policy" (Shue 2014: 74).

Socialist Planning and Workers Democracy

Socialist planning is "nothing other than radical democratization of the economy" (Loewy 2015: 27). As part of socialist planning, "Every single fact of industrial production—energy production most urgently, but also transportation, housing, trade, agriculture, manufacture of commodities, and waste production and treatment—all require gigantic systemic change and complete structural reorganization" (Williams 2010: 217).

Thus, workers', economic, or participatory democracy would constitute an integral component in a shift toward democratic eco-socialism. As Joel Kovel (2007: 143) asserts, "We call ecosocialism that society in which production is carried out by freely associated labor with consciously ecocentric means and ends." Workers' democracy entails democratic and open debate at all levels of the worksite.

Wolff (2012: 122–124) maintains that workers' self-directed enterprises (WSDEs) could play an important role in transitioning away from capitalism. Each WSDE would allow its workers to collaborate with the enterprise's board of directors in decision making and would use part of its surplus to pay taxes to various governmental sectors that have supported its operation and would also consider the impact of its operations on the environment. Enterprise managers and directors could elected so that they represent various levels of workers within an organization or directly by the entire work force. Democratic planning needs to be part and parcel of the production process, such as in deciding what goods are needed and whether they are environmentally sustainable.

Michael Albert and Robin Hahnel (1991) have developed an elaborate model of participatory economics, which they call Parecon (a shortening for "participatory economics"), that would entail a network of workplace and consumer-based councils. Parecon seeks to fulfill four key values and is compatible with the notion of democratic socialism:

- Solidarity—a solidarity economy should be based on creating solidarity among workers and consumers;
- Diversity—a good economy would take into account the wide variety of preference and choices that people display both in terms of work and consumption;
- Equity—an equity economy would orient production, consumption, and distribution toward achieving equity;
- Self-Management—a good economy will be highly democratic in that workers have strong input in decision making (Albert 2014: 3–9).

While Parecon has been criticized from various quarters for not being sufficiently sensitive to environmental factors, Hahnel argues that it will require input from environmentalists. "An active environmental movement will be necessary in a participatory economy to argue for the importance of environmental protection and restoration.... Environmentalists will have to speak up in worker and consumer councils and federations, pointing out the true benefits of environmental preservation and the magnitude of environmental degradation" (Hahnel 2007: 72).

Del Weston (2014: 179) argues that authentic or real participatory democracy will require "considerable decentralisation, scaling down and localisation of economic activities," which will also entail more labor-intensive rather than high-tech-energy-intensive production.

Meaningful Work and Shortening the Work Week

Socialism is committed to the notion of unalienated and fulfilling or meaningful work. Satisfying work contributes to positive self-esteem and a sense that one is contributing to society and one's fellow human beings.

Unemployment for most people, at least those in early adulthood and middle age, can be psychologically devastating. Even work or employment for people over the traditional retirement age of sixty to sixty-five, depending on the country in question, can be a fulfilling and meaningful activity. A shorter work week would permit everyone to be employed and thus eliminate the "industrial reserve army," which is an inherent feature of capitalist economies but should not occur in a socialist system. In his conception of eco-socialism, Sarkar maintains that there would exist no unemployment:

> Firstly, labour-intensive technologies would be preferred, not only to provide jobs, but also because such technologies reduce resource consumption and, consequently, have less negative environmental impact. Secondly, even in a steady-state economy, there would be a lot of necessary work to be done. Food, clothing, housing, and services like education, health, postal service, and so on, would have to be produced. This would demand much labour, which would be equitably distributed among all who can work. Thirdly, an eco-socialist government would pursue a policy of stabilising and then reducing the population.... Fourthly, ecologically benign technologies such as repairing, recycling, reusing, manual weeding instead of using pesticides, are all labour-intensive. (Sarkar 1999: 208)

Andre Gorz (1982: 124) asserts that an alternative to capitalism would entail "reducing what has necessarily to be done, whether enjoyable or not, to a minimum of each person's lifetime, and in extending as far as possible collective and/or autonomous activity seen as in itself." Juliet Schor (1991) some time ago wrote about the "overworked American." In her model of plentitude, she advocates what she terms "time wealth": "Millions of Americans have lost control over the basic rhythm of their daily lives. They work too much, eat too quickly, socialize too little, drive and sit in traffic for too many hours, don't get enough sleep, and feel harried too much of the time. The details of time scarcity are different across socioeconomic groups; but as a culture we have a shared experience of temporal impoverishment" (Schor 2010: 103).

What Schor says of American culture applies more or less equally to Australian culture, despite the stereotype of Aussies being a laid-back people. Despite the fact that Australian workers pioneered the eight-hour work day, albeit at a time when the work week was six days rather than five days, in the mid nineteenth century, many full-time employed Australians are now working over eight hours a day. When academics at the University of Melbourne ask each other how they are doing, a common refrain is "busy." By 2005, only 28 percent of Australian employees were working a standard week of between thirty-five and forty hours. In the case of men in full-time jobs, 42 percent were working more than forty-five hours a week, more than 30 percent were working more than fifty hours a week and nearly 15 percent were working more than sixty hours a week. Women in paid work were catching up fast on overworking: between 1978 and 2004 the proportion of women working more than fifty hours a week doubled, to more than 15 percent (Hamilton and Denniss 2005: 86).

It is difficult or impossible to say what would be the optimal work week. To some degree, this would vary from individual to individual. Marx characterized humans as "Homo Faber" or "Man the worker," but he was thinking of unalienated labor where work and play are intricately interwoven, as is often the case in foraging societies. He envisioned a society where one would be able "to hunt in the morning, fish in the afternoon, rear cattle in the evening, criticise after dinner" (Marx and Engels 1978: 160). In a sense working life would never totally end as a person has the mental and physical capacity and desire to engage in it. Thus, people should be given the option of phasing into "retirement" rather than simply going from full-time employment to full-time retirement. Work under socialism and particularly under communism will in essence contribute to human development and allow people to achieve their full potential. While some of this occurs in capitalist societies, it is restricted to a privileged few and has become increasingly hard to attain, as many academics, for instance, who feel themselves under the demands of an increasingly corporatized university structure can testify. In my own case, although I in large part became an academic to escape the corporate world in 1970, the corporate world has invaded the academy in numerous ways since then.

The Need for a Steady-State Economy

A growing number of neo-Marxian scholars as well as non-Marxian scholars have been questioning the economic growth paradigm (Higgs 2014). On the non-Marxian side, E.F. Schumacher, who maintained that "small is beautiful," argues that "infinite growth of material consumption in a finite world is an impossibility" (quoted in Bjerg 2016: 14). Leading eco-socialist John Bellamy Foster (1999: 114) asserts that economic growth has been the "chief reason for the rapid acceleration of the ecological crisis in the postwar period." On this matter, he even quotes Fred Cairncross, environmental editor for *The Economist,* who stated: "Many people hope that economic growth can be made environmentally benign. It never truly can. Most economic activity involves using energy and raw materials: that, in turn, creates waste that the planet has to absorb. Green growth is therefore a chimera" (quoted in Foster 2002: 57).

Even Simon Kuznets, the inventor of the Gross Domestic Product (GDP) metric, admitted that the "welfare of a nation can scarcely be inferred from a measure of nation income," a statement overlooked by governments and most mainstream economists around the world (quoted in Bjerg 2016: 132).

Not all Marxian scholars reject the growth paradigm per se. Some time ago, Barry Commoner (1972) maintained that an increase in economic activity does not necessarily translate into more environmental pollution. In other words, "what happens to the environment depends on how growth is achieved" (Commoner 1971: 141). He maintained that the U.S. soil conservation program during the Depression helped to restore nutrients in the soil and thereby economic growth. Conversely, the "theory of socialist economics does not appear to require that growth should continue definitely" (Commoner 1972: 281). David Pepper argues: "[A]n ecological-communist utopia requires the development of productive forces. ... Eco-socialist growth must be a rational, planned development for everyone's equal benefit, which would therefore be ecologically benign" (Pepper 1993: 197).

Schwartzman (2009: 26) maintains that an uncritical demand for zero growth on the part of environmentalists tends to isolate them from the working class around the world and thus asserts that "red greens should advocate a program of sustainable economic growth." He calls for an eco-socialist transition that would entail economic growth, culminating eventually in *solar communism* and a "steady-state with respect to material production and consumption" (Schwartzman 2014: 238). "Sustainable economic growth" would encompass processes such as solarization, demilitarization, ecosystem repair, carbon sequestration, solar-powered desalinization, and the creation of green cities and green agroecologies. Schwartzman (2014: 238) argues that the "critical metric for economic growth should be its overall ecological and health impacts with respect to artificial and natural environments, including its carbon footprint, not simply the level of material production." He maintains that solar and wind power would contribute to the decarbonization of global energy supplies and would greatly reduce the need for mining (Schwartzman 2014: 239).

Conversely, Paul Burkett maintains: "Historically, ever since Malthus, Marxists have been suspicious of any theory that posits purely natural limits to production and development. The reason is obvious: such theories tend to embody a conservative bias against all efforts to improve the human condition by fundamentally transforming class and other power relationships, or even by redistributing wealth and income" (Burkett 2006: 5–6).

In reality, this type of response has been an overreaction that has placed socialists out of touch with serious ecological considerations. A serious redistribution of the world's resources would ensure an adequate living standard for everyone on the face of the planet, but this would entail a serious discussion of how much is enough and, with the elimination of poverty, the recognition that global population would begin to dwindle, thus placing less strain on the ecosystem.

Although not a Marxist per se, Juliet Schor (2010: 170) maintains that the growth paradigm has "increasingly shown to be courting ecological disaster," noting that "perhaps simplicity is an evolutionarily superior trait." Even within capitalist economics, small mom-and-pop stores in the United States or what are termed "milk bars" in the Australian context may prosper without necessarily growing. Schor argues: "Each year it has costs and revenues, and the difference between the two is profit (or loss). The owners can use their profits to upgrade their machinery or software, or lower costs in other ways. But they don't face a growth imperative. If the business is generating a decent living, it can operate at the same size for years" (Schor 2010: 170–171).

Small businesses may for a period of time escape the growth imperative but they face the continual pressure of eventually being taken over by larger companies that may decide to enter their niche market. As Marx repeatedly noted, there is a tendency for competitive capitalism to evolve into monopoly capitalism, a process in which, metaphorically speaking, the big fish devour the little fish.

In recent years a degrowth project has emerged under the leadership of Serge Latouche and other scholars that asserts that global capitalism can be restructured so that it ceases economically expanding and reaches a steady state. According to Latouche (2012: 74), "To exit from the impasse of a growth society involves finding ways of building an alternative of voluntary sobriety and frugal abundance." However, Takis Fotopoulos views the degrowth objective within the contours of the capitalist world system as entirely utopian because: "[O]nly a society which would have first eliminated inequality could meaningfully discuss de-growth as an objective, after it has already met the basic needs of its citizens. ... But, such a society is completely incompatible with an economy based on the market economy" (Fotopoulos 2010: 120).

Obviously, there are large sectors of developed societies and smaller sectors of developing societies that need to undergo degrowth, but the abjectly poor of particularly the developing countries, but also the developed countries, such as homeless people or indigenous peoples living on reservations in North America and Australia, need to undergo some sort of growth or development in terms of access to nutritious food, decent housing and sanitation, health care, and education. Large sectors of developing countries need to undergo growth that is directed to meeting "real needs such as access to water, for, health care, education, etc." (Foster 2011: 32). Ultimately issues of growth and degrowth and development and underdevelopment are intricately interwoven with the redistribution of resources and a drastic restructuring of the social relations of production, which would be an integral component of creating democratic eco-socialism.

Herman Daly and John B. Cobb, Jr. (1990: 71) make a distinction between growth and development: "'Growth' should refer to quantitative expansion in the scale of the physical dimensions of the economic system, while 'development' should refer to the qualitative change of a physically non-growing economic system in dynamic equilibrium with the environment." Growth entails utilizing more and more resources as part and parcel of the capitalist treadmill of production and consumption. Daly and Cobb go on to argue: "Any physical subsystem of a finite and non-growing earth must itself also eventually become non-growing. Therefore, growth will become unsustainable eventually, and the term 'sustainable growth' would then be self-contradictory" (Daly and Cobb 1990: 72).

Development would entail providing people with adequate food, clothing, shelter, health care, and education. Noam Chomsky (2012: 84) maintains that the notion of growth is problematic if it entails "constant attacks on the physical environment that sustains life—like, for example greenhouse emissions, destruction of agricultural land, and so forth." Instead he advocates growth that leads to simpler lives, more livable communities, and maximization, what he terms "growth in a different direction" (Chomsky 2012: 84).

Beyond a certain point, more food, clothing, and shelter are superfluous and certainly environmentally unsustainable. How much health care is necessary would depend on each individual's physical and mental state, which are not only interwoven but highly variable. From a critical medical anthropological perspective, health can be defined as "access to and control over the basic material and nonmaterial resources that sustain and promote life at a high level of individual and group satisfaction" (Baer, Singer, and Susser 2013: 5). In a socialist society or a society seeking to construct socialism, there would be greater emphasis placed on preventive health care than curative health care.

Energy Efficiency, Renewable Energy Sources, Appropriate Technology, and Green Jobs

A crucial question is how much energy, regardless of the source, humanity needs. Given the demands of global capitalism to continually expand, under a business-as-usual scenario, humanity will need more and more energy in order to feed the treadmill of production and consumption and population growth. Energy utilization varies widely from region to region and country to country. In 2004, Asia and Oceania accounted for 31 percent of the global total annual consumption, not surprising given that this region includes the two most populated countries in the world; North America for 27 percent, a disproportionate amount given that this region accounts for only about 5 percent of the global population; Europe for 19 percent; Eurasia for 10 percent, the Middle East for 5 percent; Central and South America for 5 percent; and Africa for a mere 3 percent (Burman 2007: 20). Energy use in 2008 in Canada was 96,000 kilowatt hours per capita; 89,000 in the United States; 75,000 in Australia; 48,000 in the European Union; 19,000 in China; 4,000 in Tanzania; and 3,500 in Nepal (Chivers 2009: 103). Bear in mind, in all of these countries, there are class differences in energy utilization. Klare (2008: 33) observes that the "worldwide requirement for primary energy is expected to rise by 57 percent between 2004 and 2030," with much of the growth slated in Asia, particularly for China and India.

In a steady-state economy, energy requirements could theoretically level out or even eventually decline. According to Mark Diesendof (2014: 23), an "ecologically sustainable energy system is one that is compatible with, and preferably facilitates, the development of an ecologically sustainable and socially just society." Theoretically, his definition of an "ecologically sustainable energy system" on the surface appears to be compatible with democratic eco-socialism. In reality, Diesendorf (2010/2011: 104) appears to be a green social democrat who adheres to a program of ecological modernization as is evident in his assertion that "[s]ince technological change appears to be easier, faster and already under way, there is a case for focusing technological change while working at a more measured rate to stop growth in population and consumption per person and to redistribute wealth."

Energy efficiency is often hailed as a "cost-free tool for accelerating transition to green-energy economy," but in reality due to the Jevons paradox or the "rebound effect," increased efficiency in capitalist countries is associated with increased economic growth and consumption, thus in essence canceling out the benefits of energy savings (Owen 2012: 100). This is not to say that energy efficiency is not a desirable goal, but in order to ensure environmental sustainability, it has to be coupled with a steady-state or zero-growth economy, which would be part and parcel of a democratic eco-socialist world system.

A shift to renewable energy sources, particularly solar, wind, geothermal energy, and possibly ocean wave energy constitutes a significant component of climate change mitigation. Foster (2009) identifies two types of ecological revolutions: an eco-industrial revolution and an eco-social revolution. The first of these entails the adoption of technological innovations or ecological modernization that includes improved energy efficiency and the adoption of renewable energy sources, an approach that has been hailed by a number of corporations and governments, particularly in northern Europe and most segments of the environmental and climate movements. The second of these, which Foster favors, "draws on alternative technologies when necessary, but emphasizes the need to transform the human relations to nature and the constitution of society at its roots within the existing social relations of production. This can be accomplished only through a process of sustainable human development. This means moving decisively in the direction of egalitarian and communication forms of production, distribution, exchange, and consumption, and thus breaking with the logic of the dominant social order" (Foster 2009: 12–13).

More recently Foster has delineated two phases in a socio-ecological revolution. The "ecodemocratic phase" instigated by a broad-based radical movement would entail the following policies or goals:

- A carbon fee and dividend system
- A ban on coal-fired power plants and unconventional fossil fuels, such as tar sand oil
- A drastic shift to solar, wind, and other renewable energy sources
- A moratorium on economic growth in developed countries
- A new international climate regime modeled on the egalitarians and eco-centric principles delineated in the People's Agreement of the World People's Conference on Climate Change convened in Bolivia in 2010 (Foster 2015: 8).

The eco-socialist phase would entail "establishing more egalitarian conditions and processes for governing global society" accompanied by "requisite ecological, social, and economic planning" (Foster 2015: 8). Foster (2015: 8) believes that continuing ecological degradation and socioeconomic hardships will increase the numbers of an already-existing "environmental proletariat," drawing on so-called middle-class people and "some of the more enlightened elements of the ruling class" who forsake their class interests for the sake of humanity and the planet.

A planned centralized economy can facilitate the transition to renewable energy production. In his call for a transition to a "solar-power civilization," Schwartzman (2009: 19) concedes that fossil fuels, possibly preferably natural gas, will be needed to develop as a temporary measure for the "creature of an alternative infrastructure." Conversely, Diesendorf (2014: 75) asserts that all or most of the electricity in both the United States and Australia could be theoretically derived from three renewable energy sources, namely, concentrated solar thermal, solar photovoltaic, and wind, especially given that all of them are presently commercially available. Britain could generate most of its future electricity from wind and wave energy, and "could possibly import solar and wind power from North Africa by transmission line via continental Europe" (Diesendorf 2014: 75). While it is unclear to what degree China and India could phase out coal in the near future, both countries "have high potential for wind, PV, CST and, with large environmental and social impacts, hydro" (Diesendorf 2014: 75). Indeed, China has the highest renewable energy capacity in the world at 103.36 GW (Gigawatts or billion watts), followed by the United States at 57.99 GW, Germany at 48.86 GW, Spain at 27.78 GW, Japan at 25.96 GW, India at 18.65 GW, Italy at 16.66 GW, Brazil at 13.84 GW, and France at 9.57 GW (Dorling 2013: 25). In the period 2005–2010, China increased its renewable energy capacity by 106 percent, South Korea by 88 percent, Turkey by 85 percent, Germany by 67 percent, Italy by 45 percent, Japan by 45 percent, Brazil by 42 percent, France by 42 percent, and Spain by 42 percent. Makhijani and Ochs report: "Renewable technologies broke all growth records in recent years. In 2011, new investments in renewables topped those in conventional energy technologies for the first time in modern history. U.S. wind power capacity almost tripled and solar energy jumped ninefold since 2007. And 17.1 percent of Germany's electricity comes from renewable sources" (Makhijani and Ochs 2013: 85).

Solar photovoltaic cells and panels operate the best in sunny locations and have the potential to provide local power in remote areas, such as much of sub-Saharan Africa or even a developed society such as Australia. The silicon that is used in solar panels has tended to make them expensive but Japanese technology has produced super-thin panels that makes them more affordable. Solar thermal plants in spacious sunny areas, such as deserts, can concentrate light onto a central point from which electricity can be generated. They first came into operation in California in the 1980s but now exist in many countries. For example, in

Mongolia, some pastoral communities have their tents or yurts powered by solar panels. The Hongqiao railway station in Shanghai, the largest solar panel installation in the world, potentially can generate more than 8 million kilowatts per annum (Chambers 2011: 129). Worldwide, solar power utilization has undergone about a 40 percent increase over the past decade and some experts estimate that it could account for 10 percent of the energy supply by 2025 (Chambers 2011: 64).

Large wind farms operate very efficiently in offshore locations, such as the Baltic Sea in Europe or the Bass Strait and Southern Ocean of Australia. Reportedly Britain has the "largest offshore wind resource in Europe, capable of generating three times the nation's electricity requirements" (Godrej 2002: 134). Wind power has been on the rise in numerous countries, including Germany, China, the United States, Spain, and India. Greenpeace International and the Global Wind Energy Council project that "wind energy could supply up to 12 percent of global electricity by 2020 and more than 20 percent by 2030, given effective policies from governments" (Diesendorf 2014: 45). Geothermal energy as a renewable energy source already exists in several volcanic regions, such as Iceland, El Salvador, Kenya, the Philippines, and Costa Rica. However, it can potentially be derived from hot granite rocks far below the earth's surface. The possible impact of geothermal plants on tectonic activity, water aquifers, and wildlife should be taken into consideration (Williams 2010: 155).

While acknowledging their potential usefulness, various scholars have observed that renewable energy sources are not a panacea for mitigating climate change (McCluney 2005; Trainer 2007). The creation of solar power plants, solar panels, wind farms, and geothermal planets will require the consumption of large amounts of non-renewable energy and non-renewable resources (Sarkar and Kern 2008:20-21). In other words, renewable energy sources along with other green technologies incur *embodied* energy, the accumulated energy required to process and manufacture a product and to transport it to the site where it is utilized. Owen (2012: 187) observes that solar power plants and wind farms will require in part "vast installations, covering huge tracts of land," but theoretically this requirement can go hand-in-hand with the installation of photovoltaic panels on the tops of houses, apartment, factories, and office buildings. In essence, nonrenewable resources, such as steel and aluminum, which may rely on coal to be processed, would be required to manufacture the infrastructure for renewable energy sources such as solar and wind power. Compared with other renewable energy sources, geothermal electricity production requires a relatively high level of water consumption. The magnet generators used in the latest wind turbines require neodymium, a rare earth metal.

On the positive side, photovoltaic panels have begun to provide electricity to remote communities in developing countries that are not served by the conventional electricity grid. A downside of photovoltaic panels is that they have a life-span of only about twenty years. Another problem with solar and wind energy is their intermittency, although this would not apply to wave power. The former is available as long as the sun shines and the latter so long as the wind blows. Fortunately, the "costs of new RE sources are declining and some are projected to be competitive with fossil fuels within a few years or decades, given appropriate government policies during the transitional period" (Diesendorf 2014: 89). Off course, appropriate government policies would be much more likely under scenarios where the means of production, particularly energy production, were socialized rather than primarily privatized as is the case in most countries around the world. Conversely, while Germany's *Energiewende* or energy revolution has often received much publicity, Boisvert reports: "Despite massive construction of new capacity, electricity from renewables, especially from wind and solar, grew at a sluggish rate [in 2012]. Germany is indeed avoiding blackouts—by opening new coal- and gas-fired plants.... The spiralling cost of the renewables surge has sparked a backlash, including government proposals to slash subsidies and deployment rates. Worst of all, the *Energiewende* made no progress at all in clearing the German grid of fossil fuels or abating greenhouse emissions—nor is it likely to for at least a decade longer" (Boisvert 2013: 62).

Ultimately, the deeper question that renewable energy enthusiasts seldom ask is "how much energy is needed in the first place," particularly in developed countries (Chambers 2011: 176). A large-scale transition to solar, wind and other renewable energy sources will need to be coupled with a decline in per capita level of consumption among the affluent of the world while allowing the poor to draw on these new energy sources to achieve access to basic resources. Obviously some people in the world, particularly the poor in the developing world, desperately need access to more energy but many of affluent in both the developed and developing worlds need to reduce their energy consumption, often drastically, in order to achieve environmental sustainability and a safe climate. A shift to renewable energy sources will require an integrated approach in order to grapple for example with issue of intermittency in particular with solar and wind energy. Finally, as Makhijani and Ochs observe, "Renewable energy developments should also be in complete accord with priorities for sustainable water use to avoid large diversions of water from natural systems and to preserve scarce resources for human needs. Water scarcity already affects around 1.2 billion people globally, almost one fifth of the world, and an additional 500 million people are at risk of scarcity" (Makhijani and Ochs 2013: 97).

Aside from the matter of renewable energy sources, eco-socialism needs to grapple with developing what Wallis (2000) terms a "socialist technology." Indeed, Marx gave much attention in his numerous writings to technology as an integral part of the way that humans interact with nature and the "role of machines and machinery for the emergence of capitalism; likewise, he was interested in the technological basis of communist society" (Grundmann 1991a: 110). The component parts of a socialist technology to some extent already exist in capitalist societies but are not actively promoted by capitalism because they are not as profitable. The technology already exists to make products that endure for a long time rather than ones that are manufactured in such a way that they will break down fairly quickly, a case of built-in obsolescence. Bicycles, smaller cars, and trains, trams, and buses, as opposed to large cars, all could be part of a socialist or an appropriate technology. While apologists for capitalism often argue that it promotes continual innovations, not all innovations work for the overall benefit of humanity, as can be seen in the develop of nuclear weapons and military hardware of all sorts, ranging from tanks to jet fighters and drones. Socialist technology would entail the ability to repair things when they break down rather than disposing of them and buying a newer version of the same things. As Jones (2011: 61) observes, "Once freed from the growth imperative of capital accumulation, technology can be employed to use resources more efficiently, and thus overcome capitalism's Jevon's paradox where technological efficiencies are used to stoke the fires of even greater resource use."

A shift to democratic eco-socialism will entail the creation of "green jobs," ones that are not only environmentally sustainable but also cater to people's social, educational, and health-care needs. Ironically, global capitalism does create many jobs and working-class people often feel threatened by the loss of their jobs if there is a shift away from fossil fuels to renewable energy sources, although the latter will most definitely create new jobs, but also away from an endless treadmill of production and consumption, both of which do create jobs, both in terms of producing things and then selling them. The advertising and retail industries in themselves create numerous jobs that to a greater or lesser extent allow workers to at least meet their subsistence needs. Onaran (2010: 30) calls for "public expenditures in labour-intensive services like education, child care, nursing homes, health, and community and social services." The creation of green jobs must be accompanied by a "just transition," which means retraining displaced workers from obsolescent and environmentally destructive industries and enterprises to environmentally sustainable ones.

Sustainable Public Transportation and Travel

Referring to a society with the greatest number of cars in the world, Bohren (2009: 377) correctly argues that the car as a "symbol of the American ethos of individualism, personal freedom, and mobility, shaped twentieth-century America." In recognizing the maladaptive nature of the car in the twenty-first century, she asserts that "[c]hange is needed in the human sociocultural practices that have accompanied the adaptation to the car in order to reduce its contribution to global climate change" (Bohren 2009: 378). In his book *Ecotopia*, Ernest Callenbach (1975: 35) describes a fictional place situated in northern California, Oregon, and Washington state that has transcended cars. Aside from the question of whether such as place could exist in the modern world, as Peter Newman (2009: 108) observes, "The biggest challenge in an age of radical resource-efficiency requirements will be a way to build fast rail systems for the scattered car-dependent cities."

Environmentalists and other social activists began to challenge the pollution, health hazards, traffic congestion, urban sprawl, and fragmentation of social life resulting from motor vehicles and highways in the 1960s and 1970s, a period of social ferment on many fronts around the world (Golten et al. 1977). According to Ladd, "Anti-freeway activists joined lovers of city life, conservationists (soon to be much more numerous and known as environmentalists), urban politicians, and a growing number of transportation planners in promoting a revival of mass transit during the 1960s. The car, they believed, was reaching the limits of its usefulness, even in the suburbs" (Ladd 2008: 133).

Newman and Kenworthy (1999: 144–189) propose five policies for overcoming automobility dependency: (1) traffic calming in which speed plateaus, bottlenecks, and other strategies are employed to slow down traffic in order to make streets safer, particularly for pedestrians, cyclists, shoppers, and residents; (2) the construction of quality transit systems as well as bike and walking paths; (3) the development of "urban villages" or multimodal centers with mixed, dense land use; (4) growth management to counter the urban sprawl; and (5) increasing taxes on motor vehicle transportation. Belsky delineates the following components of a sustainable transportation system:

- "Subsidies for public transport, cycling, and affordable housing close to public transport"
- "Modernization of roads with real-time traffic management and operations"
- "Road space protected for pedestrians, cyclists, public space"

- "Bus rapid transit or rail in high-demand corridors"
- "Public-transport-oriented development"
- "Stronger governance structures for transport and land use policy, planning, and management"
- "More equitable access for the poor, disabled, young, and old" (Belsky 2012: 54).

In my view, his scheme still leaves too much wiggle room for roads.

Modern cities have evolved following, in large part, the dictates of capital with its need for manufacturing, financial, commercial, distribution, and communication centers, as well as state bureaucracies. In cities where significant public transportation infrastructure exists but has not been developed and upgraded sufficiently to discourage car use, the growing proliferation of cars also reduces the efficiency of road-based forms of public transportation. In Melbourne, for example, buses and trams are slowed greatly by congestion caused by cars. Fortunately, a few Melbourne trams operate as light-rail conveyances in some motor vehicle–free stretches between the city or central business district and various suburbs.

Newman and Jennings (2008: 45) report, "Cities that are car dependent spend between 15 and 20 percent of their wealth just on getting around, whereas transit-oriented cities spend only 5 to 8 percent of the wealth on transport." Consequently, Register (2001) maintains that cities should be designed for people, not cars. Furthermore, he proposes the notion of *pedestrian cities* in which people will not need cars and will be able to walk, cycle, or take public transportation to get around. In this light, Grongingen, a city of some 170,000 in the Netherlands, removed the roads in its central business district in 1992 and adopted various policies that promote cycling (Korten 2001: 256).

Despite the existence of massive corporate support for the ongoing use of motor vehicles, there have been some counterhegemonic efforts to resist the auto-mobilization of society by emphasizing the need for people to rely on other forms of transportation. Peter Newman and Christy Newman report: "Recent data from US and Australian cities that car use per person has been going down since 2004, and that public transport has been dramatically increasing. The first assessments of this suggest that something structural is happening, that younger people in particular are coming back into cities rather than choosing car dependence. Indeed car ownership among teenagers in the United States has dropped from fifteen million to ten million. Is it possible that a change in car culture is underway?" (Newman and Newman 2012: 360).

Unfortunately, most of the efforts to make cities greener have benefited the affluent much more than the poor and working-class people. A new urbanism that seeks to make cities more liveable and environmentally sustainable has emerged around the world and has begun to permeate urban planning. Various cities around the world, including Singapore, Hong Kong, Zurich, Copenhagen, Freiburg (Germany), Vancouver, Toronto, and Boston, are encouraging residents to rely more on public transportation, including trains, trams, and buses. While the United States remains a highly car-dependent country, there are some signs that public transit has started to exhibit a modest upswing. Lane reports: "Conventional bus service has been improved in many cities, and new modes of transit such as express bus and light rail have enhanced this serviceability and image of transit operations. Data from the American Public Transportation Association ... indicates that mass transit has been increasing in aggregate counts of unlinked passenger trips in the United States since the mid-1990s, while VMT [vehicular motor transportation] seems to have slowed to a halt and, since 2007, actually decreased" (Lane 2013: 110).

While most U.S. cities constitute a "sea of auto-dependent sprawl" in contrast to most European cities that have efficient and highly utilized public transit systems, it is important to note that the "few American cities, such as New York, San Francisco, Chicago, or Boston, that have good public transportation systems and accompanying density have per capita fuel/energy consumption and GHG emission that are not only lower than other parts of the United States, but in some cases lower than those in Europe" (Dutka 2013: 128). In these U.S. cities, people tend to live in more compact living quarters and rely less on cars for transportation than in most U.S. cities. Conversely, some European cities, such as Madrid, appear to have embraced some of the suburban sprawl characteristic of most large U.S. cities.

A global movement to make inner cities car-free has emerged in recent years. Sustainable transportation would entail many other measures, such as limiting the use of cars as much as possible, making them smaller and more energy efficient, and even banning four-wheel drives or sports utility vehicles (SUVs), except in special circumstances (such as in the outback and rugged mountainous areas) and drastically limiting air travel. Electric cars are often offered as a more environmentally sustainable form of transportation. This might be the case if they derive their power from renewable sources of energy but not necessarily if they derive their energy from coal-fired power plants. Furthermore, electric cars will not solve congestion problems and the need to build and maintain roads, which requires an enormous amount of concrete, the manufacture of which produces CO_2 emissions. While shifting from cars to public transit, particularly intercity and suburban trains and trams or light-rail systems, would serve to diminish greenhouse gas emissions, these modes of transportation are not a panacea. While there is much discussion about high-speed passenger rail substituting for cars or plane travel, according to Todorovich and Burgess (2013: 145), various studies indicate that the "direct benefits from high-speed rail in terms of overall energy and emissions may be modest." Additionally, "these analyses also neglect the indirect impacts in terms of land-use and city-entering, which may be large and are difficult to measure and attribute. It is our view that these indirect benefits may be more important than any direct reduction in energy utilization as passenger choose high-speed rail over alternative travel modes" (Todorovich and Burgess 2013: 145).

Much thought is being given to the best form of public transportation, such as train, tram, or bus, in urban areas, depending on the situation. Furthermore, there is the issue of connecting small towns and rural areas with cities.

In Europe, many villages are relatively well-connected to urban areas, but this is not generally the case in North America and Australasia. Thus, given that public transportation is often infrequent or nonexistent, most people have become dependent on cars to connect them with commercial centers, family, friends, and acquaintances. Diana Young (2001) observes that motor vehicles appear to be the most significant Western consumer item for the Anangu people in the South Australian outback in that in their remote community they have multiple purposes, including providing relatives lifts to shop and work, pick up the mail, and pay bills, and constitute an essential component of fulfilling social obligations. Likewise, the car or the pickup truck serves a similar purpose for Native Americans living on remote reservations, as any visitor to such communities will observe. Measures will need to be taken to connect rural communities to urban communities and to provide public transportation, perhaps in the form of regularly scheduled minibuses in rural areas. Furthermore, it would be possible to reinstate passenger rail service that serviced rural communities in both North America and Australia at a time in the past when their respective populations were considerably less than they are today.

For a variety of reasons, including poor scheduling and limited routes, buses are frequently underutilized. For example, as a frequent public transit user in Melbourne, while I have found that trains and trams are well-used, often requiring some passengers traveling at peak hours to stand, buses, which often serve to connect train and tram lines, are generally greatly underutilized. According to Paul Mees (2010: 38), "A bus with half a dozen passengers will be no more efficient, in greenhouse terms, than if the passengers travelled in cars at average occupancies." Conversely, various Latin American cities, such as Curitiba in Brazil, have created rapid, low-cost, high-frequency bus systems that utilize bus-only lanes and feeder buses that link with the larger bus network (Metz 2010).

Many cities also are increasing provisions for cycling and walking, including Canberra, the Australian national capital, as forms of active transport. Copenhagen has created bicycle right-of-ways and has fostered an ethos of respect for cyclists. Schiermonnikog, a national park off the northwest coast of the Netherlands, does not allow visitors to drive their private automobiles onto the island, thus meaning that "[c]yclists, local buses, taxis, pedestrians and the occasional car of one of the island's inhabitants share the streets without any problems" (Peters 2006: 128). In terms of countries, Switzerland, admittedly a small country, has the most extensive public transportation system in the world, one that connects urban and rural areas, including steep mountainous terrain. Zurich reintroduced trams "at a time when they were disappearing from many German cities, and new stretches of railway were laid when services were closing down in many other countries" (Welzer 2012: 113). Whereas the average person in EU countries makes 14.7 train trips a year, the average Swiss makes 47 train trips a year.

In capitalist societies, "time is money," and this dictates rapid movement between places. Conversely, in a more leisurely paced world based on eco-socialist principles, people might find slower train travel—although faster than presently exists in most parts of North America and Australia—to be a time to slow down by reading, chatting with fellow passengers, enjoying the passing countryside, reflecting, and even sleeping. A more sustainable form of vacationing or holidaying would entail trips much closer to home, by train or bus, if possible, rather than to distant places either by plane or car. Cheap package holidays by airplane could become things of the past (Sims 2009: 8). In addition to minimizing flying, a simpler way would also entail a disposal of or minimizing the use of private motor vehicles and reliance on alternative modes of transportation, including simple walking and cycling. Many of us, particularly those of us in the developed world, would do both humanity and other life forms on the planet a favor if we, as Katarine Alvord (2000) advices, were to divorce our cars.

Globally, cities are not only connected by highways, railway lines, and marine shipping lanes, but airplane travel, for business, social visitations, and pleasure. The growing concern about the contribution of greenhouse gases emitted by airplanes and the accompanying increase in airplane flights globally has prompted discussion about the possible revival of airships that could be powered by a hydrogen-helium mixture (Sim 2010: 92). Airships would constitute a form of slow travel given that they travel at speeds of 150 to 200 kilometers per hour. They can also "carry large loads with one-tenth the fuel of aircraft technology" (Newman 2010: 183). Furthermore, there is a need to develop airplanes that rely on solar power rather than jet fuel.

According to Gilbert and Perl (2010: 358), transoceanic ships "could make considerable use of wind power through the use of kites or solid sails." A more suitable form of vacationing would entail trips much closer to home rather than to distant places. Teleconferencing also has the potential to eliminate or reduce much air travel for the purpose of conducting business or attending conferences.

Sustainable Food Production and Forestry

Hannah Reid provides a succinct definition of "sustainable agriculture": "Sustainable agriculture is a method of farming based on providing for human needs for food, income, shelter and fuelwood. It also builds an understanding of the long-term effect of our activities on the environment. It integrates practices for plant and animal production with a focus on pest predator relationships, moisture and plants, soil health, and the chemical and physical relationship between plants and animals on the farm. Such agroecological approaches consider not just productivity but also system stability, sustainability and issues relating to equity and fairness" (Reid 2014: 45–46).

A shift in food production away from heavy reliance on meat, particularly livestock, to organic farming, vegetarianism, and even veganism would be more environmentally sustainable and an important form of climate change mitigation. At the present time, however, humanity by and large is heading in the opposite direction as the rising middle classes in the developing world develop an appetite for meat and dairy foods. Eckard et al. report: "By 2050, world demand for all meat sources is projected to increase from 269 million tonnes to 464 million tonnes, with approximately 85% of this demand coming from developing countries. Between 1997 and 2007 global populations of cattle, sheep, dairy cattle, pigs, goats and poultry increased by 4, 5, 12, 20, and 26%, respectively. Between 1973 and 2003, meat consumption per capita in developing countries increased by 160% compared to 22% in developed countries. Similarly, milk consumption increased by 55% and 7% in developing and developed countries, respectively" (Eckard et al. 2013: 145).

Factory farming as a way of raising and slaughtering animals has diffused from Europe and North America to much of the developing world, including Brazil, Malaysia, the Philippines, and Thailand (Nierenberg 2006: 24).

In 2006, factory farming produced 74 percent of the world's poultry, 50 percent of the pork, 43 percent of beef, and 68 percent of eggs (Nierenberg 2006: 26). Livestock produced in feedlots are fed grain, particularly corn and soybeans. Seventy percent of the corn harvest in the United States is fed to livestock and globally, almost 80 percent of soybeans are used to feed animals (Nierenberg 2006: 30). Animal production requires massive amounts of water and petroleum for growing, feeding, transportation, and processing. Livestock production releases methane, a powerful greenhouse gas, into the atmosphere.

While ocean fishing stocks are being depleted, humanity continues to increase its consumption of fish. Many deep sea fish, such as orange roughy and sablefish, are in danger of extinction due to overfishing. State-of-the-art technology allows fishermen to trawl fish from deep waters, often totally unregulated by international laws. Aquaculture production or fish farming has been on the increase globally, accounting now for about 50 percent of global fisheries, and relies on "wild fish further down the food chain for feeding purposes (Steffen et al. 2015: 10).

Drastic reduction of current forms of meat consumption and dairy production would greatly decrease emissions from food production and health problems as well. Weis (2013: 12) argues that there is a strong need to deindustrialize livestock production and shift away from meat diets in general as part and parcel of an effort to create a "more sustainable, just, and humane world." However, the culling of kangaroos, wild horses, camels, and rabbits in a place like Australia could theoretically serve as a more sustainable form of meat production. Raising livestock on grass would be more environmentally sustainable because it "helps preserve native grasses and control erosion, and it eliminates the need for pesticides," however, only if done in such a way that overgrazing is avoided (Nierenberg 2006: 36). Conversely, some leftists of different political stripes, Marxists, anarchists, and Fabian socialists historically "have called for the total abolition of animal exploitation and adoption of worldwide vegetarianism" (Gunderson 2011: 415).

Small-scale organic farming tends to be more fuel efficient than industrial agriculture, which relies heavily on petroleum, chemical fertilizers, and pesticides. All farming requires water, but livestock production requires much more water than does growing crops. Given that about 80 percent of the world's water is used for agriculture, organic cultures "use much less water than agribusiness, and eating locally grown foods shrinks export-oriented agribusiness and keeps water in the country" (Piper 2014: 224). Much water is used to grow grains, including corn, which in turn feeds livestock and poultry but is also used in processed foods.

There is a strong need to shift toward agroecology, which relies on the extensive knowledge of farmers of local ecosystems and seeks to transcend dependence on chemical, oil-based agriculture. Crops such as maize, wheat, sorghum, millet, and vegetables can be grown in forested areas that "provide shade, improve water availability, prevent soil erosion, and add nitrogen—a natural fertilizer—to soils" (Nierenberg 2013: 194). Ducks are used in Japan for pest control in rice paddies, thus preventing the application of expensive chemical fertilizers and pesticides. Agroforestry blends trees and shrubs with perennial crops and the production of cattle, poultry, and other animals. In the case of alley cropping, "grains or other non-woody crops are planted in strips between rows of nut, fruit, timber, or fodder trees" (Bates and Hemenway 2010: 51). The Coalition for Rainforest Nations campaigns for cash incentives to be offered to developing countries if they agree to conserve their forests (Simms 2009: 10). Permaculture, which is a contraction for "permanent agriculture," a term coined by Australians Bill Mollison and David Holmgren, seeks to "integrate concepts from organic farming, sustainable forestry, no-till management, and the village design techniques of indigenous peoples" (Bates and Hemenway 2010: 52). A shift toward vegetarianism could reverse deforestation for cattle production in the Amazon Basin with most of the meat being not consumed by Latin Americans but by Europeans and North Americans.

There is an urgent need to expand on the urban farming that already exists in many parts of the world, particularly the developing world. An estimated 90 percent of the farm produce consumed in Havana is grown either in the city or in its immediate hinterlands. Halweil and Nierenberg (2007: 572) maintain: "For cities confronted with growing waste disposal problems—which includes virtually all cities—the strongest environmental argument for local farming is the opportunity to reuse urban organic waste that would otherwise end up in distant, swollen landfills." Laws that prohibit farming in cities need to be repealed. Much urban farming can be done on rooftops, perhaps coupled with strategic placement of solar panels. A danger with urban farming is that it has the potential to serve as a reformist reform rather than a nonreformist or system-challenging reform in that it may fill in the gaps left by neoliberal state retrenchment, unless it is part and parcel of an effort to drastically restructure society, which it has the potential to do given that at some level it has rejected the capitalist food production system.

Despite the horror stories associated with the enforced collectivization of agriculture in the Soviet Union during the Stalinist era, Sarkar asserts that the notion of collective agriculture needs to be revisited for a number of reasons, including economies of scale, particularly if it is based on decentralized planning rather than centralized planning that would not account for regional variation within a country. Furthermore, "In the transition period, and more so in the low-level steady state, the quantity of machines and equipment available to agriculture as a whole would also go down, so that several family farmers (on leased land), if that were the future system, would have to share each item. They would have to buy them jointly or hire them, when needed, for some state or community organization. In either case, they would have to work out a time plan for using them" (Sarkar 1999: 227).

Collective farming would have to find a happy median between economies and diseconomies of scale, the latter of which were often present in the USSR.

Resisting the Culture of Consumption and Adopting Sustainable and Meaningful Consumption

Obviously, all humans need to consume a certain amount of food, clothing, and shelter in order to sustain themselves. Capitalism, however, converts "wants" into "needs" through voluminous and alluring advertisements and as a compensation for alienation in the workplace and everyday social life. From an eco-socialist perspective, Magdoff and Foster (2010: 26) argue that an "economic system that is democratic, reasonably egalitarian, and able to set limits on consumption will undoubtedly mean that people will live at a significantly lower level of consumption than what is sometimes referred to in wealth countries as a 'middle class' lifestyle (which has never been universalized even in those societies)." Shor (2010: 130) advocates the "art of slow spending."

One place where there has been some resistance to the culture of consumption is the new German states of united Germany, in other words, the territory of the former German Democratic Republic. Based on interviews with 20 East Germans ranging in age from twenty-seven to seventy-eight, Albinsson, Wolf, and Kopf detected consumer resistance to hyper-consumption. They report:

> [O]ur informants continue their struggle to accept the individualistic ideal and the new philosophies of hyperconsumption and throwaway-ism. They were brought up in the socialist paradigm that promised a rising standard of living through collective responsibility. Economic necessity required them to save, to repair and not to dispose of items that could be used in the future.... With the transition to a demand

economy, this necessity vanished, but many consumers still choose to integrate their cultural values and refuse to participate in wasteful consumption (Albinsson, Wolf, and Kopf 2010: 416).

Unfortunately, at least in developed societies, resistance to the culture of consumption remains confined to small niche groups. Jonathan Neale (2008: 45) warns climate activists not to talk about sacrifice to ordinary people, and one this score I am fully in agreement with his assertion that "they will find themselves without the support of ordinary people." My comments of resisting the culture of consumption are directed primarily to the affluent, even the affluent in the working class, who turn to consumerism as a compensation for alienation in the workplace and in everyday life in developed societies. Conversely, eco-socialists do not agree among themselves as to how much is enough to maintain a comfortable lifestyle. Pepper, for example, states: "The fact that in socialist development people continuously develop their needs to more sophisticated levels does not have to infringe this maxim. A society richer in the arts where people eat more varied and cleverly prepared food, use more artfully constructed technology, are more educated, have more varied leisure pursuits, travel more, have more fulfilling relationships and so on, would demand less, rather than more of Earth's carrying capacity" (Pepper 1993: 205).

It seems that Pepper wants his cake and to eat it too. While pursuing more education and having more satisfying social relationships strike me as activities that have the potential to draw people away from the culture of consumption, eating more varied foods may entail importing them from faraway places and traveling to faraway destinations, particularly in airplanes, will place a definite burden on the global ecosystem.

In reality, most people in developed societies and the more affluent sectors in developing societies will need to scale back their consumption of material goods as well as restrict the number of holidays to faraway destinations that they take. In one of his popular online commentaries, Wallerstein (2007) delineates three overarching obstacles to overcoming climate change: (1) the "interests of producers/entrepreneurs," who act as the purveyors of the capitalist treadmill of production and consumption; (2) the "interests of less wealthy nations" that are emulating the lifestyles of developed countries; and (3) the "attitudes of you and me." While it is important not to place the burden of climate change mitigation on individuals by urging them merely to become "green consumers," as Wallerstein asserts, climate change mitigation starts at the individual level, particularly for the affluent people in both developed and developing countries, but ultimately entails becoming involved in anti-systemic social movements as well.

Sustainable Trade

Over the past two centuries, global production has resulted in a tremendous cross-border trade of goods and services. The UNDP reports: "In 1800, trade accounted for 2% of world output. The proportion remained small right after the Second World War, and by 1960 it was still less than 25%. By 2011, however, trade accounted for nearly 60% of global output" (UNDP 2013: 43).

While this increased international trade has been enhanced by free trade agreements and lower transport costs, it relies heavily on oil and contributes to greenhouse gas emissions in moving goods around the world by ship or airplane as well as trucks and trains. Furthermore while developing countries, in particular China, are often criticized for their increasing greenhouse gas emissions, an appreciable amount of this due to the fact that developing countries are importing cheap resources and manufactured goods from developing countries. Peters et al. report: "Globally, in 2001, an estimated 5.3 gigatons (GT) (22% of global CO_2 emissions) of embodied CO_2 emissions were shifted around the globe due to international trade, with a general trend that developed countries were net importers of CO_2 emissions. Taking into account net trade, the European Union imported 0.6 Gt of embodied CO_2 more than it exported (13% of domestic emissions) and the USA had a net import of 0.4 Gt CO_2 (7%). In contrast, China had a net export of 0.6 Gt (CO_2) (18%) and Russia had a net export of 0.3 Gt CO_2 (22%)" (Peters et al. 2009: 381).

International aviation and marine fuels are exempt from an international taxation schemes.

The global food system has undergone a tremendous rise in "food miles"—a measurement of the distance that food travels from the site of production to the site of consumption. Vandana Shiva (2008: 104) maintains that humanity "should be reducing food-miles by eating diverse, local, and fresh foods, rather than increasing carbon pollution through the spread of corporate industrial farming, nonlocal food supplies, and processed and packaged food." In a similar vein, Sarkar (1999: 219) advocates the general principle of "as far as possible, produce locally or regionally" in order to avoid the emissions miles so often associated with long-distance trading. He admits that even in an eco-socialist world, some long-distance trade will be inevitable. For example, in the case of his home country of India, salt can only be produced in coastal regions, thus necessitating exporting it to regions in the interior. Sarkar (1999: 219) maintains: "Our needs and the required goods would have to be kept under control, so that long-distance trade can be kept within sustainable limits worldwide." There is the need for the greening of shipping, which would rely on solar and hydrogen energy–powered ships, sailing ships and "kite sails" (Simms 2009: 94–95). Also given that large quantities of products are now shipped by airplane and truck, there is a strong need to revisit railroads and waterways as less energy-intensive modes of shipping.

Sustainable Settlement Patterns and Local Communities

Modern cities have evolved following, in large part, the dictates of capital with its need for manufacturing, financial, commercial, distribution, and communication centers, as well as the administrative demands of government bureaucracies. As they have grown, cities have gobbled up precious farmland and natural areas. Overall cities are energy-intensive places on a number of counts, including in the operation of office buildings, industries, residences, shopping centers, recreational facilities, restaurants, educational institutions, hospitals, residences, transportation, highways, parking lots, airports, etc. According to Mike Davis, "Heating and cooling the urban built environment alone is responsible for an estimated 35 to 45 per cent of current carbon emissions, while urban industries and transportation contribute another 35 to 40 per cent. In a sense, city life is rapidly destroying the ecological niche—Holocene climate stability—which made its evolution into complexity possible" (Davis 2010: 41).

In reality, the ecological and carbon footprints of cities vary considerably between cities in developed and developing countries but also within cities, depending on their residential patterns (e.g., McMansions versus slum dwellings) and modes of transportation (e.g., a city with an excellent public transportation system versus a highly car-dependent one). While advocates of green cities often argue that urban density can serve to foster environmental sustainability, the reality is that: "[B]ig, dense cities are not all equal. Hong Kong is an environmental paragon [an exaggerated assertion in the larger schemes of things]; Dubai, which from a distance appears similar, is an environmental disaster" (Owen 2012: 248).

The ecological and carbon footprints of cities extend far beyond their boundaries because they rely on resources from a large hinterland that literally encompasses much of the world. Various proponents of "sustainable cities" who maintain that increasing urban density contributes to environmental sustainability downplay the "historical connections between density and economic growth" (Clement 2011: 296). Theoretically cities have the potential of becoming much greener than they presently are. Even during the early twentieth century, various socialists and anarchists pioneered efforts to make cities more liveable both socially and environmentally, such as the Karl Marx-Hof in Red Vienna and the Bauhaus housing experiments in Germany (Davis 2010: 43).

A New Urbanism that seeks to make cities more livable and environmentally sustainable has emerged around the world and has even permeated urban planning in some cities. According to Caradonna, "[T]he New Urbanists seek a return to dense, urban neighborhoods characterized by mixed-use buildings (residential, commercial, entertainment, etc.) and a vibrant sidewalk culture. They value pedestrian-only zones, abundant open spaces for leisure and social gatherings, land-use strategies that prevent sprawl, bike lanes, and integrated transportation systems that reduce reliance on the automobile. Instead of tearing down warehouse districts and industrial zones, the New Urbanists stress the need to re-use existing infrastructure and bring suburbanites back into urban areas" (Caradonna 2014: 201).

While the New Urbanism has been applied to some degree in many towns and cities, it needs to make a much stronger effort to be socially inclusive and counteract gentrification, which marginalizes low-income people. Conversely, in a democratic eco-socialist world, there would be no poor people and differences in income and wealth would not be nearly as great as is the case in capitalist societies.

The development of green cities constitutes a highly imaginative endeavor, one that will require drawing insights from numerous disciplines and fields, including architecture, building construction, urban planning, transportation planning, and last but not least the social sciences. A green or sustainable city should include medium-density housing, easy access to public transport, and minimal reliance on automobiles. Walkability should be part and parcel of the green city, which would allow people to walk as much as possible to their work sites, parks, recreational centers, theaters, shops, and eating places and contributes to a "democratized streetscape" (Agyeman 2013: 118). Some psychologists have developed the notion of *eco-psychology*, which stresses the need for people, including urban dweller, to have contact with the natural environment (Brown 2009: 162).

Eco-villages, which are increasingly found in urban and rural areas of developed and developing societies, constitute prefigurative social experiments that potentially are part and parcel of developing more sustainable settlement patterns. Urban eco-villages can reduce car dependence or eliminate it completely if closely situated to good public transportation. Furthermore, in that an eco-village is "truly a village, it has its own internal economy that enables members to work close to home, thereby reducing their transportation needs" (Liftin 2014: 44).

Arran Gare, an eco-socialist, maintains that at least in Anglophone countries the possibility of deflecting national politics away from a neoliberal agenda in the short term is improbable. For the immediate future,

> The challenge is to create a network of mutually supporting partially autonomous alternative local economic systems which can function as stepping stones for transforming the whole or society and eventually for participating in the creation of an ecologically sustainable world civilization. Such local economies have already begun to emerge.... To reduce dependence on the outside economy it is necessary to become more self-reliant. It is necessary to develop local sources of energy, to reduce the consumption of energy and in rural areas, to develop organic and other low-external-input types of agriculture. (Gare 2014: 36)

The central business districts of cities around the world now have numerous high-rise office and even apartment buildings. While there has been quite a bit of discussion on how to make buildings more environmentally sustainable through the use of green roofs and walls, fritted glazing, solar panels, more efficient lighting, "recently, under the pressure of the accelerating debate around sustainability, has the idea begun to grow that perhaps not more office buildings but rather fewer buildings, more intelligently and intensively used" (Duffy 2008: 11). Of course, in a democratic eco-socialist world, many buildings presently occupied by financial institutions, marketing firms, and department stores that facilitate the capitalist patterns of production and consumption would not be needed and could be converted to serve more socially beneficial purposes.

Cities should be easily interconnected with other cities via trains rather than auto or plane transportation. Also there is the question as to what the optimum maximum population of a city is. Some cities have become so incredibly large that it almost defies the imagination, with the world now having some twenty-five mega-cities with populations each of over ten million people: Sao Paulo has 27.6 million, Mexico City 21.6 million, and Mumbai 20.7 million people. Obviously, there is no easy answer to this question because it depends on the national context and notions of population density. I am sometimes tempted within the context of a developed society to argue that cities in which people can access other parts of the city relatively easily should not have a population of over 1.5 million. In Australia, Sydney now has a population of some five million people and Melbourne is now over four million, but both cities continue to sprawl. In contrast, the largest cities in Germany are Berlin with some 3.5 million people and Munich and Hamburg with some 2.5 million each, with many German cities having populations of 250,000 to 500,000, virtually all of which are more compact than Australian cities. Of course, the automobile has played a major role in the urban sprawl found in Australian cities, more so than has been the case for German cities.

Conclusion

The transitional steps that I have delineated constitute loose guidelines for shifting human societies or countries toward democratic eco-socialism. I do not purport that my suggested guidelines are comprehensive because undoubtedly others could be added to the list. On the issue of transitioning to a sustainable and just world, eco-anarchist Ted Trainer delineates a long list of points relating to strategies to achieve a Simpler Way, too many to repeat here. However, I have listed a few that bear discussion. Trainer (2010: 297) argues that the "Simpler Way cannot be given nor imposed by force," particularly on the part of states. However, a government that has been democratically elected and truly expresses the will of the majority of the people, at least a substantial majority, would have a right to impose regulations that would promote social parity and environmental sustainability, but that in all likelihood would be opposed by corporations or wealthy individuals. Trainer (2010: 299) advocates a "thoroughly peaceful revolution," which in theory is desirable but may not always be possible given the political situations in various countries around the world in which the capitalist class and its political allies seek to mount a violent counterrevolution. Trainer argues that in terms of achieving a Simpler Way, "[o]ur chances are not at all good" (Trainer 2010: 299). Nevertheless, he advises people to "plunge into getting those local systems going, with a view to using them to develop the crucial consciousness" (Trainer 2010: 299). Trainer (2010: 315–316) recommends a number of steps to achieving the good society. These include creating a Community Development Cooperative, setting up a community garden and workshop, producing food, forming working bees, providing local and free entertainment, establishing small family firms and co-ops, living frugally, creating an alternative money system, forming a town bank, and creating a new economy in "which all of us participate in making rational decisions about what we need and therefore are going to produce cooperatively." The new economy will theoretically gradually evolve as people opt out of the existing global economy. Trainer argues: "The new society cannot be imposed or even given. Unless it is willingly developed it will not work, and it must be learned. Communities must bumble their way to the geographies and practices that suit them in their conditions. This then means there is little choice about how to proceed: the only way to get there from here is to start now, where we live, building the new ways" (Trainer 2010: 323).

In a somewhat similar way, Graeber (2013: 283–295) asserts that he is less interested in specifically what kind of alternative economic system needs to be created than in how people decide to get there. He does suggest the following five steps for getting to some "sort of viable free society," one that avoids both economic and ecological disasters:

- A debt jubilee or a general cancellation of debts around the world
- Working less given that grim reality that the "current pace of the global work machine is rapidly rendering the planet uninhabitable"
- Shifting work from the creation of "ever more consumer products and ever more disciplined labor" to work that focuses on caretaking and teaching
- Critiquing and shifting away from the bureaucratization of social life
- Reclaiming the notion of *communism* by shifting away from viewing it as the "absence of private property arrangements" to the "original definition: 'from each according to their abilities, to each according to their needs'."

The matter is whether societies and a world system based on democratic eco-socialist principles would allow room for markets. In my vision of a democratic eco-socialist society, new left parties that rise to power in democratic elections would in time nationalize the means of production, particularly the multinational corporations and even national corporations that today have a powerful influence, both directly and indirectly on governments and mainstream political parties. I can see room for small businesses, such as restaurants, cafés, bookshops, clothing stores, experimental theaters, and so forth, preferably worker-owned and worker-managed, but possibly owned by independent entrepreneurs or families. Obviously, markets would not completely disappear but would be much diminished compared to their strong presence in capitalist societies. In contrast to reciprocity and redistribution as modes of economic exchange that are characteristic of indigenous societies, markets involve the exchange of goods and services with the use of money, predate capitalism by millennia and even existed in rudimentary form in indigenous societies, became more elaborate in precapitalist state societies, but particularly become very pronounced and hegemonic with the full-scale development of capitalism. In Schweickart's (2016: 6) conception of a socialist-oriented *economic democracy*, small businesses would "provide jobs for large numbers of people, and goods and services to even more," as they presently do. Unfortunately, most small businesses today often pay their employees low wages and provide them with minimal social benefits, if any. Such practices would have to be seriously curbed under democratic eco-socialism. Ultimately, a democratic eco-socialist society and world system, in order to prevent the emergence of large corporations, would have to create legislation that would have to restrict the size of businesses, lest the capitalist tendency for the bigger businesses to take over the smaller businesses gain ascendancy.

CHAPTER 7

Conclusion
The Future in the Balance

Under neoliberalism and particularly the economic growth of China, despite efforts on the part of the UN Framework Convention on Climate Change (UNFCCC) and twenty-one Conferences of the Parties around the world to contain global warming and associated climatic changes along with much talk about decoupling economic growth from emissions, greenhouse gas emissions have continued to increase, and at a growing rate. In light of this alarming situation, the Science and Security Board of the *Bulletin of Atomic Scientists* in consultation with its Board of Sponsors, which includes seventeen Nobel laureates, decided in 2015 to once again set its Doomsday Clock at "three minutes to midnight, two minutes closer to catastrophe than in 2014" given that "unchecked climate change, global nuclear weapons modernizations, and outsized nuclear weapons pose extraordinary and undeniable threats to the continued existence of humanity, and world leaders have failed to act with the speed or on the scale required to protect citizens from potential catastrophe" (Bulletin of Atomic Scientists 2015). On 22 January 2016, the Bulletin scientists stated that "it is *still* three minutes to midnight", adding that:

> In the past year, the international community has made some positive strides in regard to humanity's two most pressing existential threats, nuclear weapons and climate change. In July 2015, at the end of nearly two years of negotiations, six world powers and Iran reached a historic agreement that limits the Iranian nuclear program and aims to prevent Tehran from developing nuclear weaponry. And in December of last year, nearly 200 countries agreed in Paris to a process by which they will attempt to reduce their emissions of carbon dioxide, aiming to keep the increase in world temperature well below 2.0 degrees Celsius above the pre-industrial level. (Eden et al. 2016)

While the powers that be around the world are seeking to address climate change within the parameters of global capitalism, I maintain that anthropogenic climate change, particularly since the Industrial Revolution, has been by and large a byproduct of global capitalism, as Canadian journalist Naomi Klein (2014) has so forcefully argued. A world-renowned journalist, she has made a point that numerous eco-socialists have been making for some time, but one that is generally ignored in the mass media and even on the part of all too many environmentalists and climate activists. While Klein has performed an important service in identifying the proverbial "elephant in the room" in the climate change discourse, it is not entirely clearly exactly what she would like to do about capitalism, whether she would like to make it more socially just and environmentally friendly or transcend it. She maintains: "Our economic system and our planetary system are now at war. Or, more accurately, our economy is at war with many forms of life on earth, including human life. What the climate needs to avoid collapse is a contraction in humanity's use of resources; what our economic model demands to avoid collapse is unfettered expansion" (Klein 2014: 21).

Klein (2014: 89) expresses skepticism about green capitalism and ecological modernization, noting that its proponents believe that the world can "continue to function pretty much as it does now, but in which our power will come from renewable energy and all of our various gadgets and vehicles will become so much more energy-efficient that we can consume away without worrying about the impact." At least, in part, she favors a planned economy, which would include the creation of green jobs that would result in renewable energy delivery, public transportation, small-scale sustainable farming, and public ownership of certain enterprises, particularly of energy production given that it would allow communities to shift to renewable energy sources. Klein (2014: 99) praises those countries, such as the Netherlands, Austria, and Norway, that have permitted large portions of their electricity sectors to be publicly owned. She believes that the 500 million or so of the most affluent people who produce about half of all greenhouse gas emissions in the world should in particular be heavily taxed in keeping with the "polluter pays" principles (Klein 2014: 113–114).

Klein also advocates creating a guaranteed minimal income, closing tax havens, a 1 percent billionaires' tax, the slashing of the military budgets of each of the top ten military spenders by 25 percent, the implementation of a tax of $50 per metric ton of CO_2 emitted in developed societies, and the phasing out of fossil fuel subsidies as strategies for mitigating as well as adapting to climate change. She calls for "rebuilding and reinventing the very idea of the collective, the communal, the commons, the civil, and the civic after so many decades of attack and neglect" (Klein 2014: 460). However, nowhere in her eloquent treatise does Klein call for an ecological evolution along the lines advocated by eco-socialists, although she fleetingly acknowledges being intellectually indebted to John Bellamy Foster, David Harvey, and Carolyn Merchant along with an array of other scholars and writers, such as Herman Daly, Richard Heinberg, George Monbiot, Naomi Oreskes, Juliet Schor, and James Gustave Speth (see Klein 2014: 531). Nowhere in her book does she call for large-scale redistribution of the wealth, despite her advocacy of progressive taxation, and efforts to stop a small minority of people from becoming billionaires, or even multi-millionaires. In a nutshell, it appears that Klein is a very progressive green social democrat rather than a democratic eco-socialist. This, however, does not take away from her willingness to identify global capitalism as the principal driver of climate change, something that most physical scientists, economists, and even social scientists examining climate change—and climate activists for that matter—fail to do. As Foster and Clark (2015: 15) observe in their review of Klein's book, it is difficult to fault her for failing to address the issue of building a "new movement toward socialism, a society to be controlled by the associated producers" given that "[h]er aim at present is clearly confined to the urgent and strategic—if more limited—one of making the broad cases for System Change Not Climate Change." Hopefully her book will assist social democrats, liberals, and mainstream environmentalists to take the next paradigm shift toward democratic eco-socialism. As Simms (2009: 184) has observed, "global warming probably means the death of capitalism as the dominant organising framework for the global economy." In a related vein, as Kovel (2007: 258) observes, overcoming climate change or global warming will entail a shift to ecosocialism, a perspective that is largely "suppressed in normal discourse," certainly in mainstream society but all too often among some socialists who view it as a side issue.

Thus, it appears imperative to construct an alternative to global capitalism as the ultimate climate mitigation strategy, even though it will not be achieved anytime soon, if indeed ever. As Kovel so forcefully asserts,

> Global warming is not the only aspect of the ecological crisis to have reached planetary proportions, nor is it the only one with the potential to actually destroy the human species. But it definitely has the most power to seize the world's imaginations. This is because of global warming's literally spectacular quality, the way it manifestly affects other aspects of the crisis ... [and] ... for the way in which global warming puts the entire history, and the prehistory as well, or industrial capitalism into the dock. Here the leading culprits are in full view: the whole petro-apparatus, from the pusher of "automania"

to the imperial apparatus that wages endless war to keep the carbon flowing from the ground, where it belongs, to the atmosphere, where it will destroy us (Kovel 2007: 258).

In reality, many of the most threatening planetary changes are a product of anthropogenic climatic interactions with other human-induced eco-crises such as deforestation, coral reef loss, mangrove loss, wetlands loss, air pollution, and water pollution. These anthropogenic climate/environment interfaces have been termed by Merrill Singer (2010a, 2010b) "pluralea interactions," and they suggest the unintended ways in which combined human actions and their effects are impacting the earth. The term "pluralea interactions" is derived from the Latin word *plur*, meaning "many," and *alea*, meaning risks or hazards. Central to the pluralea perspective is the development of an understanding of the pathways and mechanisms through which two or more eco-crises interact to produce synergistic, magnified environmental and human impacts. For example, due to the continual need for the capitalist world system to make profits and economically grow, tropical forests in Indonesia, the Amazon Basin, and elsewhere are rapidly being cut down, removing a vital carbon sink that stores CO_2. Furthermore, the dumping of human sewage into the oceans has provided nutrients for rapid seaweed and algae growth that endangers coral reefs around the planet, in particular the Great Barrier Reef in Australia.

As humanity enters an era of dangerous climate change accompanied by tumultuous environmental and social consequences, it will have to consider alternatives that hopefully will circumvent dystopian scenarios of the order delineated earlier. Thus, I propose the imagining and creation of a democratic eco-socialist world system as real utopia, not just as a vehicle for creating a safe climate, but a more socially just, democratic, and generally sustainable world society as well. As noted earlier, democratic eco-socialism rejects the capitalist treadmill of production and consumption and its associated growth model. Instead, it recognizes that humans live on an ecologically fragile planet with limited resources that must be sustained and renewed as much as possible for future generations. While at the present time or for the foreseeable future, the notion that democratic eco-socialism may be eventually be implemented in any society, developed or developing, or in a number of societies may appear absurd, history tells us that social changes can occur very quickly once certain social structural and environmental conditions have reached a tipping point, a term that has also become popular in climate science.

As humanity lurches ever forward into the twenty-first century, our survival as a species appears to be more and more precarious, particularly given that the impact of climate change in a multiplicity of ways looms on the horizon. Global, regional, and local temperature records are repeatedly being broken around the world. In 2014, Europe witnessed its hottest year in some five hundred years and California experienced its most severe drought in the past 1,200 years. The situation can look extremely dire, particularly for the poorest people on the planet, and is one that drives all too many climate activists, along with perhaps millions of ordinary people, to despair and to feel that it is all too hard, too overwhelming to want to think about.

I often hear climate activists in Australia say that we do not have enough time to transcend global capitalism to be able to create a safe climate for humanity. Thus, they argue that climate activists need to collaborate with more supposedly progressive corporate leaders and politicians in tackling the climate crisis within the parameters of the existing global political economy. In my view, combatting climate change and global capitalism go hand-in-hand. While the more enlightened corporate elites and their political allies may permit some measures that contribute to climate change mitigation, they will certainly not consciously permit the eventual demise of global capitalism and the emergence of a democratic eco-socialist world system. Labban et al. (2013: 8) argue that climate catastrophism tends to preclude radical solutions to the climate crisis and allows the transnational capitalist class to present "capitalist expansion as the only imaginable, viable solution not only to the problem of climate change, but also to other socio-ecological problems such as food, water, and energy shortages." As I have argued and tried to demonstrate elsewhere, green capitalism and existing climate regimes are not sufficient to mitigate climate change in any serious vein (Baer 2012). How can we expect the system that created the problem to solve the problem? Indeed, some eco-socialists also argue that transcending global capitalism will be a long, drawn-out project. For instance, Christian Parenti argues: "Dealing with climate change by first achieving radical social transformation—be it as socialist or anarchist or deep-ecological/neo-primitive revolution, or a nostalgia-based *localista* conversion back to a mythical small-town capitalism—would be a very long and drawn-out, maybe even multigenerational struggle. It would be marked by years of mass education and organizing of a scale and intensity not seen in most core capitalist states since the 1960s or even the 1930s" (Parenti 2013: 51).

My own sense is that overall things will get worse, before they get better, and there is no guarantee that they will get better. Thus, in terms of the foreseeable future, I am very much in agreement with Wallerstein, who maintains: "I do not believe that our historical system is going to last much longer, for I consider it to be in a terminal structural crisis, a chaotic transition to some other system (or systems), a transition that will last twenty-five to fifty years. I therefore believe that it could be possible to overcome the self-destructive patterns of global environmental change into which the world has fallen and establish alternative patterns. I emphasize however my firm assessment that the outcome of this transition is inherently uncertain and unpredictable" (Wallerstein 2007: 382).

Wallerstein further argues that whether humanity transitions to a more egalitarian and democratic world system or into a more hierarchical and stratified world system by 2050 is highly dependent on the "political activity of everyone now and in the next twenty-five to fifty years [actually less than that now]" (Wallerstein 2007: 385). Nevertheless, while at this point in time, the capitalist world system appears to be well-entrenched, there are numerous cracks in the system, some of which I have pointed out in this book and others of which have been pointed out by numerous others (Harvey 2014; Higgs 2014; Moore 2015).

Yates (2016: 11), argues that while in recent years protest movements, ranging from the Arab Spring to workers' uprisings in China and France, and the Occupy movement and the Black Lives Matter movement in the United States, "victories have been fleeting" and all too often "protest has often devolved into ethnic and religious violence and warfare." Often it appears to be a case of two steps forward, one step backward, or one step forward and two steps backward, the latter when far-right-wing movements and right-wing populist parties are in ascendancy. Thus, it is important that progressive people keep plugging away at challenging the system in their conversations, teaching, and writing, while becoming involved in anti-systemic movements, struggling to create new left parties, pointing out alternative ways of organizing the world along democratic eco-socialist principles, and listening to critical input from other progressive perspectives, including eco-anarchism, eco-feminism, and indigenous voices, to mention only a few. Hopefully as humanity finds itself in an increasingly critical situation, counterhegemonic voices will receive a greater hearing than they do now and inspire ordinary people to become more politically involved in creating a much-needed new world.

Aside from the modest gains made at the theoretical level of the eco-socialist perspective, the development of a socialism for the twenty-first century in Latin America and the rise of new left parties in Europe indicate that a rejuvenated socialism may gradually be gaining ascendancy in the public discourse. Two cases in point are the rise of Jeremy Corbyn, a staunch social democrat, as the leader of the British Labour Party in September 2015 and the Bernie Sanders phenomenon in the 2016 Democratic Party primaries in the United States. Corbyn's election as the leader of the Labour Party resulted in a large increase in membership, with many people who had left Labour for the Green Party going back to Labour (Watson 2016: 11).

Sanders (age seventy-four) has identified himself as an "independent socialist" in his terms as the mayor of Burlington (Vermont), the sole Congressperson from Vermont in the House of Representatives, and in his term as Senator. He has been a member of Democratic Socialists of America (DSA), a group that has sought for years to move the Democratic Party to the left, with no obvious success. In reality, leftists in the United States debate among themselves as to whether Sanders is really a "democratic socialist" or a "social democrat" or classical "left liberal" (Blum 2016). Roden maintains: "Sanders is definitely not advocating a socialist revolution. His politics are broadly understood as hoping to alleviate the ills of capitalism without overthrowing capitalism. His support for some of the key aspects of the US empire is an example of this distinction" (Roden 2016: 17).

Sanders called for the rehabilitation of New Deal politics and the creation of new public insurance programs, such as paid parental leave and sickness insurance; the expansion of Social Security and Medicare; the creation of publicly funded early education and the expansion of training and job placement programs; increasing the minimum wage; and the reduction of fees for public institutions of higher education, all proposals that clearly place him more in the framework of classical social democracy rather than democratic socialism or democratic eco-socialism for that matter (Kenworthy 2016: 21).

Sanders proved to be a formidable contender against Hillary Clinton in the Democratic Party primaries. Clinton narrowly defeated Sanders in the first primary on 1 February 2016 in Iowa with 49.8 percent of the popular votes in contrast to her opponent, who won 49.6 percent of the popular vote. Sanders went on to win the primaries in Colorado (59.0 percent), Minnesota (61.7 percent), Oklahoma (51.9 percent), Vermont (86.1 percent), Kansas (67.7 percent), Maine (64.3 percent), Michigan (49.8 percent), Nebraska (57.1 percent), Alaska (81.6 percent), Hawaii (71.5 percent), Idaho (78.0 percent), Utah (79.3 percent), Washington (72.7 percent), Wisconsin (59.6 percent), and Wyoming (55.7 percent). In the end, Clinton, with the assistance of superdelegates, won the nomination with 2,842 delegates to Sanders's 1,865 delegates and with 16,914,722 (55.2 percent) popular votes to Sanders's 13,206,428 (43.1 percent) popular votes. Despite Clinton's victory, the fact that a self-proclaimed socialist and one that the U.S. mass media consistently identifies as such in a country where Barack Obama had been accused of being a "socialist" by Tea Party and other right-wingers is astounding, even though in reality he is more of a social democratic than a democratic socialist. The fact

Sanders was willing to engage in "class warfare" by calling for a "political revolution" against the "billionaire class" is rhetoric that Democratic Party candidates seldom if ever engage in in the context of generally staid campaigns for the U.S. presidency. On 1 February 2016, while Sanders's campaign was gaining momentum, Wallerstein observed in one of his biweekly commentaries on global affairs: "[Sanders] has succeeded in forcing public discussion of the income gap. He has pushed Clinton's rhetoric to the left in her attempt to recuperate political Sanders voters. Whatever the final outcome of the Democratic Party's convention, Sanders has achieved far more than almost everyone predicted at the onset of his campaign. He has, at the very least, forced a serious debate about program within the Democratic Party" (Wallerstein 2016).

The Sanders phenomenon clearly indicates that a substantial portion of the U.S. electorate has begun to question at least the neoliberal agenda, if not capitalism itself. Millions of particularly young Americans, aside from how they define the term, now identify themselves as socialists. The Sanders campaign posed a powerful challenge to neoliberalism and for a least several months opened up the political landscape in a way that it has not been opened up for decades. He was not afraid to challenge what he freely termed the "billionaire class" and the corporate media which he viewed as a threat to US democracy. In terms of combatting climate change, Sanders called for improving energy efficiency, ending government subsidies to fossil fuels corporations, taxing carbon dioxide and methane emissions, banning fracking, rejecting pipelines carrying oil from the tar sands deposits of northern Alberta, investing in renewable energy, and promoting climate justice by ensuring that "poor and minority communities share in the benefits of the clean energy revolution" (Sanders 2016: 373). If Sanders had refused to support Clinton in the wake of his defeat and accepted Jill Stein's invitation to take her place as the Green Party presidential candidate, the possibility of moving part of the U.S. political landscape even further to the left may have occurred. However, Sanders, as he said he would do if he were to lose the primary race, supported Clinton, resulting in a rapid loss of the momentum of his campaign and the potential for a reinvigorated movement for social justice in the U.S. In the wake of the final U.S. presidential race debate on 19 October 2016 between Donald Trump and Hillary Clinton, in an interview that *Democracy Now* had with Green Party presidential candidate Jill Stein, on the matter of the choice between her two rivals, stated: "So we have a very bleak reality. Everybody knows that Donald Trump is terrifying and dangerous. But are we secure with Clinton in the White House, when Clinton is telling us right now that she wants to start a war with Russia over Syria, creating a no-fly zone? It's going to be very hard not to slide into World War III with Hillary at the helm, starting off her four years with declaring war against Russia by enacting a no-fly zone" (quoted in *Green Left Weekly* 2016).

On 8 November 2016 in the U.S. presidential election, despite the overwhelming support of the mainstream media for Hillary Clinton both in the United States and much of the world, Donald Trump won the election, gaining most of the electoral votes but not the popular votes. Furthermore, the Republicans won majorities in both the House of Representatives and the Senate. Americans and much of the rest of the world now is left to wonder what to expect from a superrich businessman who has never held a political office, is a racist and misogynist, is a climate denialist who supports expansion of fossil fuel energy production, wants to double U.S. economic growth, wants to build a wall between the United States and Mexico for which he demands Mexico will pay, opposes free trade agreements, wants to impose high tariffs on imports from China, wants to reduce corporate taxes from 35 to 15 percent, wants U.S. allies, particularly ones in NATO, to pay more of the expenditure of maintaining a global military apparatus, is willing to negotiate with Putin, and asserts that he will "make America great again." From my vantage point as a dual U.S./Australian citizen residing in Australia and registered to vote in Arkansas, I cast my absentee ballot for Jill Stein, partly as a protest vote and partly knowing that it would in all probability be a "safe vote" in that Trump was predicted to win Arkansas, which he did, despite the fact that Hillary Clinton had been the First Lady of Arkansas for ten years. For me, Clinton was the lesser of two evils, perhaps barely, given her neoliberal and militarist policies in her stints as a U.S. Senator from New York and Obama's first Secretary of State in 2009–2013.

In the case of Trump, in the lead-up to the election, sociologist James Petras wrote: "Trump mouths contradictory economic statements, especially his proposals to rebuild America, while operating in the framework of an imperial system. As President of the United States, his protectionist policies will come into direct confrontation with US and global 'finance and monopoly capitalism' and will likely lead to systematic disinvestment and a disastrous economic collapse or, more likely, the Businessman-President's capitulation to the status quo" (Petras 2016: 24–25).

As a right-wing populist, Trump successfully and masterfully tapped into a nerve among millions of particularly white lower-middle-class and working-class American males and females who have been the victims of neoliberal policies and corporate globalization and feel threatened by minorities, particularly Latinos and Muslims, and women. Perhaps more so than any other U.S. election in recent times, the 2016 election reveals numerous cracks in what many describe as a declining hegemonic power in the capitalist world system. This manifested almost immediately after Trump's election when thousands of protesters took to U.S. streets to rally against the election result in numerous cities and on high school, college, and university campuses across the country. Some 1,800 protesters gathered outside the Trump International and Tower in Chicago while chanting phrases like, "No Trump! No KKK! No racist USA" (ABC News 2016). Such extensive protests following election of a U.S. presidential candidate to the presidency probably are unprecedented, at least within the borders of the United States.

Despite the certain shifts to the left in various developed countries around the world, right-wing populism has undergone an upsurge in the past decade or so, including in Greece with Golden Dawn, in Britain with the U.K. Independence Party, in France with the National Front, in Germany with Alternative in Germany, in the United States with the Trump takeover of the Republican Party and election to the U.S. presidency, and in Australia with the re-emergence of the One Nation Party under the leadership of Pauline Hanson.

To a large degree, this book has been an exercise in the anthropology of the future. I have tried to make the case that humanity needs over the course of the twenty-first century to shift from the reigning capitalist world system to an alternative world system based on social justice, democratic processes, environmental sustainability, and a safe climate, a world system that can fairly be termed democratic eco-socialism. I have also posed not a blueprint per se, but some loose guidelines, for getting from the existing world system to an alternative one, a "real utopia" so to speak. In referring to Michael Albert's Parecon plan, Greaeber maintains:

"[S]uch models can only be thought experiments. We cannot really conceive the problems that will arise when we start actually trying to build a free society. What now seem likely be the thorniest problems might not be problems at all; others that never even occurred to us might prove devilishly difficult. There are innumerable factors" (Graeber 2013: 284).

Indeed, this characterization could well apply to the vision of democratic eco-socialism and guidelines to achieve it delineated in this book. Conversely, perhaps my propositions are not unrealistic given that others have somewhat similar visions for the future. For example, John Bodley (2012), in his book *Anthropology and Contemporary Human Problems*, has a concluding chapter on "The Future" in which he envisions a *sustainable planetary society*. He maintains that such a new society:

> will be brought about by popularly based non-violent political action to redistribute social power and decentralize decision making in or-

> der to promote human economic and social rights, freedom, social equity, and sustainability. In this transformation global corporations would lose much of their immortality, omniscience, and omnipotence, just as monarchs lost their divinity at the beginning of the Commercial Age. Totalitarian, plutocratic, and aristocratic systems would be replaced by genuine systems.... In the Eco-social World social power would be decentralized, local economies would have priority over the global, knowledge and information would be more important than machine-technology, and the dominant ideology would be dominant humanist. (Bodley 2012: 323)

What Bodley alludes to is the often-avoided recognition that while the various and problematic socialist-oriented societies that have emerged are faulted for being undemocratic to varying degrees, in the end the most egregious violator of democracy, liberty, and equality is the capitalist world system, and particularly the United States as its leading actor at the nation-state level. By looking up from the bottom to the top, the worst and most punishing features of capitalism become clear. Add the ecological dimension to the critique of capitalism, and the threat of this system to life on Earth becomes increasingly apparent.

Needless to say, to a large degree Bodley's vision of a sustainable planetary society is consistent with the notion of *global democratic eco-socialism* or what world systems theorists Terry Boswell and Christopher Chase-Dunn term *global democracy*. Adopting a comparative world system perspective, Chase-Dunn and Lerro (2014: 364–365) present, like a few other neo-Marxian scholars whose scenarios I mentioned earlier, three possible future scenarios. In their first scenario, they portray another "round of US economic centrality based on comparative advantage in generative sectors and another round of US political leadership (hegemony instead of supremacy)" (Chase-Dunn and Lerro 2014: 363). In the worst-case scenario, Chase-Dunn and Lerro (2014: 363–366) see the United States continuing to serve as a hegemonic but not clearly supreme power in which it undergoes hegemonic decline accompanied by hegemonic rivalry among various core states and rising semi-peripheral states, deglobalization, resource wars, ecological disasters, and the appearance of deadly epidemic diseases. In what I regard to be a real utopian scenario very similar to what I term a democratic eco-socialist world system, they see the rise of global democratic and sustainable commonwealth characterized by "capable, democratic, multilateral, and legitimate global governance strongly supported by progressive transnational social movements, global parties, and semi-peripheral democratic socialist regimes with help of important movements and parties in the core and the periphery. This new global polity would democraticize global governance, reduce global inequalities, and make the global human economy more sustainable" (Chase-Dunn and Lerro 2014: 362–363).

Even if a socialist world government or a democratic eco-socialist world system ever comes to fruition, much thought that is beyond the purview of this book will have to be given to how to respect the diversity found among the billions of people on the planet, one that includes national, ethnic, cultural, religious, and ideological differences. Democratic eco-socialism is a vision that still needs much fleshing out and my book is a modest effort to start and extend the dialogue as to how to create an alternative world system based on social justice and parity, democratic processes, environmental sustainability, and safe climate, but one that respects human diversity on a multiplicity of levels. As Amin (2008b: 262) observes, the "independence and sovereignty of states, nations and peoples must be respected and a polycentric international system must be built on this basis." While acknowledging that a working universal global society does not exist in the early twenty-first century and is not visible on the immediate horizon, Martinelli (2005: 249) asserts that a "cosmopolitan ethic is emerging among significant among significant minorities in the contemporary world," which is committed to common aims, such as peace, human dignity, social justice, cultural pluralism, and environmental sustainability" and is based on "four basic types of consciousness: the anthropological consciousness that recognizes unity in our diversity, the ecological consciousness that recognizes our singular human nature within the biosphere, the civic consciousness of our common responsibilities and solidarity, and the dialogical consciousness that refers both to the critical mind and to the need for mutual understanding."

Humanity is obviously at a crossroads—between business-as-usual, a shift to some variant of green capitalism that has gained much support among people somewhere left-of-center, and finally an eco-socialist vision that unfortunately remains very muted at this point in time. As Loewy so aptly observes,

> There will be no radical transformation unless the forces committed to radical socialist and ecological programmes become hegemonic, in the Gramscian sense of the word. In one sense, time is on our side, as we work for change, because the global situation of the environment is becoming worse and worse, and the threats are becoming closer and closer. But on the other hand time is running out, because in some years—no one can say how much—the damage may be irreversible. There is no reason for optimism: the entrenched ruling elites of the system are incredibly powerful, and the forces of radical opposition are still small. But they are the only hope that capitalism's 'destructive progress' will be halted. (Loewy 2006: 307)

Anthropologists and other social scientists can play a critical role in contributing their analytical skills and insights to humanity's struggle for a better future—one in which we as a species learn to live in reasonable harmony with each other and with nature. Perhaps more than any other issue, climate change allows critical social scientists to identify the contradictions of the existing capitalist world system and to contemplate the creation of an alternative world system based on democratic eco-socialist principles and to apply them in keeping with the notion of praxis—the merger of theory and action.

Anthropology as a discipline that seeks to make holistic cross-cultural and temporal observations must collaborate with the other social sciences, such as sociology, political science, and human geography, as well as with the natural sciences. It has been very good at looking at human societies in the present and in the past, the immediate past and even the distant past. With a few exceptions, the social sciences have been rather weak in attempting to look at the future and new imaginings, but the seriousness of anthropogenic climate change, the ecological, and the human condition in general within the context of the capitalist world system will inspire anthropologists and social scientists to give more serious consideration to future scenarios and be part of the larger project of developing a critical, and future-directed, social science. In this regard, they can reinvigorate the work of W. Warren Wagar, a historian who doubled as a social scientist in his efforts to envision future scenarios, including by resorting to science fiction in order to do so.

Anthropologists and other social scientists can play a small but vital role in contributing their analytical skills and insights to a much larger struggle for the future—one that already exists in various anti-systemic movements, leftist political parties, and even left-wing governments, and one in which we as a species learn to live in relative harmony with each other and nature. Perhaps more than any other issue, climate change allows critical social scientists to illuminate the contradictions of the existing capitalist world system and to contemplate the creation of an alternative world system based on democratic eco-socialist principles. Anthropology is a discipline that has the potential to present the "big picture" while drawing on observations of "small pictures" in different societies and cultures around the world. It has been very good at looking at human societies in the present and past, both the immediate past and distant past. With a few exceptions, anthropology has been rather weak in attempting to look at the future, but hopefully new imaginings, prompted by the seriousness of anthropogenic climate change, will inspire anthropologists to engage with the anthropology of the future and be part of the larger project of developing a critical, and future-directed social science.

REFERENCES

ABC News. 2016. "Donald Trump: Protesters Take to US Streets to Rally against Election Results. http://www.abc.net.au/news/2016-11-10/protests-break-out-in-response-to-trump-election/8013024.
Agger, Ben. 1979. *Western Marxism: An Introduction, Classical and Contemporary Sources.* Santa Monica, CA: Goodyear Publishing Company.
Agyeman, Julian. 2013. *Introducing Sustainabilities: Policy, Planning, and Practice.* London: Zed Books.
Ali, Tariq. 2009. *The Idea of Communism.* Seagull Books.
———. 2013. "Introduction: Stalinist Legacy." In *The Stalinist Legacy: Its Impact on Twentieth-Century Politics.* Ed. Tariq Ali. Chicago: Haymarket Books. 9–29.
Albert, Michael. 2014. *Realizing Hope: Life beyond Capitalism.* London: Verso.
Albert, Michael, and Robin Hahnel. 1991. *The Political Economy of Participatory Economics.* Princeton, NJ: Princeton University Press.
Albinsson, Pia A., Marco Wolf, and Dennis A. Kopf. 2010. "Anti-consumption in East Germany: Consumer Resistance to Hyperconsumption." *Journal of Consumer Behavior* 9: 412–425.
Alexander, Lisa, et al. 2013. "Summary for Policymakers." In *Climate Change 2013—The Physical Science Basis: Working Group I Contribution to the Fifth Assessment Report of the Intergovernmental Panel on Climate Change.* Cambridge, U.K.: Cambridge University Press, 1–36.
Alexander, Samuel, Ted Trainer, and Simon Ussher. 2012. "The Simpler Way: A Practical Action Plan for Living More on Less." Simplicity Institute Report 12A. www.simplerway.org.
Alexander, Samuel, and Simon Ussher. 2011. "The Voluntary Simplicity Movement: A Multi-national Survey Analysis in Theoretical Context." Simplicity Institute Report 11a. www.simplerway.org.
Althusser, Louis. 2014. *On the Reproduction of Capitalism: Ideology and Ideological State Apparatuses.* London: Verso
Altvater, Elmar. 2006. "The Social and Natural Environment of Fossil Capitalism." In *Socialist Register 2007 Socialist Register 2007—Coming to Terms with Nature,* ed. Leo Panich and Colin Leys. London: Verso, 37–59.
Alvord, Katharine. 2000. *Divorce Your Car! Ending the Love Affair with the Automobile.* Gabriola, BC: New Society Publishers.

Amin, Samir. 1985. *Delinking: Towards a Polycentric World*, trans. Michael Wolfers. London: Zed Books.
———. 2007. "Towards the Fifth International?" In *Global Political Parties*, ed. Katarina Sehm-Patomaeki and Marko Ulvila. London: Zed Books, 123–143.
———. 2008a. *The World We Wish to See: Revolutionary Objectives in the Twenty-First Century*. New York: Monthly Review Press.
———. 2008b. "The Defense of Humanity Requires the Radicalisation of Popular Struggles." In *Socialist Register 2009: Violence Today—Actually-Existing Barbarism*, ed. Leo Panith and Colin Leys. London: Merlin Press, 260–272.
———. 2009. "Capitalism and the Ecological Footprint." *Monthly Review* (November): 19–22.
———. 2013. "China 2013." *Monthly Review* 64, no. 10: 14–33.
———. 2014. "Contra Hardt and Negri: Multitude or Generalized Proletariatization?" *Monthly Review* (November): 25–36.
———. 2015. "Contemporary Imperialism." *Monthly Review* (July–August): 23–36.
Anderson, James. 2006. "Afterword: Only Sustain ... the Environment, 'Antiglobalization,' and the Runaway Bicycle." In *Nature Revenge: Reclaiming Sustainability in an Age of Corporate Globalization*, ed. Josse Johnston, Michael Gismondi, and James Goodman. Toronto: Broadview, 245–273.
Anderson, Kevin. 2014. "Revisiting Lenin's Hegel Notebooks, 100 Years Later." *Socialism and Democracy* 28, no. 1: 142–152.
Anderson, Perry. 2010. "Two Revolutions: Rough Notes." *New Left Review* no. 61: 59–66.
Anderson, Robert. 1976. *The Cultural Context: An Introduction to Cultural Anthropology*. Minneapolis: Burgess Publishing Company.
Anderson, Tim. 2015. "From Havana to Quito: Understanding Economic Reform in Cuba and Ecuador." *Journal of Australian Political Economy*, no. 76: 103–128.
Angus, Ian, and Simon Butler. 2011. *Too Many People? Population, Immigration and the Environmental Crisis*. Chicago: Haymarket Books.
Antilla, Liisa. 2005. "Climate of Scepticism: U.S. Newspaper Coverage of the Science of Climate Change." *Global Environmental Change* 15: 338–352.
Appardurai, Arjun. 2013. *The Future as Cultural Fact: Essays on the Global Condition*. London: Verso.
Arrighi, Giovanni. 2007. *Adam Smith in Beijing: Lineages of the Twenty-First Century*. London: Verso.
Arrighi, Giovanni, Terence K. Hopkins, and Immanuel Wallerstein. 1989. *Anti-Systemic Movements*. London: Verso.
Assadourian, Erik. 2012. "The Path to Degrowth in Overdeveloped Countries." In *State of the World 2012: Moving Toward Sustainable Prosperity*, ed. Linda Starke. Washington, DC: Island Press, 22–37.
Azzellini, Dario. 2010. "Constituent Power in Motion: Ten Years of Transformation in Venezuela." *Socialism and Democracy* 24, no. 2: 8–31.

Badiou, Alan. 2007. "One Divides Himself into Two." In *Lenin Reloaded*, ed. Sebastian Budgen, Stathis Kouvelakis, and Slavoj Ziziek. Durham, NC: Duke University Press, 7–17.

———. 2010. *The Communist Hypothesis*. London: Verso.

Badiou, Alain, and Marcel Gauchet. 2016. *What Is to Be Done? A Dialogue on Communism, Capitalism, and the Future of Democracy*, trans. Susan Spitzer. London: Polity.

Baer, Hans A. 1976a. "The Effect of Technological Innovation on Hutterite Culture." *Plains Anthropologist* 21: 187–197.

_____. 1976b. "The Levites of Utah: The Development of and Recruitment to a Millenarian Sect: The Levites." Unpublished PhD Dissertation, Department of Anthropology, University of Utah.

_____. 1987. *Recreating Utopia in the Desert: A Sectarian Challenge to Modern Mormonism*. Albany: State University of New York Press.

_____. 1998. *Crumbling Walls and Tarnished Ideals: An Ethnography of East Germany Before and After Unification*. Lanham, MD: University Press of America.

———. 2008. "Global Warming as a By-product of the Capitalist Treadmill of Production and Consumption: The Need for an Alternative Global System." *Australian Journal of Anthropology* 19: 58–62.

———. 2012. *Global Capitalism and Climate Change: The Need for an Alternative World System*. Lanham, MD: AltaMira Press.

_____. 2014. "The Australian Climate Movement: A Disparate Response to Climate Change and Climate Politics in a Not So 'Lucky Country'." In *Routledge Handbook of the Climate Change Movement*, ed. Matthias Dietz and Heiko Garrelts. Routledge, 147–162.

Baer, Hans A., and Merrill Singer. 2009. *Global Warming and the Political Ecology of Health: Emerging Crises and Systemic Solutions*. Walnut Creek, CA: Left Coast Press.

———. 2014. *The Anthropology of Climate Change: An Integrated Critical Perspective*. London: Earthscan Routledge.

———. 2015a. "Towards an Anthropology of the Future: Visions of a Future World in the Era of Climate Change." In *Environmental Change and the World's Futures: Ecologies, Ontologies and Mythologies*, ed. Jonathan Paul Marshall and Linda H. Connor. London: Routedge, 17–32.

———. 2015b. "Al Gore and the Climate Reality Project Down Under: The Upmarket of the Climate Movement." *Practicing Anthropology* 37, no. 1: 10–14.

Baer, Hans A., Merrill Singer, Debbi Long, and Pamela Erickson. 2016. "Rebranding Our Field? Toward an Articulation of Health Anthropology. *Current Anthropology* 57: 494–510.

Baer, Hans A., Merrill Singer, and Ida Susser. 1997. *Medical Anthropology and the World System: A Critical Perspective*. South Hadley, MA: Bergin & Garvey.

———. 2003. *Medical Anthropology and the World System: A Critical Perspective*. 2nd ed. Westport, CT: Praeger.

———. 2013. *Medical Anthropology and the World System: A Critical Perspective*. 3rd ed. Westport, CT: Praeger.

Bahro, Rudolf. 1978. *The Alternative in Eastern Europe*. London: Verso.

———. 1982. *Socialism and Survival*. London: Heretic Books.

Baran, Paul, and Paul Sweezy. 1966. *Monopoly Capital: An Essay on the American Economic Order*. New York: Monthly Review Press.

Barker, Colin, Laurence Cox, John Krinsky, and Alf Gunvald Nilsen. 2013. "Marxism and Social Movements: An Introduction." In *Marxism and Social Movements,* ed. Colin Barker, Laurence Cox, John Kinsky, and Alf Gunvald Nilsen. Leiden: Brill, 1–37.

Bates, Albert, and Toby Hemenway. 2010. "From Agriculture to Permaculture." In *2010 State of the World: Transforming Cultures—From Consumerism to Sustainability,* ed. Linda Starke and Lisa Mastny. New York: W.W. Norton & Company, 47–53.

Beal, Tim. 2005. *North Korea: The Struggle against American Power*. London: Pluto Press.

Buechler, Steven M. 2014. *Critical Sociology*, 2nd ed. Boulder, CO: Paradigm Publishers.

Beck, Ulrich. 1992. *Risk Society: Towards a New Modernity*. London: Sage.

———. 2000. *The Risk Society and Beyond: Critical Issues for Social Theory*. London: Sage.

———. 2007. *World at Risk*. London: Polity.

———. 2010. "Climate for Change, or How to Create a Green Modernity?" *Theory, Culture & Society* 27, no. 2–3: 254–266.

Becker, Marc. 2012. "Building a Plurinational Ecuador: Complications and Contradictions." *Socialism and Democracy* 26, no. 3: 72–92.

Bell, Karen. 2016. "Green Economy of Living Well? Assessing Divergent Paradigms for Equitable Eco-social Transition in South Korea and Bolivia." *Journal of Political Ecology* 23: 7192.

Bello, Walden. 2009. *Food Wars*. London: Verso.

Belsky, Eric S. 2012. "Planning for Inclusive and Sustainable Urban Development." In *State of the World 2012: Moving Toward Sustainable Prosperity,* ed. Linda Starke. Washington, DC: Island Press, 38–65.

Bencievenni, Marcella. 2006. "The New World Order and the Possibility of Change: A Critical Analysis of Hardt and Negri's Multitude." *Socialism and Democracy* 20, no. 2: 23–43.

Benedict, Ruth. 1971. "The Growth of Culture." In *Man, Culture, and Society,* ed. Harry L. Shapiro. New York: Oxford University Press, 223–236.

Benton, Ted. 1989. "Marxism and Natural Limits: An Ecological Critique and Reconstruction." *New Left Review*, no. 178: 51–86.

Berkhout, Frans. 2014. "Anthropocene Futures." *The Anthropocene Review* 1, no. 2: 154–159.

Bettelheim, Charles. 1976. *Class Struggles in the USSR, First Period: 1917-1923*. New York: Monthly Review Press.

———. 1978. *Class Struggles in the USSR, Second Period: 1923-1930*. New York: Monthly Review Press.

Bjerg, Ole. 2016. *Parallax of Growth: The Philosophy of Ecology and Economy*. London: Polity.

Black, William R. 2007. "Sustainable Solutions for Freight Transfer." In *Globalized Freight Transport*, ed. T.T. Leinbach and C. Capineri. Aldershot, U.K.: Edward Elgar, 189–216.

Blum, William. 2016. "Is Bernie Sanders a 'Socialist'?" *Counterpunch*, 5 February.

Bodley, John H. 2003. *The Power of Scale: A Global History Approach*. Armonk, NY: M.E. Sharpe.

———. 2008. *Anthropology and Contemporary Human Problems*. 5th ed. Walnut Creek, CA: AltaMira Press.

———. 2012. *Anthropology and Contemporary Human Problems*. 6th ed. Walnut Creek, CA: AltaMira Press.

———. 2013. *The Small Nation Solution: How the World's Smallest Nations Can Solve the World's Biggest Problems*. Lanham, MD: AltaMira Press.

———. 2015. *Victims of Progress*. 6th ed. Lanham, MD: Rowman & Littlefield.

Boggs, Carl. 1995. *The Socialist Tradition: From Crisis to Decline*. New York: Routledge

Bohren, Lenora. 2009. "Car Culture and Decision-Making: Choice and Climate Change." In *Anthropology & Climate Change: From Encounters to Actions*, ed. Susan A. Crate and Mark Nuttal. Walnut Creek, CA: Left Coast Press, 370–379.

Boisvert, Will. 2013. "Green Energy Bust in Germany." *Dissent*. Summer, 62–70.

Bonds, Eric, and Liam Downey. 2012. "'Green' Technologies and Ecologically Unequal Exchange: The Environmental and Social Consequences of Ecological Modernization in the World-System." *Journal of World System Research* 18: 167–186.

Bonini, Astra. 2012. "Complementary and Competitive Regimes of Accumulation." *American Sociological Review* 18: 50–68.

Bookchin, Murray. 1991. *The Ecology of Freedom*. 2nd ed. Montreal: Black Rose Books.

Boothroyd, Rachel. 2016. "Maduro Announces New Economic Measures amid Crisis." *Green Left Weekly*. 23 February, 18.

Bosteels, Bruno. 2014. *The Actuality of Communism*. London: Verso.

Boswell, Terry, and Christopher Chase-Dunn. 2000. *The Spiral of Capitalism and Socialism*. Boulder, CO: Lynne Riener.

Brandal, Nik, Oivind Bratberg, and Dag Einar Thorsen. 2013. *The Nordic Model of Social Democracy*. New York: Palgrave Macmillan.

Brie, Michael. 2007. "On the Difficulties of Speaking about the GDR." *Socialism and Democracy* 9, no. 1: 37–57.

Bronner, Stephen Eric. 2011. *Socialism Unbound: Principles, Practices, and Prospects*. 2nd ed. New York: Columbia University Press.

———. 2014. *Moments of Decision: Political History and the Crises of Radicalism*. London: Bloomsbury.

Brown, Lester R. 2008. *Plan B 3.0: Mobilization to Save Civilization.* New York: W.W. Norton & Company.

———. 2009. *Plan B 4.0: Mobilizing to Save Civilization.* New York: W.W. Norton & Company.

Brownhill, Leigh. 2010. "Earth Democracy and Ecosocialism: What's in a Name?" *Capitalism, Nature, Socialism* 21, no. 1: 96–99.

Buechler, Steven M. 2014. *Critical Sociology.* 2nd ed. Boulder, CO: Paradigm Publishers.

Bulletin of Atomic Scientists. 2015. "Three Minutes and Counting." http://thebulletin.org/three-minutes-and-counting7938.

Burgmann, Verity, and Hans A. Baer. 2012. *Climate Politics and Climate Movement in Australia.* Melbourne: Melbourne University Press.

Burkett, Paul. 2006. *Marxism and Ecological Economics: Toward a Red and Green Political Economy.* Leiden: Brill.

———. 2014. *Marx and Nature: A Red and Green Perspective.* Chicago: Haymarket Books.

Burman, Stephen. 2007. *The State of the American Empire: How the USA Shapes the World.* London: Earthscan.

Burrant, Stephen R. 1988. "The Society and Its Environment." In *East Germany: A Country Study*, ed. Stephen R. Washington, DC: U.S. Government Printing Office, 61–115.

Buxton, Julia. 2016. "Venezuela after Chavez." *New Left Review* no. 99: 5–25.

Callenbach, Ernest. 1975. *Ecotopia: The Notebooks and Reports of William Weston.* New York: Bantam Books.

Callinicos, Alex. 2003. *An Anti-Capitalist Manifesto.* Cambridge, UK: Polity.

Campbell, Al. 2016. "'Updating Cuba's Economic Model': Socialism, Human Development, Markets and Capitalism." *Socialism and Democracy* 30, no. 1: 1–29.

Caradonna, Jeremey L. 2014. *Sustainability: A History.* Oxford, U.K.: Oxford University Press.

Carneiro, Robert L. 1981. "Leslie A. White." In *Totems and Teachers: Perspectives on the History of Anthropology*, ed. Sydel Silverman. New York: Columbia University Press, 209–251.

———. 2004. "The Political Unification of the World: Whether, When, and How—Some Speculations." *Cross-Cultural Research* 38, no. 2: 162–177.

Castells, Manuel. 2012. *Networks of Outrage and Hope: Social Movements in the Internet Age.* London: Polity.

Chambers, Neil B. 2011. *Urban Green: Architecture for the Future.* New York: Palgrave Macmillan.

Chase-Dunn, Christopher. 1982. "Socialist States in Capitalist World-Economy." In *Socialist States in the World-System*, ed. Christopher Chase-Dunn. Beverly Hills, CA: Sage, 21–55.

———. 2005. "Social Evolution and the Future of World Society." *Journal of World-Systems Research* 11, no. 2: 171–192.

———. 2010. "Globalization from Below: Toward a Democratic Global Commonwealth." *Journal of Globalization Studies* 1, no. 1: 46–57.

———. 2013. "Five Linked Crises in the Contemporary World System." *Journal of World-System Research* 1, no. 2: 175–180.

Chase-Dunn, Christopher, and Ellen Reese. 2007. "The World Social Forum—A Global Party in the Making?" In *Global Political Parties*, ed. Katarina Sehm-Patomaeki. London: Zed Books, 53–91.

Chase-Dunn, Christopher and Bruce Lerro. 2014. *Social Change: Globalization from the Stone Age to the Present.* Boulder, CO: Paradigm Publishers.

Chattopadhyay, Paresh. 2010. "The Myth of Twentieth-Century Socialism and the Continuing Relevance of Karl Marx." *Socialism and Democracy* 24, no. 3: 23–45.

Childs, David. 1983. *The GDR: Moscow's German Ally.* London: George Allen & Unwin.

———. 2001. *The Fall of the GDR: Germany's Road to Unity.* Harlow, Essex: Longman.

Chirot, Daniel. 1986. *Social Change in the Modern Era.* San Diego: Harcourt Bruce Jovanovich Publishers.

Chivers, Danny. 2009. "Climate Choices." In *People First Economics*, ed. David Ranson and Vanessa Baird. Oxford, U.K.: World Changing, 103–209.

Chomsky, Noam. 2012. *Occupy.* Brooklyn: Zuccotti Park Press.

Christoff, Peter, ed. 2014. *Four Degrees of Global Warming: Australia in a Hot World.* London: Earthscan Routledge.

Clement, Matthew Thomas. 2011. "The Town-Country Antithesis and the Environment: A Sociological Critique of a "Real Utopian" Project." *Organization & Environment* 24: 292–311.

Cliff, Tony. 1974. *State Capitalism in Russia.* London: Pluto.

Cohen, Stephen F. 1973. *Bukharin and the Bolshevik Revolution: A Political Biography, 1888-1938.* New York: Vintage Books.

———. 2009. *Soviet Fates and Lost Alternatives: From Stalinism to the New Cold War.* New York: Columbia University Press.

Collins, Randall. 1995. "Prediction in Macrosociology: The Case of the Soviet Collapse." *American Journal of Sociology* 100: 1553–1593.

Collins, Samuel Gerald. 2008. *All Tomorrow's Cultures: Anthropological Engagements with the Future.* New York: Berghahn Books.

Commoner, Barry. 1972. *The Closing Circle: Confronting the Environmental Crisis.* London: Cape.

———. 1990. *Making Peace with the Planet.* New York: Pantheon Books.

Connell, Raewyn. 2011. *Confronting Equality: Gender, Knowledge and Global Change.* Sydney: Allen & Unwin.

Connolly, Kate. 2007. "Wall Remembered: Germans Hanker after the Barrier." *The Guardian*, 8 November, 20.

Connor, Linda. 2012. "Experimental Publics: Activist Culture and Political Intelligibility of Climate Change Action in the Hunter Valley, Southeast Australia." *Oceania* 82: 228–249.

Crate, Susan A. 2011. "Climate and Culture: Anthropology in the Era of Contemporary Climate Change." *Annual Review of Anthropology* 40: 175–194.

Crate, Susan A., and Mark Nuttall, eds. 2009. *Anthropology and Climate Change: From Encounters to Actions.* Walnut Creek, CA: Left Coast Press.

———. 2016. *Anthropology and Climate Change: From Encounters to Actions.* 2nd ed. London: Routledge.

Crutzen, Paul, and Eugene Stoermer. 2000. "The Anthropocene." *Global Change Newsletter* 41: 17–18.

Cumings, Bruce. 2004. *North Korea: Another Country.* New York: The New Press.

Daly, Herman. 2015. "Economics for a Full World." *Great Transition Initiative.* http://greattransition.org/publication/economics-for-a-full-world.

Daly, Herman E., and John B. Cobb, Jr. 1990. *For the Common Good.* London: Green Point.

Daniela, Danna. 2014. "Population Dynamics and World-System Analysis." *Journal of World Systems Theory* 20, no. 2: 207–208.

Dauverge, Peter. 2005. "Dying of Consumption: Accidents or Sacrifices of Global Mortality?" *Global Environmental Politics* 5, no. 3: 35–47.

David-West, Alzo. 2013. "Eric Fromm and North Korea: Social Psychology and the Political Regime." *Critical Sociology* 40, no. 4: 575–600.

Davis, Mike. 2010. "Who Will Build the Ark?" *New Left Review,* no. 61 (Jan/Feb): 29–46.

Dealeage, J. 1989. "Eco-Marxist Critique of Political Economy." *Capitalism, Nature, Socialism* no. 3: 15–31.

Dean, Jodi. 2016. *Crowds and Party.* London: Verso.

Delina, Laurence L., and Mark Diesendorf. 2013. "Is Wartime Mobilisation a Suitable Policy Model for Rapid National Climate Mitigation?" *Energy Policy* 58: 371–380.

Dennis, Mike. 1988. *German Democratic Republic: Politics, Economics and Society.* London: Pinter.

Derber, Charles. 2010. *Greed to Green: Solving Climate Change and Remaking the Economy.* Boulder, CO: Paradigm Publishers.

———. 2011. *Marx's Ghost: Midnight Conversations on Changing the World.* Boulder, CO: Paradigm Publishers.

Deutscher, Isaac. 1967. *The Unfinished Revolution: Russia 1917-1967.* London: Oxford University Press.

———. 1971. *Marxism in Our Time.* San Francisco: Ramparts Press.

Di Cauter, Lieven. 2008. "The Mad Max Phase." In *Urban Climate Change Crossroads,* ed. Richard Plunz and Maria Paola Sutto, 111–117. Farnham, England: Ashgate.

DIE LINKE. 2011. *Das Programme der Partei DIE LINKE: Resolution of the Party Congress of the Party THE LEFT.* Erfurt, Germany, 21–23 October.

Diesendorf, Mark. 2010/2011. "Strategies for Radical Climate Change Mitigation." *Journal of Australian Political Economy* no. 66: 98–117.

———. 2014. *Sustainable Energy Solutions for Climate Change*. Sydney: UNSW Press.
Dietz, Matthias, and Heiko Garrelts, eds. 2014. *Routledge Handbook of the Climate Change Movement*. London: Routledge.
Dilworth, Craig. 2010. *Too Smart for Our Own Good*. Cambridge, U.K.: Cambridge University Press.
Dingsdale, Alan, and Denes Loczy. 2001. "The Environmental Challenge of Societal Transition in East Central Europe." In *East Central Europe and the Former Soviet Union: Environment and Society*, ed. David Turnock. London: Arnold, 187–199.
Dixon, Chris. 2014. *Another Politics: Talking Across Today's Transformative Movements*. Berkeley: University of California Press.
Doran, Peter T., and Maggie Kendall Zimmerman. 2009. "Examining the Scientific Consensus on Climate Change." *Eos* 90, no. 3: 22–23.
Dorling, Danny. 2013. *Population 10 Billion: The Coming Demographic Crisis and How to Survive It*. London: Constable.
Douzinas, Costas and Slavoj Zizek, eds. 2010. *The Idea of Communism*. London: Verso.
Dove, Colin, ed. 2014. *The Anthropology of Climate Change: A Historical Review*. Malden, MA: Wiley Blackwell.
Dryzek, John S., Richard B. Norgaard, and David Schlosberg, eds. 2011. *Oxford Handbook of Climate Change and Society*. Oxford, U.K.: Oxford University Press.
Dryzek, John S., Richard B. Norgaard, and David Schlosberg. 2013. *Climate-Challenged Society*. Cambridge, U.K.: Cambridge University Press.
Duffy, Frank. 2008. *Work and the City*. London: Black Dog Publishing.
Dutka, Projjal. 2013. "Taking the Car Out of the Carbon: Mass Transit and Emission Avoidance." In *Transport Beyond Oil: Policy Choices for a Multimodal Future*, ed. John L. Renne and Billy Fields. Washington, DC: Island Press, 126–140.
Eagleton, Terry. 2011. *Why Marx Was Right*. New Haven, CT: Yale University Press.
Eckard, Richard, Matthew Bell, Karen Christie, and Richard Rawnsley. 2013. "Livestock." In *Living in a Warmer World: How a Changing Climate Will Affect Our Lives*, ed. Jim Salinger. Melbourne: CSIRO Publishing, 144–157.
Eckhardt, Giana, Russell Belk, and Timothy M. Devinney. 2010. "Why Don't Consumers Consume Ethically?" *Journal of Consumer Behaviour* 9: 426–436.
Eden, Lynn, Robert Rosner, Rod Ewing, Lawrence M. Krauss, Sivan Kartha, Thomas R. Pickering, Raymond T. Pierrehumbert, Ramamurti Rajaraman, Jennifer Sims, Richard C. J. Somerville, Sharon Squassoni, and David Titley. 2016. "It Is Still Three Minutes to Midnight." *Bulletin of Atomic Scientists*. 22 January. http://thebulletin.org/it-still-three-minutes-midnight9107.
el-Ojeili, Chamsy. 2014. "Anarchism as the Contemporary Spirit of Anti-capitalism? A Critical Survey of Recent Debates." *Critical Sociology* 40: 451–468.

Engels, Frederich. 1969 [original 1845]. *The Condition of the Working Class in England.* London: Grenada.

Ergas, Christina. 2010. "A Model of Sustainable Living: Collective Identity in an Urban Ecovillage." *Organization & Environment* 23, no. 1: 32–54.

Escobar, Arturo. 2010. "Latin America at a Crossroad: Alternative Modernizations, Post-liberalism, or Post-development?" *Cultural Studies* 24, no. 1: 1–66.

Eurostat. 2012. *Statistics by the Theme.* http://ec.europa.eu/eurostat/data/browse-statistics-by-theme.

Farber, Samuel. 2011. *Cuba since the Revolution of 1959: A Critical Assessment.* Chicago: Haymarket Books.

Fenelon, James V. 2012. "Indigenous Peoples, Globalization and Autonomy in World-Systems Analysis." In *Routledge Handbook of World-Systems Analysis.* Hoboken, NJ: Taylor and Francis.

Ferguson, Peter. 2015. "The Green Economy Agenda: Business as Usual or Transformational Discourse." *Environmental Politics* 24: 17–37.

Fertl, Duroyan. 2014. "Twenty-Five Years after Fall of Berlin Wall, Socialists to Head Government in Eastern State." *Green Left Weekly.* 4 September, 18.

Feshman, Murray and Alfred Friendly, Jr. 1992. *Ecocide in the USSR: Health and Nature Under Siege.* New York: Basic Books.

Fidler, Richard. 2014. "Bolivia Leads Climate Change Fight." *Green Left Weekly.* 22 October, 18.

Filtzer, Donald. 2014. "Privilege and Inequality in Communist Society." In *The Oxford Handbook of the History of Communism,* ed. Stephen A Smith. Oxford, U.K.: Oxford University Press, 504–521.

Fiske, Shirley J. 2009. "Global Change Policymaking from Inside the Beltway: Engaging Anthropology." In *Anthropology & Climate Change: From Encounters to Actions,* ed. Susan A. Crate and Mark Nuttall. Walnut Creek, CA: Left Coast Press, 277–291.

———. 2012. "The Anthropocene? Planet Earth in the Age of Humans." *Anthropology News,* 16 October.

Flacks, Richard. 2013. "Where Is It Likely to Lead?" *Sociological Quarterly* 54: 202–206.

Flaherty, Patrick. 1991. "Stalinist Power Structures and Military." *Socialism and Democracy* 7, no. 3: 141–156.

Flassbeck, Heiner, and Costas Lapavitsas. 2015. *Against the Troika: Crisis and Austerity in the Eurozone.* London: Verso.

Foster, John Bellamy. 1999. *The Vulnerable Planet.* New York: Monthly Review Press.

———. 2000. *Marx's Ecology.* New York: Monthly Review Press.

———. 2002. *Ecology against Capitalism.* New York: Monthly Review Press.

———. 2005. "Organizing Ecological Revolution." *Monthly Review* 61, no. 5: 1–23.

———. 2009. *The Ecological Revolution: Making Peace with the Planet.* New York: Monthly Review Press.

———. 2011. "Capitalism and Degrowth—An Impossibility Theorem." *Monthly Review* 62, no. 8: 26–33.
———. 2013. "James Hansen and the Climate-Change Exit Strategy." *Monthly Review* 64(9): 1–19.
———. 2014a. *The Theory of Monopoly Capitalism: An Elaboration of Marxian Political Economy.* New ed. New York: Monthly Review Press.
———. 2014b. "Paul Burkett's *Marx and Nature* Fifteen Years After." *Monthly Review* December: 56–62.
———. 2015. "Marxism and Ecology: Common Fonts of a Great Transition." *Great Transition Initiative.* October. http://www.greattransition.org/publication/marxism-and-ecology.
Foster, John Bellamy, and Brett Clark. 2015. "Crossing the River of Fire: The Liberal Attack on Naomi Klein and This Changes Everything." *Monthy Review* February, 1–17.
———. 2016. "Marxism and the Dialectics of Ecology." *Monthly Review* 68, no. 5: 1–17.
Foster, John Bellamy, Brett Clark, and Richard York. 2010. "Capitalism and the Curse of Energy Efficiency: The Return of the Jevons Paradox. *Monthly Review* 62, no. 6: 1–12.
Fotopoulos, Takis. 1997. *Towards an Inclusive Democracy: The Crisis of the Growth Economy and the Need for a New Liberatory Project.* New York: Cassell.
———. 2010. "The De-growth Utopia: The Incompatibility of De-growth within an Internationalised Market Economy." In *Eco-Socialism as Politics: Rebuilding the Basis of Our Modern Civilization,* ed. Q. Huan. New York: Springer, 103–121.
Frank, Andre Gunder. 1990. "Civil Democracy: Social Movements in Recent Social History." In *Transforming the Revolution: Social Movements and the World-System,* ed. Samir Amin, Giovanni Arrighi, Andre Gunder Frank, and Immanuel Wallerstein. New York: Monthly Review Press, 139–180.
French, Paul. 2014. *North Korea: State of Paranoia.* London: Zed Press.
Friedlingstein, P., et al. 2014. "Persistent Growth of CO_2 Emissions and Implications for Reaching Climate Targets." *Nature Geoscience* 7: 709–715.
Ganti, Tejaswini. 2014. "Neoliberalism." *Annual Review of Anthropology* 43: 89–104.
Gare, Arran. 2014. "Creating an Ecological Socialist Future." *Capitalism, Nature, Socialism* 11, no. 2: 23–40.
Gates, Becky A. 1988. "The Economy." In *East Germany: A Country Study,* ed. Stephen R. Burrant. Washington, DC: U.S. Government Printing Office, 119–159.
Gautier, Catherine. 2008. *Oil, Water, and Climate: An Introduction.* Cambridge, U.K.: Cambridge University Press.
Geall, Sam. 2013. "China's Environmental Journalists: A Rainbow Confusion." In *China and the Environment: The Green Revolution,* ed. Sam Geall and Isabel Hilton. London: Zed Books, 15–39.
Gerth, Karl. 2010. *As China Goes So Goes the World: How Chinese Consumers Are Transforming Everything.* New York: Hill & Wang.

Giacomini, Terran, and Terisa Turner. 2015. "The 2014 People's Climate March and Flood Wall Street Civil Disobedience: Making the Transition to a Post-fossil Capitalist, Commoning Civilization." *Capitalism, Nature, Socialism* 26, no. 2: 27–45.

Giddens, Anthony. 2009. *The Politics of Climate Change*. London: Polity.

———. 2014. *Turbulent and Mighty Continent: What Future for Europe?* London: Polity.

Gilbert, Richard, and Anthony Perl. 2010. "Transportation in the Post-carbon World." In *The Post-Carbon Reader: Managing the 21st Century's Sustainability Crises*, ed. Richard Heinberg and Daniel Lercy. Healdsburg, CA: Watershed Media, 347–360.

Gilding, Paul. 2011. *The Great Disruption: How the Climate Crisis Will Transform the Global Economy*. London: Bloomsbury.

Gill, Stephen. 2007. "The Globalization of Party Politics." In *Global Political Parties*, ed. Katarina Sehm-Patomaeki and Marko Ulvila. London: Zed Books, 114–122.

Giri, S. 2012. "Capitalism Expands but the Discourse Is Radicalized: Wither '21st Century Venezuelan Socialism'?" *Critical Sociology* 31, no. 1: 21–36.

Glaser, Daryl. 2007. "Marxist Theory, Marxist Practice." In *Twentieth-Century Marxism: A Globalist Perspective*, ed. Daryl Glaser and David M. Walker. London: Routledge, 196–210.

Glassman, Ronald M. 1995. *The Middle Class and Democracy in Socio-Historical Perspective*. Leiden: E.J. Brill.

Glickson, Andrew. 2013. "Fire and Human Evolution: The Deep-Time Blueprints of the Anthropocene." *Anthropocene* 3: 89–92.

Godrej, Dinyar. 2002. *The No-Nonsense Guide to Climate Change*. Oxford, U.K.: New Internationalist.

Goldfrank, Walter L. 1987. "Socialism or Barbarism? The Long-Run Fate of the Capitalist World Economy." In *America's Changing Role in the World System*, ed. Terry Boswell and Albert Bergesen. New York: Praeger, 85–92.

Golten, T. E. and B.R. eds. 1977. *The End of the Road: A Citizen's Guide to Transportation Problem Solving*. Washington, DC: National Wildlife Federation.

Goodal, Chris. 2010. *How to Live a Low-Carbon Life*. 2nd ed. London: Earthscan.

Goodkind, Daniel, Loraine West, and Peter Johnson. 2011. "A Reassessment of Mortality in North Korea, 1993-2008." Paper presented at the annual meeting of the Population Association of America, 31 March–2 April, Washington, DC.

Gore, Al. 2006. *An Inconvenient Truth: The Planetary Emergency of Global Warming and What We Can Do About It*. London: Rodale Books.

———. 2007. *Earth in the Balance: Ecology and Human Spirit* 3rd ed. Boston: Houghton Mifflin.

———. 2009. *Our Choice: A Plan to Solve the Climate Crisis*. London: Bloomsbury.

Gore, Al. 2013. *The Future*. London: WH Allen.

Gorz, Andre. 1973. *Socialism and Revolution*. Garden City, NY: Anchor.

———. 1980. *Ecology as Politics*. Boston: South End Press.

———. 1982. *Farewell to the Working Class: An Essay on Post-Industrial Socialism*. London: Pluto Press.

———. 1994. *Capitalism, Socialism, Ecology*. London: Verso.

Gorz, Bronislaw, and Kurek. 2001. "Sustainable Agriculture." In *East Central Europe and the Former Soviet Union: Environment and Society*, ed. David Turnock. London: Arnold, 200–215.

Gottlieb, Roger S. 1992. *Marxism 1844-1990: Origins, Betrayal, Rebirth*. New York: Routledge.

Gowdy, John. 1999. "Hunters-Gatherers and the Mythology of the Market." In *The Cambridge Encyclopedia of Hunters and Gatherers*, ed. Richard B. Lee and Richard Daly. Cambridge, U.K.: Cambridge University Press, 391–398.

Graeber, David. 2005. "The Globalization Movement: Some Points of Clarification." In *The Anthropology of Development and Globalization: From Classical Political Economy of Contemporary Neoliberalism*, ed. Mac Edelman and Angelique Haugerud. Malden, MA: Blackwell Publishing, 169–172.

———. 2013. *The Democracy Project: A History, a Crisis, a Movement*. New York: Spiegel & Grau.

Green Left Weekly. 2016. "Stein Slams 'Trump's Psychosis and Clintons Distortions'." 25 October, 18.

Grigsby, Mary. 2004. *Buying Time and Getting By: The Voluntary Simplicity Movement*. Albany: State University of New York Press.

Grisaffi, Thomas. 2013. "'All of Us Are Presidents': Radical Democracy and Citizenship in the Chapare Province, Bolivia." *Critique of Anthropology* 33, no. 1: 47–65.

Gross, Bertram. 1980. *Friendly Fascism: The New Face of Power in America*. Boston: South End Press.

Grundmann, R. 1991a. "The Ecological Challenge to Marxism." *New Left Review* no. 187: 103–120.

———. 1991b. *Marxism and Ecology*. New York: Oxford University Press.

Gulick, John. 2011. "The Long Twentieth Century and Barriers to China's Hegemonic Accession." *Journal of World-Systems Research* 17, no. 1: 4–38.

Gunderson, Ryan. 2011. "The Metabolic Rifts of Livestock Agriculture." *Organization & Environment* 24: 404–422.

Haggard, Stephan, and Marcus Noland. 2007. *Famine in North Korea: Markets, Aid, and Reform*. New York: Columbia University Press.

Hahnel, Robin. 2007. "Eco-localism: A Constructive Critique." *Capitalism, Nature, Socialism* 18, no. 2: 62–72.

———. 2016. "Participatory Economics & the Next System." *The Next System Project*. http://nextsystemproject.org.

Hall, Gillette and Harry Anthony Patrino. 2012. "Introduction." In *Indigenous Peoples, Poverty, and Development*. Gillette Hall and Harry Anthony Patrino, eds. Cambridge, UK: Cambridge University Press, 117.

Halweil, Brian, and Danielle Nierenberg. 2007. "Farming the Cities." In *2007 State of the World: Our Urban Future*, ed. Linda Starke. New York: W.W. Norton & Company, 48–66.
Hamilton, Clive, and Richard Denniss. 2005. *Affluence: When Too Much Is Never Enough.* Sydney: Allen & Unwin.
Hammond, Allen. 1998. *Which World? Scenarios for the 21st Century.* Washington, DC: Island Press.
Hammond, John L. 2012. "Social Movements and Struggles for Socialism." In *Taking Socialism Seriously,* ed. Anatole Anton and Richard Schmitt. Lanham, MD: Lexington Books, 213–247.
Hannerz, Ulf. 2015. Writing Futures: Anthropologist's View of Global Scenarios. *Current Anthropology* 26: 797–818.
Hardt, Michael, and Antonio Negri. 2009. *Commonwealth.* Cambridge, MA: Harvard University Press.
Harnecker, Marta. 2015. *A World to Build: New Paths toward Twenty-First Socialism.* New York: Monthly Review Press.
———. 2016. "Social Movements and Progressive Governments: Rebuilding a New Relationship in Latin America." *Monthly Review.* January, 25–34.
Harris, Marvin. 1992. "Distinguished Lecture: Anthropology and the Theoretical and Paradigmatic Significance of the Collapse of Soviet and East European Communism." *American Anthropologist* 92: 295–305.
Harrison, Mark. 2014. "Communism and Economic Modernization." In *The Oxford Handbook of the History of Communism,* ed. Stephen A. Smith. Oxford, U.K.: Oxford University Press, 387–406.
Harsch, Donna. 2014. "Communism and Women." In *The Oxford Handbook of the History of Communism,* ed. Stephen A. Smith. Oxford, U.K.: Oxford University Press, 488–504.
Hart-Landsberg, Martin. 1998. *Korea: Division, Reunification, and U.S. Foreign Policy.* New York: Monthly Review Press.
Hart-Landesberg, Martin, and Paul Burkett. 2005. *China and Socialism: Market Reforms and Class Struggle.* New York: Monthly Review Press.
Harvey, David. 2012. *Rebel Cities: From the Right to the City to the Urban Revolution.* London: Verso.
———. 2014. *Seventeen Contradictions and the End of Capitalism.* London: Profile Books.
Haug, Frigga, and the Dialektikfrauen. 2012. "The Politics of Die LINKE—A Politics of Time." *Socialism and Democracy* 26, no. 1: 143–155.
Hawken, Paul, Amory Lovins, and L. Hunter Lovins. 1999. *Natural Capitalism: Creating the Next Industrial Revolution.* London: Little, Brown.
Haynes, Mike. 2002. *Russia: Class and Power 1917-2000.* London: Bookmarks Publications.
Heinberg, Richard. 2004. *Powerdown: Options and Actions for a Post-Carbon World.* Gabriola Island, BC: New Society Publishers.

———. 2011. *The End of Growth: Adapting to Our New Economic Reality*. Gabriola Island, BC: New Society Publishers.

———. 2015. *Afterburn: Society Beyond Fossil Fuels*. Gabriola Island, BC: New Society Publishers.

Henderson, Gail E., and Myron S. Cohen. 1984. *The Chinese Hospital: A Socialist Work Unit*. New Haven, CT: Yale University Press.

Hertsgaard, Mark. 2011. *Hot: Living Through the Next Fifty Years on Earth*. Boston: Houghton Mifflin.

Herzfeld, Michael. 2016. "The Hypocrisy of European Moralism: Greece and the Politics of Cultural Aggression—Part 2." *Anthropology Today* 32, no. 2: 10–13.

Higgs, Kerryn. 2014. *Collision Course: Endless Growth on a Finite Planet*. Cambridge, MA: MIT Press.

Hodges, Donald C. 1981. *The Bureaucratization of Socialism*. Amherst: University of Massachusetts.

Hoffman, John. 1984. *The Gramscian Challenge: Coercion and Consent in Marxist Political Theory*. Oxford, U.K.: Basil Blackwell.

Hofrichter, Richard. 2000. "Introduction: Critical Perspectives on Human Health and the Environment." In *Reclaiming the Environmental Debates: The Politics of Health in a Toxic Culture*. Richard Hofrichter, ed. Pp. 1:15. Cambridge, MA: MIT Press.

Hogwood, Patricia. 2012. "Political (Re)learning and Consumer Culture in Post-GDR Society." *German Politics* 21: 1–16.

Hollender, Rebecca. 2015. "Post-growth in the Global South: The Emergence of Alternatives to Development in Latin America." *Socialism and Democracy* 29, no. 1: 73–101.

Holloway, John. 2005. *Change the World without Taking Power: The Meaning of Revolution Today*. New ed. London: Pluto Press.

———. 2010. *Crack Capitalism*. London: Pluto Press.

Holmes, Les. 1997. *Post-Communism: An Introduction*. Durham, NC: Duke University Press.

Holmgren, David. 2009. *Future Scenarios: How Communities Can Adapt to Peak Oil and Climate Change*. White River Junction, VT: Chelsea Green.

Hopkins, Rob. 2008. *The Transition Handbook: Creating Local Sustainable Communities Beyond Oil Dependency*. Sydney: Finch Publishing.

Hornborg, Alf. 2001. *The Power of the Machine: Global Inequalities of Economy, Technology, and Environment*. Walnut Creek, CA: AltaMira Press.

Horvat, Branko. 1982. *The Political Economy of Socialism: A Marxist Social Theory*. Armonk, NY: M.E. Sharpe.

Hou, Yanjie, and Jie Xu. 2012. "Socialism and Ecological Crises: A View from China." *Journal of Sustainable Development* 5, no. 4: 126–131.

Huan, Qingzhi. 2010a. "Eco-socialism in an Era of Capitalist Globalisation: Bridging the West and the East." In *Eco-Socialism as Politics: Rebuilding the Basis of Our Modern Civilization*, ed. Quigzhi Huan. New York: Springer, 3–12.

———. 2010b. "Growth Economy and Its Ecological Impacts upon China: An Eco-socialist Analysis." In *Eco-socialism as Politics: Rebuilding the Basis of Our Modern Civilization.* ed. Quigzhi Huan. New York: Springer, 191–203.

———. 2016. "Socialist Eco-Civilization and Social-Ecological Transformation." *Capitalism, Nature, Socialism* 27, no. 2: 51–66.

Huberman, Leo and Paul M. Sweezy. 1969. *Socialism in Cuba.* New York: Monthly Review Press.

Hui, Wang. 2009. *The End of the Revolution: China and the Limits of Modernity.* London: Verso.

Humphrys, Elizabeth. 2013. "Organic Intellectuals in the Australian Global Justice Movement: The Weight of 9/11." In *Marxism and Social Movements,* ed. Colin Barker, Laurence Cox, John Kinsky, and Alf Gunvald Nilsen. Leiden: Brill, 357–375.

Hunter, Helen-Louise. 1999. *Kim Il-song's North Korea.* Westport, CT: Praeger.

Iglesias, Pablo. 2015. *Politics in a Time of Crisis: Podemos and the Future of a Democratic Europe.* London: Verso.

Jackson, Tim. 2009. *Prosperity Without Growth.* London: Earthscan.

Jacques, Peter, Riley Dunlap, and Mark Freeman. 2008. "The Organization of Denial: Conservative Think Tanks and Environmental Scepticism." *Environmental Politics* 17, no. 3: 349–385.

Jacques, Peter J., and Jessica R. Jacques. 2012. "Monocropping Cultures into Ruin: The Loss of Food Varieties and Cultural Diversity." *Sustainability* 4: 2970–2997.

Jameson, Fredric. 2005. *Archaeologies of the Future: The Desire Called Utopia and Other Science Fictions.* London: Verso.

Jami, Dahr. 2010. "Tar Sands—A Toxic Nightmare." *Green Left Weekly,* 23 June, 16.

Jasanoff, Sheila. 2010. "A New Climate for Society." *Theory, Culture & Society* 27, no. 2–3: 233–253.

Jessop, Bob. 2016. *The State: Past, Present, Future.* London: Polity.

Johnston, Hank. 2011. *States & Social Movements.* London: Polity.

Jones, Andrew W. 2011. "Solving the Ecological Problems of Capitalism: Capitalist and Socialist Possibilities." *Organization & Environment* 24: 54–73.

Jorgenson, Andrew K., Brett Clark, and Jeffrey Kentor. 2010. "Militarization and the Environment: A Panel Study of Carbon Dioxide Emissions and the Ecological Footprints of Nations, 1970–2000." *Global Environmental Politics* 10, no. 1: 7–29.

Kagarlitsky, Boris. 1995. *The Mirage of Modernization.* New York: Monthly Review Press.

———. 2002. *Russia Under Yeltsin and Putin: Neo-liberal Autocracy,* trans. Renfrey Clarke. London: Pluto Press.

———. 2004. "The Russian State in the Age of American Empire." In *Socialist Empire: The Empire Reloaded,* ed. Leo Panitch and Colin Leys. London: Merlin Press, 271–283.

———. 2008. *Empire of the Periphery: Russia and the World System*, trans. Renfrey Clarke. London: Pluto Press.
Kaldor, Mary. 1999. "Transnational Civil Society." In *Human Rights in Global Politics*, ed. T. Dunne and N. Wheeler. Cambridge, U.K.: Cambridge University Press, 195–213.
Kath, Elizabeth. 2010. *Social Relations and the Cuban Health Miracle*. New Brunswick, NJ: Transaction Books.
Kaup, Brent Z. 2015. "Reiterated Problem Solving in Neoliberal and Counter-neoliberal Shifts: The Case of Bolivia's Hydrocarbon Sector." *Theory and Society* 44: 445–470.
Keller, Edmond J. 1987. "Afro-Marxist Regimes." In *Afro-Marxist Regimes: Ideology and Public Policy*, ed. Edmond J. Keller and Donald Rothchild. Boulder, CO: Lynne Rienner Publishers, 1–21.
Kelly, Raymond C. 2000. *Warless Societies and the Origins of War*. Ann Arbor: University of Michigan Press.
Kennemore, Amy, and Gregory Weeks. 2011. "Twenty-First Century Socialism? The Elusvive Search for a Post-neoliberal Development Model in Bolivia and Ecuador." *Bulletin of Latin American Research* 30: 267–281.
Kenworthy, Lane. 2016. "Social Democracy." *The Next System Project*. http//:thenextsystem.org.
Klare, Michael T. 2008. *Rising Powers, Shrinking Earth: The New Geopolitics of Energy*. New York: Metropolitan Books.
Klein, Naomi. 2007. *The Shock Doctrine: The Rise of Disaster Capitalism*. New York: Metropolitan Books.
———. 2014. *This Changes Everything: Capitalism vs. the Climate*. London: Allen Lane.
Kolko, Gabriel. 2006. *After Socialism: Reconstructing Critical Social Thought*. London: Routledge.
Korten, David. 2001. *When Corporations Rule the World*. 2nd ed. San Francisco: Kumerian Press.
Kovel, Joel. 2007. *The End of Nature: The End of Capitalism or the End of the World?*. 2nd ed. London: Zed Books.
Kovel, Joel, and Michael Loewy. 2001. *An Ecosocialist Manifesto*. September. https://greenfrombelow.wordpress.com/an-eco-socialist-manifesto/.
Kuran, Timur. 1995. "The Inevitability of Future Revolutionary Surprises. *American Journal of Sociology* 100: 1528–1551.
Labban, Mazen, David Correia, and Matt Huber. 2013. "Apocalypse, the Radical Left and the Post-colonial condition." *Capitalism, Nature, Socialism* 24, no. 1: 6–8.
Ladd, Brian. 2008. *Autophobia: Love and Hate in the Automotive Age*. Chicago: University of Chicago Press.
Lane, Bradley. 2013. "Public Transportation as a Solution to Oil Dependence." In *Transport Beyond Oil: Policy Choices for a Multimodal Future*, ed. John L. Renne and Billy Fields. Washington, DC: Island Press, 107–125.

Lane, David. 1976. *The Socialist Industrial State*. London: George Allen & Unwin.
———. 1981. *Leninism: A Sociological Interpretation*. Cambridge, U.K.: Cambridge University Press.
Lankov, Andrei. 2013. *The Real North Korea: Life and Politics in the Failed Stalinist Utopia*. Oxford, U.K.: Oxford University Press.
Laquer, Walter. 1985. *Germany Today: A Personal Report*. Boston: Little, Brown & Co.
Latouche, Serge. 2012. "Can the Left Escape Economism?" *Capitalism, Nature, Socialism* 23, no. 1: 74–78.
Le Blanc, Paul. 2006. *Marx, Lenin, and the Revolutionary Experience: Studies of Communism and Radicalism in the Age of Globalization*. London: Routledge.
Leahy, Terry. 2011. "The Gift Economy." In *Life Without Money: Building Fair and Sustainable Economics*, ed. Anitra Nelson and Frans Timmerman. London: Pluto Press, 111–135.
Lebowitz, Michael. 2012. *The Contradictions of Real Socialism: The Conductor and the Conducted*. New York: Monthly Review Press.
———. 2014. "Proposing a Path to Socialism: Two Papers for Hugo Chavez." *Monthly Review* March, 1–19.
Leech, Garry. 2012. *Capitalism: A Structural Critique*. London: Zed Books.
Lenin, Vladamir I. 1964. "The Tasks of the Proletariat in the Present Revolution." In *Collected Works, 24*. Moscow: Progress Publishers.
Lever-Tracy, Constance, ed. 2010. *Routledge Handbook of Climate Change and Society*. London: Routledge.
Levins, Richard. 2010. "How to Visit a Socialist Country." *Monthly Review* 61, no. 11: 1–27.
Lewin, Moshe. 2005. *The Soviet Century*, ed. Gregory Elliott. London: Verso.
Li, Minqi. 2008. *The Rise of China and the Demise of the Capitalist World-Economy*. New York: Monthly Review Press.
———. 2009. "Capitalism, Climate Change and the Transition to Sustainability: Alternative Scenarios for the US, China and the World." *Development and Change* 40: 1039–1061.
———. 2016. *China and the 21st Century*. London: Pluto Press.
Lichtheim, George. 1970. *A Short History of Socialism*. London: Flamingo.
Liftin, Karen T. 2014. *Eco-Villages: Lessons for Sustainable Community*. London: Polity.
Linden, Marcel van der. 2007. *Western Marxism and the Soviet Union*. Chicago: Haymarket Books.
Little, Amanda. 2009. *Power Trip: From Oil Wells to Solar Cells—Our Ride to the Renewable Future*. London: Harper Press.
Loewenstein, Antony. 2015. *Disaster Capitalism: Making a Killing out of Catastrophe*. London: Verso.
Loewy. Michael. 2006. "Eco-Socialism and Democratic Planning." In *Coming to Terms with Nature*, ed. Leo Panitch and Colin Leys. London: Merlin Press, 294–309.

Loewy, Michael. 2015. *Ecosocialism: A Radical Alternative to Capitalist Catastrophe.* Chicago: Haymarket.

Lohmann, Larry. 2006. "Carbon Trading: A Critical Conversation on Climate Change, Privatisation and Power." *Development Dialogue* no. 28. Uppsala, Sweden: Dag Hammarskjoeld Centre.

Lorimer, Doug. 1997. *The Collapse of 'Communism' in the USSR: Its Causes and Significance.* Chippendale, NSW: Resistance Books.

Lovelock, James. 2006. *The Revenge of Gaia.* London: Penguin Books.

———. 2009. *The Vanishing Face of Gaia: A Final Warning.* Camberwell, VIC: Allen Lane.

———. 2014. *A Rough Ride to the Future.* London: Allen Lane.

Lynas, Mark. 2007. *Six Degrees: Our Future on a Hotter Planet.* London: Fourth Estate.

Lynch, Michael. 2004. *The People's Republic of China, 1949-76.* London: Hodder Education.

Lynd, Staughton, and Andrej Grubacic. 2008. *Wobblies & Zapatistas: Conversations on Anarchism, Marxism and Radical History.* Oakland, CA: PM Press.

Madarasz, Jeannette. 2006. *Working in East Germany: Normality in a Socialist Dictatorship, 1961-1979.* Basingstoke, U.K.: Palgrave Macmillan.

Magdoff, Fred. 2013. "Global Resource Depletion: Is Population the Problem?" *Monthly Review* 64, no. 8: 13–28.

———. 2014. "An Ecologically Sound and Socially Just Economy. *Monthly Review* 66, no. 4: 23–34.

Magdoff, Fred, and John Bellamy Foster. 2010. "What Every Environmentalist Needs to Know about Capitalism." *Monthly Review* 61, no. 10: 1–30.

———. 2011. *What Every Environmentalist Needs to Know About Capitalism: A Citizen's Guide to Capitalism and Environment.* New York: Monthly Review Press.

Makhijani, Shakuntala, and Alexander Ochs. 2013. "Renewable Energy's Natural Resource Impacts." In *State of the World 2013: Is Sustainability Still Possible?*, ed. Linda Starke. Washington, DC: Island Press, 84–98.

Malinowski, Bronislaw. 1941. "War—Past, Present, and Future." In *War as a Social Institution,* ed. J.L. Clarkson and T.C. Cochran. New York: Columbia University Press, 21–31.

Malm, Andreas. 2014. "Tahrir Submerged? Five Theses on Revolution in the Era of Climate Change." *Capitalism Nature Socialism* 25(3): 28–44.

———. 2016. *Fossil Capital: The Rise of Steam Power and the Roots of Global Warming.* London: Verso.

Mamdani, Mahomood. 1972. *The Myth of Population Control: Family, Caste, and Class in an Indian Village.* New York: Monthly Review Press.

Mandel, 1981. "The Laws of Motion in the Soviet Economy." *Review of Radical Political Economics* 13, no. 1: 35–39.

———. 1992. *Trotsky as Alternative.* London: Verso.

Marcuse, Peter. 1990. "Letter from the German Democratic Republic." *Monthly Review* 42, no. 3: 30–62.

———. 1991. *Missing Marx: A Personal and Political Journal of a Year in East Germany, 1989-1990*. New York: Monthly Review Press.

Marshall, Jonathan Paul, and Linda H. Connor, eds. 2015. *Environmental Change and the World's Futures: Ecologies, Ontologies and Mythologies*. London: Earthscan from Routledge.

Martinelli, Alberto. 2005. "From World System to World Society." *Journal of World Systems Research* 11, no. 2: 241–260.

Martinez-Fernandez, Luis. 2014. *Revolutionary Cuba: A History*. Gainesville: University Press of Florida.

Marx, Karl. 1975. "Critical Notes on the King of Prussia and Social Reform." In *Early Writings*. Harmondsworth, England: Penguin.

———. 1977. *Economic and Philosophic Manuscripts*. Moscow: Progress Publishers.

Marx, Karl, and Frederich Engels. 1978. "The German Ideology." In *The Marx-Engels Reader*. 2nd ed. New York: Norton.

Mathews, Freya. 1991. *Ecological Self*. London: Routledge.

McAdams, A. James. 1985. *East Germany and Détente: Building Authority after the Wall*. Cambridge, U.K.: Cambridge University Press.

McCarthy, John D., and Mayer N. Zald. 1977. "Resource Mobilization and Social Movements: A Partial Theory." *American Journal of Sociology* 82: 1112–1241.

McClintock, Nathan. 2014. "Radical, Reformist, and Garden-Variety Neoliberal: Coming to Terms with Urban Agriculture's Contradictions." *Local Environment* 19, no. 2: 147–171.

McCluney, Ross. 2005. "Renewable Energy Limits." In *The Final Energy Crisis*, ed. Andrew McKillop and Sheila Newman. London: Pluto Press 153–175.

McCormack, Gavan. 2008. "North Korea: Birth Pangs of a New Northeast Asian Order." *Arena Journal*, no. 29/30: 81–100.

McLaughlin, Andrew. 1990. "Ecology, Capitalism, and Socialism." *Capitalism, Nature, Socialism* 6, no. 1: 69–102.

McLellan, David. 1983. *Karl Marx: The Legacy*. London: BBC.

McKay, Joanna. 2012. "The Berlin Land Election 2011." *German Politics* 21: 228–238.

McMichael, Philip. 2012. *Development and Change: A Global Perspective* (5th edition). Los Angeles: Sage.

McNeill, J.R., and Peter Engelke. 2014. *The Great Acceleration: An Environmental History of the Anthropocene since 1945*. Cambridge, MA: Belknap Press of Harvard University Press.

McQuaig, Linda. 2004. *Crude, Dude: War, Big Oil and the Fight for the Planet*. Toronto: Doubleday Canada.

McWilliams, Wayne C., and Harry Piotrowski. 2009. *The World since 1945: A History of International Relations*. 7th ed. Boulder, CO: Lynne Rienner Publishers.

Mead, Margaret. 1978. "The Contribution of Anthropology to the Science of the Future." In *Cultures of the Future*, ed. Magorah Maruyama and Arthur M. Harkins. The Hague: Mouton, 3–6.

Mees, Paul. 2010. *Transport for Suburbia: Beyond the Automobile.* London: Earthscan.
Meinshausen, M., N. Meinshausen, and W. Hare. 2009. "Greenhouse-Gas Emission Targets for Limiting Global Warming to 2°C. *Nature* 458, no. 7242: 1158–1162.
Mellor, Mary. 2012. "Co-operative Principles for a Green Economy." *Capitalism, Nature, Socialism* 25, no. 2: 108–110.
Merchant, Carolyn. 1996. *Earthcare: Women and the Environment.* New York: Routledge.
———. 2005. *Radical Ecology: The Search for a Liveable World.* 2nd ed. New York: Routledge.
Metz, Bert. 2010. *Controlling Climate Change.* Cambridge, UK: Cambridge University Press.
Metz, David. 2008. *The Limits to Travel: How Far Will You Go?* London: Earthscan.
Meyer, Thomas, with Lewis Hinchman. 2007. *The Theory of Social Democracy.* London: Polity.
Mies, Maria. 2010. *The Village and the World: My Life, Our Times.* Melbourne: Spinifex Press.
Mies, Maria, and Veronika Bennholdt-Thomsen. 1999. *The Subsistence Perspective: Beyond the Globalised Economy.* Zed Books.
Milanovic, Branko. 2012. "Global Inequality Recalculated and Updated: The Effect of New PPP Estimates on Global Inequality and 2005 Estimates." *Journal of Economic Inequality* 10: 1–18.
Miliband, Ralph. 1994. *Socialism for a Skeptical Age.* London: Courage Press.
Minnerup, Gunter. 1989. "The October Revolution in East Germany." *Labour Focus on Eastern Europe* 3: 5–9.
———. 1990. "Kohl Hijacks East German Revolution." *Labour Focus on Eastern Europe.* No. 3: 3–4.
Mintz, Sidney W. 1985. *Sweetness and Power: The Place of Sugar in Modern History.* New York: Viking.
Molyneux, John. 1978. *Marxism and the Party.* Chicago: Haymarket Books.
Moore, Jason W. 2015. *Capitalism in the Web of Life: Ecology and the Accumulation of Capital.* London: Verso.
———. 2016. "Anthropocene or Capitalocene? Nature, History, and the Crisis of Capitalism." In *Anthropocene or Capitalocene: Nature, History, and the Crisis of Capitalism,* ed. Jason W. Moore. Oakland, CA: PM Press, 1–13.
Morgan, Lewis Henry. 1964. *Ancient Society,* ed. Leslie A. White. Cambridge, MA: Harvard University Press.
Morris, Emily. 2014. "Unexpected Cuba." *New Left Review,* no. 88 (July–August): 4–45.
Moschonas, Gerassimos. 2002. *In the Name of Social Democracy—The Great Transformation: 1945 to the Present,* trans. Gregory Elliott. London: Verso.
Mumford, Lewis. 1967. *Technics & Civilization.* New York; Harcourt, Brace & World.

Munckton, Stuart. 2015. "Why Syriza is Greek for Hope." *Green Left Weekly*, January 30, https://www.greenleft.org.au/content/why-syriza-greek-hope.

Murdoch, William H. 1980. *The Poverty of Nations: The Political Economy of Hunger and Population*. Baltimore: Johns Hopkins University Press.

Murphy, Pat. 2008. *Plan C: Community Survival Strategies for Peak Oil and Climate Change*. Gabriola Island, BC: New Society Publishers.

Murphy, Pat, and Faith Morgan. 2013. "Cuba: Lessons from a Forced Decline." In *State of the World 2013: Is Sustainability Still Possible?*, ed. Linda Starke. Washington, DC: Island Press, 332–342.

Myers, Norman, and Jennifer Kent. 2004. *The New Consumers: The Influence of Affluence on the Environment*. Washington, DC: Island Press.

Naroll, Raoul. 1967. "Imperial Cycles and World Order." *Peace Research Society: Papers* 8: 83–101.

NASA. 2016. "Carbon Ddioxide—Llatest Mmeasurement: September 2016." Retrieved 5 October 2016 at http://climate.nasa.gov/vital-signs/carbon-dioxide/.

———. 2017. "NASA, NOAA Data Show 2016 Warmest Year on Record Globally." Jan. 19, 2017. RELEASE 17-006. https://www.nasa.gov/press-release/nasa-noaa-data-show-2016-warmest-year-on-record-globally

National Public Radio. 2012. *Poverty in the U.S. by the Numbers*.

Navarro, Vincente. 1982. "The Limits of the World System Theory in Defining Capitalist and Socialist Formations." *Science and Society* 46: 77–90.

Neale, Jonathan. 2008. *Stop Global Warming: Change the World*. UK: Bookmark Publications.

Nelson, Anitra. "Money versus Socialism." 2011. In *Life without Money: Building Fair and Sustainable Economies*, ed. Anrita Nelson and Frans Timmerman. London: Pluto, 23–46.

Nelson, Anitra, and Frans Timmerman. 2011. "Use Value and Non-market Socialism." In *Life without Money: Building Fair and Sustainable Economies*, ed. Anrita Nelson and Frans Timmerman. London: Pluto, 1–20.

Newell, Peter, and Matthew Patterson. 2010. *Climate Capitalism: Global Warming and the Transformation of the Global Economy*. Cambridge, U.K.: Cambridge University Press.

Newman, Peter. 2009. "Transport Opportunities: Towards a Resilient City." In *Opportunities Beyond Carbon: Looking for a Sustainable World*, ed. John O'Brien. Melbourne: Melbourne University Press, 98–115.

———. 2010. "Responding to Oil Vulnerability and Climate Change in Our Cities." In *Managing Climate Change: Papers from the Greenhouse 2009 Conference*, ed. Imogen Jubb, Paul Holper, and Wenjce Cai. Melbourne: CSIRO Publishing, 177–183.

Newman, Peter, and Jeffrey Kentworthy. 1999. *Sustainability and Cities: Overcoming Automobile Dependency*. Washington, DC: Island Press.

Newman, Peter, and Isabella Jennings. 2008. *Cities as Sustainable Ecosystems: Principles and Practices*. Washington, DC: Island Press.

Newman, Peter, and Christy Newman. 2012. "Car." In *Sociology: Antipodean Per-

spectives, ed. Peter Beilharz and Treveor Hogan. Sydney: Oxford University Press, 358–363.

Nichols, Dick. 2015. "European Elites Declare War on Syriza." *Green Left Weekly,* 11 February. www.greenleft.org.au.

Nierenberg, Danielle. 2006. "Rethinking the Global Meat Industry." In *2006 State of the World,* ed. Linda Starke. New York: W.W. Norton & Company, 24–40.

———. 2013. "Agriculture: Growing Good—And Solutions." In *State of the World 2013: Is Sustainability Still Possible?,* ed. Linda Starke. Washington, DC: Island Press, 190–200.

Nhanenge, Jytte. 2011. *Eco-Feminism: Towards Integrating the Concerns of Women, Poor People, and Nature into Development.* Lanham, MD. University Press of America.

Nugent, Stephen. 2012. Anarchism Out West: Some Reflections on Sources. *Critique of Anthropology* 32: 206–216.

O'Connor, James. 1988. "Capitalism, Nature, Socialism: A Theoretical Introduction." *Capitalism, Nature, Socialism* 1: 11–38.

———. 1989. "The Political of Ecology and Capitalism." *Capitalism Nature Socialism* 3: 93–127.

———. 1998. *Natural Causes: Essays in Ecological Marxism.* New York: Guilford Press.

Oldfeld, Jonathan D., and Denis J.B. Shaw. 2001. "The State of the Environment in the CIS." In *East Central Europe and the Former Soviet Union,* ed. David Turnock. London: Arnold, 173–186.

Onaran, Oezlem. 2010. "The Crisis of Capitalism in Europe, West and East." *Monthly Review* 62, no. 5: 18–33.

Oppelland, Torsten. 2012. "Rituals of Commeration—Identity and Conflict: The Case of Die Linke in Germany." *German Politics* 21: 429–443.

Organisation for Economic Co-operation and Development (OECD). 2011. *Divided We Stand: Why Inequality Keeps Rising—An Overview of Growing Income Inequalities in OECD Countries, Main Findings.* Paris: OECD Publishing.

Oswald, Franz. 2002. *The Party That Came Out of the Cold: The Party of Democratic Socialism in United Germany.* Westport, CT: Praeger.

Owen, David. 2012. *The Conundrum: How Scientific Innovation, Increased Efficiency, and Good Intentions Can Make Our Energy and Climate Problems Worse.* Melbourne: Scribe.

Oxfam. 2015. *Wealth: Having It All and Wanting More.* www.oxfam.org.

———. 2016. "An Economy for the 1%: How Privilege and Power in the Economy Drive Extreme Inequality and How this Can Be Stopped." *210 Oxfam Briefing Paper,* 18 January. www.oxfam.org.

Padgett, Stephen, and William E. Paterson. 1991. *A History of Social Democracy in Postwar Europe.* London: Longman.

Panitch, Leo. 2001. *Renewing Socialism: Democracy, Strategy, and Imagination.* Boulder, CO: Westview Press.

Parenti, Christian. 2013. "A Radical Approach to the Climate Crisis." *Dissent.* Summer, 51–57.

———. 2014. "Climate Change: What Role for Reform?" *Monthly Review*. April, 49–51.

Parenti, Michael. 1997. *Blackshirts & Red: Rational Fascism & the Overthrow of Communism*. San Franciso: City Light Books.

Parr, Adrian. 2013. *The Wrath of Capital: Neoliberalism and Climate Change Politics*. New York: Columbia University Press.

Paskal, Cleo. 2010. *Global Warring: How Environmental, Economic, and Political Crises Will Redraw the World Map*. New York: Palgrave Macmillan.

Patico, Jennifer. 2016. "Spinning the Market: The Moral Alchemy of Everyday Talk in Postsocialist Russia." *Critique of Anthropology* 29: 205–224.

Patton, David F. 2011. *Out of the East: From PDS to Left Party in Unified Germany*. Albany: State University of New York Press.

———. 2013. "The Left Party at Six: The PDS-WASG Merger in Comparative Perspective." *German Politics* 22: 219–234.

Paz, Juan Valdes. 2007. "The Cuban Political System in the 1990s: Continuity and Change." *Socialism and Democracy* 15, no. 1: 98–110.

Peace, William J. 2004. *Leslie A. White: Evolution and Revolution in Anthropology*. Lincoln: University of Nebraska Press.

Pels, Peter. 2015. "Modern Times: Seven Steps toward an Anthropology of the Future." *Current Anthropology* 56: 779–796.

Pepper, David. 1993. *Eco-Socialism: From Deep Ecology to Social Justice*. London: Routledge.

———. 2010. "On Contemporary Eco-socialism." In *Eco-Socialism as Politics: Rebulding the Basis of Our Modern Civilization*, ed. Q. Huan. New York: Springer, 33–44.

Perez, Christina. 2008. *Caring for Them from Birth: The Practice of Community-Based Cuban Medicine*. Lanham, MD: Lexington Books.

Peters, Glen P., Gregg Marland, Edgar G. Hertwich, Laura Saikku, Aapo Rautiainen, and Peka E. Kauppi. 2009. "Trade, Transport, and Sinks Extend the Carbon Dioxide Responsibility of Countries: An Editorial Essay." *Climatic Change* 97: 379–388.

Peters, Peter Frank. 2006. *Time, Innovation and Mobilities: Travel in Technological Cultures*. London: Routledge.

Petras, James. 2011. "Latin America's Twenty-First Century Socialism." In *21st Century Socialism: Reinventing the Project*, ed. Henry Veltmeyer. Halifax: Nova Scotia: Fernwood Publishing.

———. 2016. *The End of the Republic and the Delusion of Empire*. Atlanta: Clarity Press.

Pieke, Frank N. 2014. "Anthropology, China, and the Chinese Century." *Annual Review in Anthropology* 43: 123–138.

Piper, Karen. 2014. *The Price of Thirst: Global Water Inequality and the Coming Chaos*. Minneapolis: University of Minnesota Press.

Poston, Dudley L. Jr., and Leon F. Bouvier. 2010. *Population and Society: An Introduction to Demography*. Cambridge, U.K.: Cambridge University Press.

Price, Andy. 2012. *Recovering Bookchin: Social Ecology and the Crises of Our Time.* Grenmarsvegan, Norway: New Compass Press.
Price Waterhouse Coopers. 2012. *Low Carbon Economy Index 2012: Too Late for 2°C.* http://www.pwc.com/gx/en/services/sustainability/publications/low-carbon-economy-index.html.
Princen, Thomas, Jack P. Manno, and Pamela Martin. 2013. "Keep Them in the Ground: Ending the Fossil Fuel Era." In *State of the World 2013: Is Sustainability Still Possible?*, ed. Linda Starke. Washington, DC: Island Press, 161–171.
Randers, Jorgen. 2012. *2052: A Global Forecast for the Next Forty Years.* White River Junction, VT: Chelsea Green Publishing.
Randers, Jorgen, and Paul Gilding. 2010. "The One Degree War Plan." *Journal of Global Responsibility* 1, no. 1: 170–188.
Raskin, Paul. 2016. *Journey to Earthland: The Great Transition to Planetary Civilization.* Boston: Tellus Institute.
Raskin, Paul, Tariq Banuri, Gilberto Gallopin, Pablo Gutman, Al Hammond, Robert Kates, and Rob Swart. 2002. *Great Transition: The Promise and the Lure of Times Ahead.* Stockhom: Global Scenario Report.
Rasmus, Jack. 2016. *Looting Greece: A New Financial Imperialism Emerges.* Atlanta: Clarity Press.
Ratner, Carl. 2015. "Neoliberal Co-optation of Leading Co-op Organizations, and a Socialist Counter-politics of Cooperation." *Monthly Review.* February, 18–30.
Register, Richard. 2001. *Ecocities: Building Cities in Balance with Nature.* Berkeley, CA: Berkeley Hills Books.
Reid, Hannah. 2014. *Climate Change and Human Development.* London: Zed Books.
Riner, Reed D. 1998. The Future of a Sociocultural Problem: A Personal Ethnohistory. *American Behavioral Scientist* 42: 347–364.
Robbins, Richard H. 2011. *Global Problems and the Culture of Capitalism.* 5th ed. Boston: Prentice-Hall.
Roberts, J. Timmons and Bradley C. Parks. 2007. *A Climate of Injustice: Global Inequality, North-South Politics, and Climate Policy.* Cambridge, MA: MIT Press.
Rocktroem, J. et al. 2009. "Planetary Boundaries: Exploring the Safe Operating Space for Humanity. *Ecology and Society* 14(2): 32.
Robinson, Andrew, and Simon Tormey. 2012. "Beyond the State: Anthropology and 'Actually-Existing–Anarchism'." *Critique of Anthropology* 32: 143–157.
Robinson, William I. 2014. *Global Capitalism and the Crisis of Humanity.* Cambridge, U.K.: Cambridge University Press.
Roden, Duncan. 2016. "Bernie Sanders' Campaign Targets the 'Billionaire Class'." *Green Left Weekly.* 9 February, 17.
Roemer, John. 1994. *A Future for Socialism.* London: Verso.
Rogatyuk, Denis. 2014. "Left Unity Creates a New Party for the Left: Interview with Kate Hudson." *Green Left Weekly,* 29 October, 20.
Rogers, Chris. 2014. *Capitalism and Its Alternatives.* London: Zed Books.

Rogner, H.-H., D. Zhou, R. Bradley. P. Crabbé, O. Edenhofer, B. Hare, L. Kuijpers, M. Yamaguch. 2007. "Review of the Last Three Decades." In *Climate Change 2007: Mitigation. Contribution of Working Group III to the Fourth Assessment Report of the Intergovernmental Panel on Climate Change,* ed. B. Metz, O.R. Davidson, P.R. Bosch, R. Dave, L.A. Meyer. Cambridge, U.K.: Cambridge University Press.

Roper, Brian S. 2013. *The History of Democracy: A Marxist Interpretation.* London: Pluto Press.

Rosa, Eugene A., and Thomas Dietz. 2010. "Global Transformations: Passage to a New Ecological Era. In *Human Footprints on the Global Environment,* ed. Eugene A. Rose, Andreas Dielmann, Thomas Dietz, and Carlo Jaeger. Cambridge, MA: MIT Press, 1–45.

Rossman, Peter. 1990. "The Paradox of the East German Revolution in East Germany." *New Politics* 3: 65–73.

Rothkopf, David. 2012. *Power, Inc.: The Epic Rivalry between Big Business and Government—and the Reckoning that Lies Ahead.* New York: Farrar, Straus and Giroux.

Ruddiman, William F. 2005. *Plows, Plagues and Petroleum: How Humans Took Control of Climate.* Princeton, NJ: Princeton University Press.

Ruiz, Pollyanna. 2014. *Articulating Dissent: Protest and the Public Sphere.* London: Pluto Press.

Rundle, Guy. 2015. "Greece: Openings to a Future." *Arena Magazine: The Australian Magazine of Left Political, Social and Cultural Commentary* no. 134, 5–8.

Ruyle, Eugene L. 1978. "A Socialist Alternative for the Future." In *Cultures of the Future,* ed. Magorah Marayuma and Arthur M. Harkins. The Hague: Mouton, 613–627.

Ryle, Martin. 1988. *Ecology and Socialism.* London: Radius.

Sachs, Jeffrey D. 2008. *Common Wealth: Economics for a Crowded Planet.* New York: Allen Lane.

Salleh, Ariel. 1997. *Ecofeminism as Politics: Nature, Marx and the Postmodern.* London: Zed Books.

———. 2011. "The Value of a Synergistic Economy." In *Life without Money: Building Fair and Sustainable Economics,* ed. Anitra Nelson and Frans Timmerman. London: Pluto Press, 94–110.

Sanders, Bernie. 2016. *Our Revolution: A Future to Believe In.* London: Profile Books.

Sanderson, Stephen K. 1995. *Social Transformations: A General Theory of Historical Development.* Oxford, U.K.: Blackwell.

———. 2010. *Revolutions: A Worldwide Introduction to Social and Political Contention.* 2nd ed. Boulder, CO: Paradigm Publishers.

———. 2015. *Modern Societies: A Comparative Perspective.* Boulder, CO: Paradigm Publishers.

Sanderson, Stephen K., and Arthur S. Alderson. 2005. *World Societies: The Evolution of Human Social Life.* Boston: Pearson.

Sarkar, Saral. 1999. *Eco-Socialism or Eco-Capitalism? A Critical Analysis of Humanity's Fundamental Choices.* Zed Books.

———. 2010. "Prospects for Eco-socialism." In *Eco-Socialism as Politics: Rebuilding the Basis of Our Modern Civilization,* ed. Q. Huan. New York: Springer, 207–222.

Sarkar, Sarkar and Bruno Kern. 2008. *Eco-Socialism or Barbarism: An Up-to-Date Critique of Capitalism.* Cologne: Initiative Eco-Socialism.

Satterhwaite, David. 2009. "The Implications of Population Growth and Urbanization for Climate Change." *Environment and Urbanization* 21: 545–567.

Saunois, Tony. 2015. "A New Turn in Latin America?" *Socialism Today.* May, 5–11.

Sayer, Andrew. 2016. *Why We Can't Afford the Rich.* Bristol, U.K.: Polity Press.

Scharf. C. Bradley. 1984. *Politics and Change in East Germany: The Evolution of Socialist Democracy.* Boulder, CO: Westview Press.

Schiavoni, C., and W. Camacaro. 2009. "The Venezuelan Effort to Build a New Food and Agriculture System." *Monthly Review* 61, no. 3: 129–141.

Schmitt, Richard. 2012. "Beyond Capitalism and Socialism." In *Taking Socialism Seriously,* ed. Anatole Anton and Richard Schmitt. Lanham, MD: Lexington Books. 187–211.

Schor, Juliet. 1991. *The Overworked American: Unexpected Decline of Leisure.* New York: Basic Books.

———. 2010. *Plenitude: The New Economics of True Wealth.* Melbourne: Scribe.

Schwartzman, David. 2009. "Ecosocialism or Ecocatastrophe?" *Capitalism, Nature, Socialism.* 20, no. 1: 6–33.

———. 2014. "Is Zero Economic Growth Necessary to Prevent Climate Catastrophe?." *Science and Society* 78: 235–240.

Schwartzman, David, and Quincy Saul. 2015. "An Ecosocialist Horizon for Venezuela: A Solar Communist Horizon for the World." *Capitalism, Nature, Socialism* 26, no. 3: 14–30.

Schweickart, David. 2012. "But What Is Your Alternative? Reflections on Having a 'Plan'." In *Taking Socialism Seriously,* ed. Anatole Anton and Richard Schmitt. Lanham, MD: Lexington Books, 47–66.

———. 2016. "Economic Democracy: An Ethically Desirable Socialism that is Economically Viable." The Next System Project. http://nextprojectsystem.org.

Scott-Cato, Molly, and Jean Hillier. 2010. "How Could We Study Climate-Related Social Innovation? Applying Deleuzean Philosophy to Transition Towns." *Environmental Politics* 19: 869–887.

Sears, Alan. 2014. *The Next New Left: A History of the Future.* Halifax: Ferwood Publishing.

Selenyi, Ivan. 2015. "Capitalisms after Communism." *New Left Review,* no. 96: 39–51.

Sernau, Scott. 2011. *Social Inequality in a Global Age.* 3rd ed. Los Angeles: Sage.

Shearman, David, and Joseph Wayne Smith. 2007. *The Climate Change Challenge and the Failure of Democracy.* Westport, CT: Praeger.

Sherman, Howard J. 1972. *Radical Political Economy.* New York: Basic Books.
———. 1995. *Reinventing Marxism.* Baltimore: Johns Hopkins University Press.
Sherman, Howard J., and James L. Wood. 1989. *Sociology: Traditional and Radical Perspectives.* New York: Harper & Row, Publishers.
Shiva, Vandana. 2005. *Earth Democracy: Justice, Sustainability, and Peace.* Cambridge, MA: South End Press.
———. 2008. *Soil Not Oil: Environmental Justice in an Age of Climate Crisis.* Boston, MA: South End Press.
Shiva, Vandana, Maria Mies, and Ariel Salleh. 2014. *Ecofeminism.* London: Zed Books.
Shue, Henry. 2014. *Climate Justice: Vulnerability and Protection.* Oxford, U.K.: Oxford University Press.
Silber, Irwin. 1994. *Socialism: What Went Wrong? An Inquiry into the Theoretical and Historical Sources of the Socialist Crisis.* London: Pluto Press.
Sim, Stuart. 2010. *The Carbon Footprint Wars: What Might Happen If We Retreat from Globalization?* Edinburgh: Edinburgh University Press.
Simms, Andrew. 2009. *Ecological Debt: Global Warming and the Wealth of Nations.* 2nd ed. London: Pluto Press.
Singer, Merrill. 2010a. "Eco-nomics: Are the Planet-Unfriendly Features of Capitalism Barriers to Sustainability?" *Sustainability* 2, no. 1: 127–144.
———. 2010b. "Atmospheric and Marine Pluralea Interactions and Species Extinction Risks." *Journal of Cosmology* 8: 1832–1837.
Singer, Merrill, and Hans A. Baer. 2007. *Introducing Medical Anthropology: A Discipline in Action.* Lanham, MD: AltaMira.
———. 2012. *Introducing Medical Anthropology: A Discipline in Action.* 2nd ed. Lanham, MD: AltaMira.
Smith, Stephen A. 2014. "Towards a Global History of Communism." In *The Oxford Handbook of the History of Communism,* ed. Stephen A. Smith. Oxford, U.K.: Oxford University Press, 21–52.
Smith, Murray E.G. and Joshua D. Dumont. 2011. "Socialist Strategy, Yesterday and Today: Notes on Classical Marxism and the Contemporary Radical Left." In *21st Century Socialism: Reinventing the Project.* ed. Henry Veltmeyer. Halifax, Nova Scotia: Fernwood Publishing.
Solty, Ingar. 2008. "The Historic Significance of the New German Left Party." *Socialism and Democracy* 22, no. 1: 1–34.
Sousanis, John. 2011. "World Vehicle Population Tops 1 Billion Units." *Wards Auto.* 15 August. http://wardsauto.com/news-analysis/world-vehicle-population-tops-1-billion-units.
Spehr, Christoph. 2012. "Die Linke Today: Fears and Desires." *Socialist Register 2013: The Question of Strategy,* ed. Leo Panitch, Greg Albo, and Vivek Chibber. London: Merlin Press, 159–173.
Sperling, Daniel and Deborah Gordon. 2010. *Two Billion Car: Driving Toward Sustainability.* New York: Oxford University Press.
Speth, James Gustave. 2008. *The Bridge at the Edge of the World: Capitalism, the*

Environment, and the Crossing from Crisis to Sustainability. New Haven, CT: Yale University Press.

———. 2011. "Beyond the Growth Paradigm: Creating a Unified Progressive Politics." *Great Transitions Initiative,* March.

———. 2012. *America the Possible: Manifesto for a New Economy.* New Haven, CT: Yale University Press.

Srnicek, Nick, and Alex Williams. 2015. *Inventing the Future: Postcapitalism and a World without Work.* London: Verso.

Stand, Kurt. 2012. "Ambivalences, Contradictions, Choices: The Legacy of GDR Socialism." *Socialism and Democracy* 26, no. 1: 58–64.

Starr, Amory. 2000. *Naming the Enemy: Anti-Corporate Movements Confront Globalization.* London: Zed Books.

———. 2005. *Global Revolt: A Guide to the Movements against Globalization.* London: Zed Books.

Steffen, Will, Wendy Broadgate, Lisa Deutsch, Owen Gaffney, and Cornelia Ludwig. 2015. "The Trajectory of the Anthropocene: The Great Acceleration." *The Anthropocene Review* 2, no. 1: 1–18.

Steinfeld, H., P. Gerber, T. Wassenaar, V. Castel, M. Rosales, and C. de Hart. 2006. *Livestock's Long Shadow: Environmental Issues and Options.* Rome: FAQ.

Stilwell, Frank. 1992. *Understanding Cities and Regions.* Leichhardt, NSW: Pluto Australia.

———. 2000. *Changing Track: A New Political Economic Direction for Australia.* Sydney: Pluto Press.

Stokes, Raymond G. 2000. *Constructing Socialism: Technology and Change in East Germany 1945-1990.* Baltimore: Johns Hopkins University Press.

Streeck, Wolfgang. 2016. *How Will Capitalism End?* London: Verso.

Suvin, Darko. 2016. "What Is to Be Done? A First Step." *Socialism and Democracy* 30, no. 1: 105–127.

Sweezy, Paul. 1980. *Post-Revolutionary Society.* New York: Monthly Review Press.

Szymanski, Albert. 1979. *Is the Red Flag Still Flying? The Political Economy of the Soviet Union Today.* London: Zed Books.

Tabb, William. 2014. "What If? Politics and Post-capitalism." *Critical Sociology* 40: 501–509.

Tammilehto, Olli. 2012. "On the Prospect of Preventing Global Climate Catastrophe Due to Rapid Social Change." *Capitalism, Nature, Socialism* 22, no. 1: 79–92.

Tanuro, Daniel. 2010. *Green Capitalism: Why It Can't Work.* London: Merlin Press.

Taylor, Graeme. 2008. *Evolution's Edge: The Coming Collapse and Transformation of Our World.* Garbriola Island, BC: New Society Publishers.

Taylor, Peter J. 1996. *The Way the Modern World Works: World Hegemony to World Impasse.* New York: John Wiley & Sons.

Thatcher, Ian D. 2007. "Rosa Luxemburg and Leon Trotsky Compared." In *Twentieth-Century Marxism: A Globalist Introduction,* ed. Daryl Glaser and David M. Walker. London: Routledge, 30–45.

Tickell, Oliver. 2008. *Kyoto2: How to Manage the Global Greenhouse*. London: Zed Books.
The Next System Project. n.d. http://thenextsystem.org/.
Therborn, Goeran. 2008. *From Marxism to Post-Marxism?*. London: Verso.
———. 2012. "Class in the 21st Century." *New Left Review*, no. 78: 5–29.
Todorovich, Petra, and Edward Burgess. 2013. "High-Speed Rail and Reducing Oil Dependence." In *Transport Beyond Oil: Policy Choices for a Multimodal Future*, ed. John L. Renne and Billy Fields. Washington, DC: Island Press, 141–160.
Tokar, Brian. 2010. "Toward a Movement for Climate Justice." www.social-ecology.org.
Touissant, Eric. 2010. "Is Bolivia Heading for Andean-Amazonian Capitalism?" *Upsidedownworld*. 11 March. http://upsidedownworld.org.
Touraine, Alain. 1974. *The Post-Industrial Society: Tomorrow's Social History: Classes, Conflicts and Culture in the Programmed Society*. London: Wildwood House.
Trainer, Ted. 2007. *Renewable Energy Cannot Sustain a Consumer Society*. Sydney: UNSW Press.
———. 2009. "The Transition Towns Movement: Its Huge Significance and a Friendly Criticism—(We) Can Do Better." 30 July. http://candobetter.net/node//1439.
———. 2010. *The Transition to a Sustainable and Just World*. Canterbury NSW: Envirobook.
———. 2014. "An Anarchism for Today: The Simpler Way. 18 October. http://simplicitycollective.com/ted-trainer-and-the-simpler-way.
Tucker, Robert, ed. 1978. *The Marx-Engels Reader: Second Edition*. New York: W.W. Norton.
Turner, Terisa E., and Leigh S. Brownhill. 2001. "'Women Never Surrendered': The Mau Mau and Globalization from Below in Kenya 1980-2000." In *There is an Alternative: Subsistence and Worldwide Resistance to Corporate Globalization*, ed. Veronika Bennholdt-Thomsen, Nicholas Faraclas, and Claudia von Werlhof. London: Zed Books, 106–132.
United Nations Development Programme (UNDP). 2013. *Human Development Report 2013: The Rise of the South: Human Progress in a Diverse World*. New York: United Nations Development Programme.
———. 2016. *Human Development Report 2016: Human Development for Everyone*. New York: United Nations Development Programme.
United Nations Environment Programme. 2009. "Green New Deal: An Update for the G20 Pittsburgh Summit." September. www.unep.org/pdf/G20_policy_Final.pdf.
United Nations Population Fund. 2010. *State of the World Population 2010*. New York: UNPF.
Varoufakis, Yanis. 2013. *The Global Minotaur: America, Europe and the Future of the Global Economy*. London: Zed Books.

Vu, Tuong. 2014. "Workers under Communism: Romance and Reality." In *The Oxford Handbook of the History of Communism,* ed. Stephen A. Smith. Oxford, U.K.: Oxford University Press, 471–487.

Wagar, W. Warren. 1991. *The Next Three Futures: Paradigms of Things to Come.* New York: Greenwood Press.

———. 1992. *A Short History of the Future.* Chicago: University of Chicago Press.

Wainright, Joel and G. Mann. 2012. "Climate Leviathan." *Antipode* 45: 1–22.

Wall, Derek. 2010a. *The Rise of the Worldwide Ecosocialist Movement.* Pluto Press.

———. 2010b. "Ecosocialism, the Left, and the U.K. Greens." *Capitalism, Nature, Socialism* 21, no. 3: 109–115.

Wallerstein, Immanuel. 1979. *The Capitalist World-Economy: Essays.* New York: Cambridge University Press.

———. 1984. *The Politics of the World Economy: The States, the Movements, and the Civilizatons.* Cambridge, UK; Cambridge University Press.

———. 1990. Antisystemic Movements: History and Dilemmas." In *Transforming the Revolution: Social Movements and the World System,* ed. Samir Amin, Giovanni Arrighi, Andre Gunder Frank, and Immanuel Wallerstein. New York: Monthly Review Press, 13–53.

———. 1998. *Utopistics: Or Historical Choices of the Twenty-First Century.* New York: New Press.

———. 2004. *World-Systems Analysis: An Introduction.* Durham, NC: Duke University Press.

———. 2007. "Commentary No. 205: Climate Disasters: Three Obstacles to Doing Anything." Fernand Braudel Center. 15 March. http://www.binghamton.edu/fbc/archive/205en.htm.

———. 2008. "Remembering Andre Gunder Frank." *Monthly Review* 60, no. 1: 50–61.

———. 2011. "Structural Crisis in the World System: Where Do We Go from Her?" Monthly Review, March, 31–39.

———. 2013. "Structural Crisis, or Why Capitalists May No Longer Find Capitalism Rewarding." In *Does Capitalism Have a Future?,* ed. Immanuel Wallerstein, Randall Collins, Michael Mann, Georgi Derluguian, and Craig Calhoun. Oxford, U.K.: Oxford University Press, 9–35.

———. 2014. "Antisystemic Movements, Yesterday and Today." *Journal of World Systems Research* 20, no. 2: 158–172.

———. 2015. "Cuba and the United States Resume Relations: Happy New Year." *Commentary* No. 392. 1 January. http://iwallerstein.com/cuba-and-the-united-states-resume-relations-happy-new-year/.

———. 2016. "A Left Electoral Strategy? France and the United States." *Commentary* no. 418, 1 February. http://www.binghamton.edu/fbc/commentaries/archive-2016/418en.htm.

Wallis, Victor. 2000. "'Progress' or Progress? Defining a Socialist Technology." *Socialism and Democracy* 14, no. 1: 45–61.

---. 2008. "On Marxism, Socialism, and Ecofeminism: Continuing the Dialogue." *Capitalism, Nature, Socialism* 19, no. 1: 107–111.

---. 2009. "Economic/Ecological Crisis and Conversion." *Socialism and Democracy* 23, no. 2: 94–101.

---. 2010. "Socialism and Technology: A Sectoral Overview." In *Eco-Socialism as Politics: Rebuilding the Basis of Our Modern Civilization*, ed. Q. Huan. New York: Springer, 45–61.

Wang, Zhihe. 2012. "Ecological Marxism in China." *Monthly Review*, February, https://monthlyreview.org/2012/02/01/ecological-marxism-in-china/.

Wang, Zhihe, Huili He, and Meijun Fan. 2014. "The Ecological Civilization Debate in China: The Role of Ecological Marxism and Constructive Postmodernism—Beyond the Predicament of Legislation. *Monthy Review* Nov.: 37–59.

Ward, Peter. D. 2010. *The Flooded Earth: Our Future in a World without Ice Caps.* New York: Basic Books.

Watson, Judith. 2016. "Peace Is an Ecosocialist Issue: Some Experiences from Local UK Politics and Suggestions for Global Action." *Capitalism, Nature, Socialism* 27: 9–20.

Webber, Jeffrey R. 2011. *From Rebellion to Reform in Bolivia: Class Struggle, Indigenous Liberation, and the Politics of Evo Morales.* Chicago: Haymarket Books.

---. 2012. *Red October: Left-Indigenous Struggles in Modern Bolivia.* Chicago: Haymarket Books.

Weil, Robert. 2006. "'To Be Attacked by the Enemy Is a Good Thing': The Struggle over the Legacy of Mao Zedong and the Chinese Socialist Revolution." *Socialism and Democracy* 20(2): 19–53.

Weiner, D.R. 1988. *Models of Nature: Ecology, Conservation, and Cultural Revolution in Soviet Russia.* Bloomington: Indiana University Press.

Weis, Tony. 2013. *The Ecological Hoofprint: The Global Burden of Industrial Livestock.* London: Zed Books.

Welzer, Harald. 2012. *Climate Wars: Why People Will Be Killed in the Twenty-First Century,* trans. Patrick Camiller. London: Polity.

Wescott, Roger W. 1978. "'The Anthropology of the Future' as an Academic Discipline." In *Cultures of the Future,* ed. Magorah Maruyama and Arthur M. Harkins. The Hague: Mouton, 509–523.

Weston, Del. 2014. *The Political Economy of Global Warming: The Terminal Crisis.* London: Routledge.

Whettenhall, Roger. 1965. "Public Ownership in Australia." *Political Quarterly*, no. 36: 426–440.

White, Leslie A. 1949. "Ethnological Theory." In *Philosophy for the Future: The Quest of Modern Materialism,* ed. R.W. Sellars, V.J. McGill, and M. Farber. New York: Macmillan, 357–384.

---. 1959. *The Evolution of Culture: The Development of Civilization to the Fall of Rome.* New York: McGraw-Hill.

---. 1975. *The Concept of Cultural Systems.* New York: Columbia University Press.

———. 2008. *Modern Capitalist Culture,* ed. Burton J. Brown, Benjamin Urish, and Robert L. Carneiro. Walnut Creek: CA: Left Coast Press.
White, Leslie A., with Beth Dillingham. 1973. *The Concept of Culture.* Minneapolis: Burgess.
Williams, Chris. 2010. *Ecology & Socialism: Solutions to Capitalist Ecological Crisis.* Chicago: Haymarket Books.
Wilpert, Gregory. 2005. "Venezuela: Participatory Democracy or Government as Usual?" *Socialism and Democracy* 19, no. 1: 7–32.
Wlodzierz, Wesolowski. 1967. "Marx's Theory of Class Domination." In *Marx and the Western World,* ed. Nicholas Lobkowicz. 103–131.
Wolf, Frieder Otto. 2015. "The Radical Left in Germany." *Socialism and Democracy* 29, no. 3: 81–92.
Wolff, Richard. 2012. *Democracy at Work: A Curve for Capitalism.* Chicago: Haymarket Books.
Woo, Jongseok. 2014. "Kim Jong-il's Military-First Politics and Beyond: Military Control Mechanisms and the Problem of Power Succession." *Communist and Post-Communist Studies* 47: 117–125.
World Bank. 2012. *Turn Down the Heat: Why a 4ºC Warmer World Must Be Avoided.* Washington, DC: International Bank for Reconstruction and Development/ World Bank.
Wright, Olin Erik. 2010. *Envisioning Real Utopias.* London: Verso.
Wright, Olin Erik and Joel Rogers. 2011. *American Society: How It Really Works.* New York: WW. Norton & Co.
Xu, Zhun. 2013. "The Political Economy of Decollectivization in China." *Monthly Review* 65, no. 1: 17–36.
Yates, Michael D. 2016. "Measuring Global Inequality." *Monthly Review* 68, no. 6: 1–13.
Yih, Katherine. 1990. "The Red and the Green." *Monthly Review* 42, no. 5: 16–27.
Young, Diana. 2001. "The Life and Death of Cars: Private Vehicles on the Pitjantjatjara Lands, South Australia." In *Car Cultures,* ed. Daniel Miller. Oxford, U.K.: Berg, 35–57.
Zatlin, Jonathan R. 2007. *The Currency of Socialism: Money and Political Culture in East Germany.* Cambridge, U.K.: Cambridge University Press.
Zee News. 2016. "September 2016 Warmest in 136 years Says NASA." http:// zeenews.india.com/environment/september-2016-warmest-in-136-years-says-nasa_1941307.html.
Ziegler, Charles E. 1987. *Environmental Policy in the USSR.* Amherst: University of Massachusetts Press.
Zimbalist, Andrew, and Howard J. Sherman. 2015. *Competing Economic Systems: A Political-Economic Approach.* St. Louis: Elsevier Science.

INDEX

advertising, 17, 94, 122, 236
advertisements, 245
Afro-Marxism, 47
Agricultural Revolution, 7, 23, 206
agroecology, 244
airplanes, 41, 31, 33, 106, 241–242
airships, 241–242
Allende, Salvador, 206
Althusser, Louis, 206
anarchism, 1, 145, 150, 178, 181
Anthropocene, 115–116
    futures, 115–117
anthropology of climate change, vii, 7, 151–152, 170
anthropology of the future, 7–9, 16–17, 171, 263, 266
Anderson, Robert, 5
appropriate technology, 13, 192, 231, 236
Arab Spring, 176, 184–185, 259
Australian Labor Party, 98, 180, 220
automobile(s), 18, 32–33, 64, 106, 239–241, 249. *See* cars
Autonomists, 177

Badiou, Alan, 51, 134, 136
Beck, Ulrich, 12, 104–106
Berlin Wall, 6, 46, 69
blueprint (s), 13, 43, 129, 134, 202, 205, 169
Bodley, John H., 2, 16, 170
Boeing Corporation, 4
Boliva, 13, 153, 158–165, 168, 171, 196, 232

Bolivarian Revolution, 155
Bolshevik Revolution, 43, 52, 128, 136, 201
Bolsheviks, 43, 50, 62
Brandt, Willy, 97
Brazilian Landless Workers Movement, 181
Brezhnev, Leonid, 57
Brown, Lester R., 108
Bruderhof, 191
bureaucratic centralism, 54
Bush, George W., 31, 74, 166, 185, 187

capitalism
    climate, 106
    consumer, 17
    disaster, 40
    green, 8, 42, 106, 117, 119, 177, 188, 255, 258, 265
    social, 135
    state, 12, 16, 53, 78–79, 81–82, 106, 128, 159
    sustainable, 118
    welfare, 96–100 (*see* social democracy)
capitalist world system, 15–42
carbon footprint, 228, 248
Carbon Price Mechanism, 220–221
cars, 8, 23, 32–33, 90, 106, 198, 114, 117, 126, 168, 191, 193, 236–240. *See* automobile
Castells, Manuel, 197–198
Castro, Fidel, 76–78

Castro, Raul, 166
Chase-Dunn, Christopher, 56, 82, 152, 264
Chavez, Hugo, 153, 155–157
China, 61–65, 88—90, 95–96, 167–168
Chomsky, Noam, 185, 230
Chou En Lai, 64
climate catastrophism, 258
climate change mitigation, 246
climate denialists, 30
climate emergency mobilization model, 12, 113–115
Climate Reality Project, 188
climate science, 7, 29, 105, 117, 139, 155, 190, 254
climate skeptics, 30
Clinton, Hillary, 260–261
Cold War, 9, 66
collectivization, 43, 53, 55–57, 63, 85, 91, 244
communism, 45
Communist Manifesto, 45
Conference of the Parties (COP), UN, 41, 254
contradiction, 15
Corbyn, Jeremy, 259–260
corporations, 1, 10, 16, 30, 32, 36, 64, 105–109, 112, 115, 119, 124, 126, 149, 159, 174, 179–181, 188, 196–197, 206, 219, 223–224, 232, 251–253, 261, 264
Correa, Rafael, 162
cosmopolitanism, 105
Council for Marxist Anthropology, 170
Council for Mutual Economic Assess (COMECON), 57, 67, 72, 76–77, 168
counterhegemony, 259
countermovement, 175
cracks in the system, 195–198
Cuba, 76–78, 90, 123, 163–167
  health care system, 165–166
  Special Period, 163–164
Cuban Energy Revolution, 164
Cuban Missile Crisis, 57
Cuban Revolution, 47
culture of consumption, 13, 65, 68, 152, 205, 245–246

Daly, Herman, 108–109, 230, 256
deforestation, 7, 29, 30, 34, 95, 112, 114, 244, 257
degrowth project, 229–230
deindustrialization, 180
delinking, 13, 167–168
democratic centralism, 51
Democratic People's Republic of Korea, 49, 71–76. *See* North Korea
Deng Xiaoping, 64
development
  appropriate, 147
  human, 227
  sustainable, 27, 39, 99, 104, 112–113, 164, 178
  sustainable human, 156
development state, 100
dictatorship of the proletariat, 16, 44
DIE LINKE, 207–210
Doomsday Clock, 254
dystopia, 11, 37–41

Eagleton, Terry, 207
Earth democracy, 1, 142–143
eco-anarchism, 137–145, 150–151, 259. *See* anarchism
eco-authoritarianism, 39
eco-civilization, 140
eco-fascism, 39–40
eco-feminism
  socialist, 140–143
eco-Marxism, 137
eco-psychology, 249
eco-villages, 19, 2493
Ecosocialist Horizons, 144
Ecosocialist International Network (EIN), 143

Ecosocialist Manifesto, 143
ecological footprint, 171
ecological Marxism, 140
economic democracy, 1, 3, 109, 126, 131, 135, 143, 153, 191, 203, 252
economic growth, 8, 11, 15, 22, 25–26, 57, 64, 67, 86–87, 92, 94, 96, 111–113, 121, 128, 138, 153, 177, 214, 218, 223–224, 227–228, 231–232, 248, 254, 262
economy
    centrally planned, 86
    command, 86
    gift, 150
    global, 106
    moneyless, 150
    postgrowth, 109
    second, 87
    steady state, 108–110, 227–230
Ecuador, 13, 153, 162–163, 168, 184
embodied energy, 234
emissions taxes, 217–221
emissions trading, 42, 117–118, 128, 218–220
Energiewende, 235
energy efficiency, 13, 231–232
Engels, Friedrich, 26, 43–44, 170
environmental degradation, 1, 11, 16–17, 26, 29, 34–35, 58, 64, 92–93, 106, 113, 119, 141, 170, 199, 224–225
EU Emissions Trading Scheme, 42, 219
European Central Bank, 213–214
Eurozone, 213
extractivism, 162

farming
    collective, 244–245
    factory, 243
    organic, 243
    urban, 244
fascism, 130
February Revolution, 50

Federal Republic of Germany, 6, 49, 66, 85, 97
fishing, 244
foraging societies, 2, 227. *See* hunting and gathering societies
fossil fuels, 7
Foster, John Bellamy, 105
Fotopoulos, Takis, 145
Fourth International, 80
Framework Convention on Climate Change (FCCC), 42, 254
Fuel Revolution, 31

Gang of Four, 79
German Democratic Republic, 6, 12, 49, 65–71, 85–88
Giddens, Anthony, 221
Gilding, Paul, 112
Gillard, Julia, 220
Glasnost, 58
global democracy, 1, 131, 152, 196, 264
Global Marshall Plan, 117
Global Scenario Group, 12, 104–105
Gorbachev, Michel, 57
Gramsci, Antonio, 175, 181, 265
Great Famine, 63
Great Transition, 124–126
green jobs, 13, 204–205, 231, 236, 255
Great Leap Forward, 62
Green New Deal, 12
Green Peace, 187
Green Revolution, 33
Greens
    German, 99
    Sweden, 99
    U.K., 143, 215
    U.S., 187
greenhouse gas emissions, 9, 17, 21, 25–26, 28, 32–36, 64, 95–96, 106, 112–114, 118, 122, 131, 152, 161, 205, 218–219, 239, 247, 254–255
Gore, Al, 12, 117–120

growth paradigm, 13, 42, 99, 120, 138, 146, 227–229
Guevara, Che, 76–78

Hansen, James, 219
Harvey, David, 176
health anthropology, vii
Heinberg, Richard, 110–111
Herzfeld, Michael, 212
historical/dialectical materialism, 129
Holocene, 27
Holmgren, David, 121–124. *See* permaculture paradigm
hominines, 2
Honecker, Erich, 69
horizontalism, 177, 199
human nature, 169
Humboldt University, 6, 49
Hunting and gather societies, 2, 169. *See* foraging societies
Hutterites, 5, 191

immigrants, 25, 214
indigenous societies, 25, 147–148, 169–170, 252
Industrial Revolution, 7, 23, 29–31, 206
Intergovernmental Panel on Climate Change (IPCC), 8, 28–29, 107
International Monetary Fund (IMF), 212

Jackson, Tim, 112–113
James, Fredric, 129
Jevons paradox, 29, 113, 120, 231

Kampuchea, 48
Keynesianism, 97–98
Khrushchev, Nikita, 57, 92
Kim Il Sung, 72–73
Kim Jong-il, 73
Kim, Jong-nam, 74
Kim Jong-un, 74
Klein, Naomi, 255–256

Korean Workers' Party, 72
Kovel, Joel, 191
Kyoto Protocol, 218

La Via Campesina, 181–182
Latouche, Serge, 229
Leahy, Terry, 150–151
Left Opposition, 54
Left Unity (U.K.), 215
legitimation crisis, 86
Lenin, Vladimir Ilyich, 41, 43, 46, 50–54, 59–60, 62
Leninist regimes, 48–49, 107
Levites of Utah, 5
livestock production, 34
Lovelock, James, 38–39
Luxemburg, Rosa, 51
Lynas, Mark, 37–38

Maduro government, 172
Malinowski, Bronislaw, 9
Mao, Mao Tse Sung, 41, 48, 62–64, 72, 95, 101, 167
Marcuse, Peter, 70
Marx, Karl, 3, 16, 18, 26, 29, 43–45, 50, 62, 119, 129, 137, 170, 201
mass media, 255
materialism
    cultural, 6
    historical, 6
McCarthyism, 170
Mead, Margaret, 9
medical anthropology, 6
Mensheviks, 50–53
Merkel, Angela, 22, 211, 213
metabolic rift, 26
Miliband, Ralph, 132
military expenditure, 36
military-industrial complex, 35
modernity, 105
modernization
    in China, 64
    ecological, 12, 105
    reflexive, 105

Modrow, Hans, 69
Mondragon Cooperative Corporation, 191–192
Morales, Evo, 158–161
Morgan, Lewis Henry
motor vehicle(s), 31–33, 205, 237–238, 240–241
movement(s). *See* countermovement
   anti-systemic, 151, 173
   climate movement, 13, 34, 41, 174, 179, 186–189, 232
   environmental, 186–190
   global justice or anti-corporate globalization, 183–185
   indigenous, peasant, and ethnic rights, 181–182
   labor, 179–181
   peace, 185–186
   progressive, 175
   reactionary, 175
   social, 173
   women's, 182–183
Mumford, Lewis, 31
Movement Toward Socialism, 158–159

Nader, Ralph, 187
NASA (National Aeronautics and Space Administration), 29
neoliberalism, 16
neo-Marxism, 6, 13
new class societies, 81–84
New Economic Plan, 53
New Global Left, 128
New left parties, 13, 206–217
Next System Project, xii
New Urbanism, 249
New Zealand, 168
North American Free Trade Agreement (NAFTA), 181
nuclear holocaust, 9

Obama, Barrack, 166, 260
Occupy Wall Street, 184, 190, 197, 200

ocean acidification, 28
October Revolution, 50
Organisation of Latin American Solidarity, 77

Parecon, 139, 225, 263
Paris Commune, 45
Party of Democratic Socialism (PDS), 70
peak oil, 31
Pedestrian cities, 238
Perestroika, 58–59, 69, 94, 133
permaculture paradigm, 121–124. *See* David Holmgren
permanent revolution, 55, 154
Plan B, 108
Plan C, 108
planetary boundaries, 26
pluralea interactions, 257
political economy of ecology, 137
population growth, 16
postcapitalism, 9
postcolonialism, 9
postmodernism, 10
postsocialism, 9
poststructuralism, 9
post-revolutionary societies, xii, 3, 12, 84–96
poverty, 22–26, 36, 74, 85–86, 104, 108–109, 118, 126, 147, 157, 161–162, 170, 192, 198–199, 213, 223–224, 229
prefigurative social experiments, 190–195
profit-making, 16–17
Provisional Government, Russia, 51
public ownership of the means of production, 13, 221–223
Putin, Vladimir, 61

radical democracy, 1, 3, 131, 176
radical ecology, 137, 142
Randers, Jorge, 111
Raskin, Paul, 104, 124–126

real utopia, 2, 14, 126, 169, 203, 257, 263–264
rebound effect, 231
Red Guards, 63
reforms
　nonreformist, 201
　reformist, 201
　system-challenging, 3
renewable energy, 12–13, 96, 108, 115–116, 120, 122, 126, 139, 152, 161, 168, 192, 204–205, 220–221, 232–236, 255, 261
Republic of Korea, 49. See South Korea, 49
resource wars, 35–37
revolution, 10, 13, 45, 201–202
risk society, 105
Rockstroem, Johan, 26
Roemer, John, 133
Rudd, Kevin, 220
Ruddiman, William, 7
Russia, 45, 50
Russian Social Democrats, 50

Salleh, Ariel, 141–142
Sanders, Bernie, 269–272
Saudi Arabia, 32
Scholar-activist, 197
shipping, sea, 33
Shiva, Vandana, 142–143
Simpler Way, 251–252
Singer, Merrill, xii, 3, 6, 257
social democracy, 96–100. See welfare capitalism
　Nordic model, 98–99
Social Democratic Party, Germany, 66
social equality, 1, 13, 18, 22, 25, 44, 90, 109, 152, 205, 223–224
social inequality, 11–13, 21–22, 89–90, 101, 109, 113, 118, 126, 204, 223
social media, 197–198
socialism
　actually existing, 78–79
　democratic, 3–4, 133–136

　eco-socialism, 136–140, 152–153
　for the 21st century, 153–155
　market, 133–134
　petro-socialism, 156
　reconceptualising, 131–143
　state, 78–79
Socialist Alliance (Australia), 144, 217
socialist democracy, 43–45, 56, 78, 166
socialist ecology, 137
Socialist International, 100
socialist-oriented societies, 3, 12, 127–172
socialist planning, 224
Socialist Unity Party of Germany, 6
socialist world government, 265
solar communism, 228
Soviet Union (USSR), 43, 50–61, 86–87, 89–95
Speth, James Gustav, 109–110
Stalinism, 50, 62, 80, 101
state capitalism, 16, 53
statism, 134
statist societies, 83–84
Stockholm Resilience Center, 26
subsistence perspective, 148, 183
Susser, Ida, xii, 3
sustainable food production, 13, 242–245
sustainable forestry, 13
sustainable public transportation, 13, 237–239
Syzmanski, Albert, 79–80

Taylor, Graeme, 120–121
tipping point, 27, 29, 155, 207, 257
Trainer, Ted, 146–147, 251—252
transition, between capitalism and socialism, 79–81
Transition Town movement, 194–195
transnational state, 206
treadmill of production and consumption, 11, 15, 108, 119, 230–231, 236, 246, 257

Trotsky, Leon, 43
Trump, Donald, 262–263
Tsipras, Alexis, 213–214
Turnbull, Malcolm, 220

United Nations, 9
    Security Council, 61
United Nations Conference on
    Sustainable Development, 27
United Nations Development
    Programme, 223

Venezuela, 153, 155–158
voluntary simplicity movement, 192

Wagar, W. Warren, 103, 266
Wall, Derek, 203–204

Wallerstein, Immanuel, 11
wars
    civil war, 1918–1920, 52
    interstate, 37
    resource, 35–37
Wende, 69
workers' democracy, 3, 44, 224–225
world state, 106
World Trade Organization (WT), 64
World's Peoples Conference on
    Climate Change and the Rights of
    Mother Earth, 161
White, Leslie A., 5, 9–11, 116
World Social Forum, 179
Wright, Erik Olin, 2

Zaptistas, 174

www.ingramcontent.com/pod-product-compliance
Lightning Source LLC
Chambersburg PA
CBHW070910030426
42336CB00014BA/2353